TECHNIQUES OF CHEMISTRY

ARNOLD WEISSBERGER, *Editor*

VOLUME VII

MEMBRANES IN SEPARATIONS

TECHNIQUES OF CHEMISTRY

ARNOLD WEISSBERGER, *Editor*

VOLUME I

PHYSICAL METHODS OF CHEMISTRY, in Five Parts
(INCORPORATING FOURTH COMPLETELY REVISED AND AUGMENTED
EDITION OF PHYSICAL METHODS OF ORGANIC CHEMISTRY)
Edited by Arnold Weissberger and Bryant W. Rossiter

VOLUME II

ORGANIC SOLVENTS, Third Edition
John A. Riddick and William S. Bunger

VOLUME III

PHOTOCHROMISM
Edited by Glenn H. Brown

VOLUME IV

ELUCIDATION OF ORGANIC STRUCTURES BY PHYSICAL
AND CHEMICAL METHODS, Second Edition, in Three Parts
Edited by K. W. Bentley and G. W. Kirby

VOLUME V

TECHNIQUE OF ELECTROORGANIC SYNTHESIS, in Two Parts
Edited by Norman L. Weinberg

VOLUME VI

INVESTIGATION OF RATES AND MECHANISMS OF
REACTIONS, Third Edition, in Two parts
PART I: *Edited by Edward S. Lewis*
PART II: *Edited by Gordon G. Hammes*

VOLUME VII

MEMBRANES IN SEPARATIONS
Sun-Tak Hwang and Karl Kammermeyer

TECHNIQUES OF CHEMISTRY

VOLUME VII

MEMBRANES IN SEPARATIONS

SUN-TAK HWANG
Professor
AND
KARL KAMMERMEYER
Professor
Department of Chemical and Materials Engineering
University of Iowa
Iowa City, Iowa

A WILEY-INTERSCIENCE PUBLICATION

JOHN WILEY & SONS

New York • London • Sydney • Toronto

QD
61
.T4
v. 7

Library of Congress Cataloging in Publication Data

Hwang, Sun-Tak, 1935-
  Membranes in separations.

  (Techniques of chemistry; v. 7)
  Includes bibliographies.
  1. Separation (Technology) 2. Membranes (Technology) I. Kammermeyer, Karl, 1904-   joint author.
II. Title. III. Series.

QD61.T4 vol. 7 [QD63.S4] 540'.28s [544] 74-22418
ISBN O-471-93268-X

Printed in the United States of America

10 9 8 7 6 5 4 3 2 1

INTRODUCTION TO THE SERIES

Techniques of Chemistry is the successor to the Technique of Organic Chemistry Series and its companion—Technique of Inorganic Chemistry. Because many of the methods are employed in all branches of chemical science, the division into techniques for organic and inorganic chemistry has become increasingly artificial. Accordingly, the new series reflects the wider application of techniques, and the component volumes for the most part provide complete treatments of the methods covered. Volumes in which limited areas of application are discussed can easily be recognized by their titles.

Like its predecessors, the series is devoted to a comprehensive presentation of the respective techniques. The authors give the theoretical background for an understanding of the various methods and operations and describe the techniques and tools, their modifications, their merits and limitations, and their handling. It is hoped that the series will contribute to a better understanding and a more rational and effective application of the respective techniques.

Authors and editors hope that readers will find the volumes in this series useful and will communicate to them any criticisms and suggestions for improvements.

ARNOLD WEISSBERGER

Research Laboratories
Eastman Kodak Company
Rochester, New York

PROLOGUE, IN THE FORM OF A PERSONAL LETTER

TO KARL KAMMERMEYER*

Dear Karl,

When you asked me to write a largely historical "Prologue" to your book, you referred to my intimate association with the developments in the physical chemistry of membranes and stated that "what we would like to see is the personal touch," and further added: "It would be fascinating, especially to the younger people in the field, to get some historical perspective. For instance, how did you ever get interested in membranes and then what were some of the exciting events in your early work?" I gladly agreed to write such a prologue for which I can claim indeed some unusual qualifications: personal acquaintanceship with many of the old classical masters of "membranology" and active work on the physical chemistry, particularly the electrochemistry, of membranes which has occupied more than 80% of the last 45 years of my life. Also, writing this Prologue gave me a most welcome incentive to look systematically into the historical development of the physical chemistry of membranes, a topic about which I hope now to publish at length in the near future.

I use here the form of a personal letter, which frees me of many restrictions customary in scientific writing.

When I studied chemistry in my home town at the University of Vienna, Austria, in the first half of the 1920s one did not hear much about membranes. You got just a few bits of casual information. "Membranes" were not an organized field, a far cry from today's situation. In physical chemistry you learned that M. Traube developed in the 1860s "semipermeable" copperferrocyanide precipitation membranes which were used

* The author of this letter is Dr. Karl Sollner, recently retired Chief of the Section on Electrochemistry and Colloid Physics, National Institute of Arthritis, Metabolism and Digestive Diseases, National Institutes of Health, Bethesda, Maryland 20014.

in 1877 by the "botanist" Pfeffer to measure the osmotic pressure of sucrose solutions, and that these measurements were in turn utilized by the great van't Hoff (1887) and Raoult (1884) as the basis for the classical theory of solutions. It also was impressed on you that the measurement of osmotic pressures by means of semipermeable membranes for the determination of molecular weights is a most difficult procedure, for practical purposes essentially useless.

In analytical chemistry word spread that one can prepare some very dense filters called "membrane filters" and "ultrafilters," which are useful with very fine precipitates. Although I was a teaching assistant in analytical chemistry from 1922 to 1926, I never saw this type of filter. In chemical technology you learned that parchment paper is used in the sugar and the dyestuff industries to remove by "dialysis" impurities, mainly salts. Only by chance did I hear that biologists were deeply interested in the "permeability" of living membranes. Thus, when I received a Ph.D. in 1926 the idea had never occurred to me that membranes and membrane effects per se could be worthwhile research topics.

After receiving my degree, I looked long and frantically for a position. For young scientists in Central Europe the job situation in those days was infinitely worse, and chronically so, than that which has confronted young American Ph.D.'s in the last few years. Soon I realized that in order to find a position as a chemist I must obtain postdoctoral training and that specializing in colloid chemistry might be a good bet. After considerable difficulties (postdoctorate fellowships at that time were virtually nonexistent) early in 1927 I arranged that I would join in the fall (at my family's expense) the laboratory of Professor Herbert Freundlich at the Kaiser Wilhelm-Institut (now Max Plank-Institut) for Physical Chemistry in Berlin, who was then the best authority on colloid chemistry in Germany. The interim period I used by working out a laboratory course in colloid chemistry for the Institute of Physical Chemistry of the Technical University of Vienna. In doing this I became somewhat acquainted with the use of membranes in dialysis, electrodialysis, and ultrafiltration. More importantly, I also familiarized myself thoroughly with the literature on colloids, paying particular attention to the 1922 and 1923 editions of Freundlich's "Kapillarchemie" and the 1920 and 1925/1927 editions of Zsigmondy's "Kolloidchemie."

Freundlich's book, for several decades a kind of "Bible" in colloid chemistry, impressed me by its emphatically physicochemical, not merely descriptive, approach. I was fascinated by his then unsurpassed presentation of the electrical aspect of colloids, colloid stability, electrokinetics and electrocapillarity, but found myself confused by a section on "Anomalous Osmosis, Electrostenolysis and the Becquerel Phenomenon." Freundlich's reviews of the mechanisms of these effects which had been suggested in the literature amazed me by their haziness and by their lack of a rigorous physicochemical reasoning.

In the fall of 1927 I started work in Professor Freundlich's laboratory and finished during the next ten months three experimental studies of a diversified nature. The nearly daily contact with Professor Freundlich was most stimulating and rewarding. The general atmosphere at the institute under the directorship of the Nobel laureate Fritz Haber of ammonia synthesis fame was one of intellectual fermentation with the emphasis on originality, and a great deal of independence for the younger scientists. By good luck I had fallen into a hothouse for the development of budding scientists about which I was to read thirty years later in the Encyclopedia Britannica: "This research institute, which he [Haber] headed until 1933, became the finest of its kind in the world and many mature chemists from all nations came there to work."

Thus I was highly pleased when toward the end of my first year at this institute I was awarded one of the few then available and therefore highly coveted "Haber fellowships," established by Professor Haber from his income as a director of the I. G. Farben Industries.

Now a few words about how I got into work on membranes: It occurred in a somewhat roundabout way, which I think now was significantly influenced by a strong boyhood interest in electricity. In the fall of 1928, when I looked for a sharply defined problem, I recalled the several aforementioned poorly understood electrical membrane effects, and decided to attempt to clarify the mechanism of the so-called electrocapillary Becquerel phenomenon, first described 60 years earlier by A. C. Becquerel (1867) as follows: when a cracked test tube or porous clay diaphragm filled with a Na_2S solution is placed into a solution of a copper or silver salt, CuS and Ag_2S precipitates are formed where the two solutions meet, and after some time beautiful copper or silver crystals appear at the crack or diaphragm on the side of the heavy-metal solution. The same effect is observed if solutions of Na_2Se or Na_2Te instead of Na_2S are used. The cracked test tubes and diaphragms, as was shown later, act only as convenient mechanical supports for the heavy-metal sulfide and other precipitation membranes.

The first modern attempt to explain this effect from the point of view of the newly established ion theory of electrolyte solutions was made in 1890 by Wilhelm Ostwald (1890), one of the founding fathers of physical chemistry. Ostwald's ideas, reported approvingly in Freundlich's book, did not make sense to me. Another explanation suggested by F. Braun (1891), however, seemed to be on the right track, but his view needed further clarification and reformulation in terms of modern electrochemistry.

The sulfides, selenides, and some other precipitates at which the Becquerel effect occurs are all metallic conductors, which when precipitated from aqueous solutions are highly porous. Accordingly short-circuited electrochemical cells of the type:

Soln. 1	Membrane CuS (solid)	Soln. 2
Na_2S		$CuSO_4$
	Na_2SO_4 soln. in pore	

can form readily in which the heavy-metal sulfides act as electronically conducting electrodes. Cations and anions may be discharged continuously if the emf which arises in a particular cell surpasses the decomposition potential. Thus, the Bequerel effect turned out to be a rather trivial phenomenon and without wider significance. Nevertheless, I was elated by having solved this long open problem and greatly savored Haber's complimentary remarks when I presented my result in a seminar which I recall as a highlight of my budding career.

With great difficulty I persuaded Professor Freundlich to join me as coauthor on the Becquerel-effect paper (1928). I would have felt embarrassed to criticize my much beloved and admired master, and furthermore I felt that he had made a substantial contribution not only by leading me through his book to this problem but also by his interest in and encouragement of my effort. At that time I did not dream that this paper, which in my eyes was of a strictly electrochemical nature, would turn out to be the first step toward a lifetime's work on membranes. Similarly, I did not realize that Ostwald's aforementioned paper, aside from his obviously nonacceptable explanation of the electrocapillary Becquerel phenomenon, contains some most ingenious speculations concerning the electromotive properties of "semipermeable" membranes, which at that time already had been used by several, to me then still unknown, investigators as the seminal ideas of the rational electrochemistry of membranes.

Next, I tried my hand on another of the poorly understood electrical effects described in Freundlich's book, namely "Electrostenolysis" and analogous "electrostenolytic effects," the formation of metallic precipitates at membranes and diaphragms on the passage of current, an effect which has been used as the basis of several farfetched and hazy speculations. I demonstrated that this phenomenon can be due to two entirely different mechanisms according to the experimental conditions (1929).

Electrostenolysis, like the electrocapillary Becquerel effect, also turned out to be readily explainable within the framework of conventional electrochemistry, in other words, trivial. Nevertheless, I felt rather pleased when more than 40 years later I read with reference to my electrostenolysis paper of 1929:

"The subject of the electrolytic processes at membranes has been capably reviewed and discussed by Sollner and thus, there is no need for further treatment here."*

While trying to solve the riddle of electrostenolysis I immersed myself totally for many months in the rather widely scattered and never in toto systematically reviewed literature on the physical chemistry, particularly the electrochemistry of membranes. Fortunately, the literature on membranes, though already rather voluminous, was forty-five years ago still rather small compared with today's literature on this subject. It was then not too difficult even for a newcomer to identify the papers of the rather limited number of investigators who had made the major contributions. One can only marvel at the intuitive insight and experimental skill of the great masters and admire their ability to draw essentially correct, general, and far-reaching conclusions from limited experimental data, and it seems appropriate to review here briefly some of the most outstanding pre-1930 papers on the physical chemistry of membranes.

Thomas Graham, the father of systematic physicochemical membrane studies, described in 1829 the inflation (to the point of bursting) of water-wet (pig?) bladders containing some air, which were inserted into a jar filled with CO_2. He correctly explained this effect as the consequence of the solubility of the CO_2 in the water in the "capillary canals" of these membranes, and its diffusion through it followed by the release of the CO_2 into the air inside the bladder. In 1854 Graham also observed and clearly explained the process, which is referred to today as "pervaporation" and in 1866 reported extensively "On the absorption and dialytic separation of bases by colloidal septa" (1866). Exner (1874 and Stefan (1878) demonstrated that the rates of the movements of different gases across liquid (soap) lamellae are strictly proportional to the solubility of the various gases in water times their diffusion velocities. These and numerous similar papers were carefully reviewed by Waitz (1908) in an article that is an excellent guide to certain phases of the older literature. Lhermit in 1855 and Nernst in 1890 considered membrane systems consisting of three liquid phases and

* A. Brenner and J. L. Sligh, Jr., J. Electrochem. Soc.: Electrochemical Science, 117, 602 (1970).

proved that only those substances that are soluble in the middle phase, the membrane, can pass across it. For solid membranes, such as sheets of rubber, which act as solvents, the same was demonstrated by Flusin (1908).

The basic ideas concerning the *electrochemistry of liquid membranes*, particularly their electromotive behavior in concentration cells, were presented in a rather casual manner, as by-products of certain investigations directed toward other ends, which were carried out by the later Nobel laureates Walter Nernst and Fritz Haber and their collaborators.

In papers that deal mainly with the polarization of liquid–liquid phase boundaries under the influence of a direct current, Nernst and Riesenfeld came in 1902 to the following conclusion: the phase-boundary potential arising at the interface between two mutually sparingly soluble liquid phases, which are in a state of distribution equilibrium with respect to an electrolytic solute, is independent of the absolute concentrations of the latter. Accordingly, in a concentration cell $A^+L^-c_1$ |liquid membrane| $A^+L^-c_2$ in which the electrolyte A^+L^- is present in the membrane near the phase boundaries at the concentrations kc_1 and kc_2 (k being the distribution coefficient) the two phase-boundary potentials are equal and opposite. The measurable EMF of such cells is the liquid junction potential arising within the membrane due to the difference in the concentrations of A^+L^- within the membrane.

An entirely different electromotive behavior of certain liquid–liquid phase boundaries, namely, an electromotive response to changes in the hydrogen-ion concentration in reasonably good conformity with the Nernst equation for concentration cells, was observed by Haber and Klemensiewicz (1909) in the course of an investigation that dealt mainly with the phase-boundary potentials at the interfaces between electrolytically conducting solid electrolytes and aqueous solutions having a common ion. The same paper also describes the glass-membrane electrode and presents its theory. The work on liquid membranes was later continued by Beutner (1912), a former student of Haber.

The papers of Nernst and Haber were difficult reading, at least for me, and things became much clearer when I found Michaelis' systematic, clear presentation of these matters in his book *Hydrogen Ion Concentration* (1926), which is a classic and still worthwhile reading.

Donnan's famous theoretical paper on the membrane equilibrium and membrane potential (1911), because of its generality (it applies equally to liquid and to porous membranes) and its rigorous mode of presentation, was somewhat awe-inspiring; it and subsequent experimental work created the feeling that everything that could be said about these problems had been expounded by Donnan. Professor Donnan was, indeed, a man of superior intellectual capacity as I could observe firsthand after joining his laboratory at University College, London, several years later.

The papers dealing with porous membranes were much more numerous than those on liquid membranes. Many of them were not membrane studies per se but described the use of such membranes for specific purposes, and thus contributed but little to the understanding of the basic physical chemistry of membranes, which was my primary interest. The outstanding examples of the use of porous membranes for the measurement of osmotic pressures were the careful experimental investigations by Morse and his school (1914) and by Berkeley and Hartley (1906–1916). There was also a considerable body of literature dealing with the use of membranes in dialysis, beginning with a paper by Graham (1861), and in ultrafiltration. Early papers by Bigelow and Gamberling and by Brown (1915) gave much useful information on the preparation of collodion membranes. The outstanding systematic investigation on various types of ultrafilters, however, originated in the laboratory of the Nobel laureate R. Zsigmondy (1918–1929). The technological literature on ultrafiltration and dialysis, as exemplified, for example, by a review of Hebler (1927), did not provide much of basic interest.

As to the nonelectrolyte permeability of membranes, considerably denser than those used ordinarily in dialysis and ultrafiltration, it was known that, aside from the porosity of the membrane, the deciding factor is the molecular weight (or size) of the permeants. With electrolytes, the situation was rather confused. The literature on these questions was reviewed briefly by Collander in three papers (1924–1926), on the permeability of copper ferrocyanide and of collodion membranes of graded porosities. In these systematic masterly studies, carried out with minimal facilities, Collander established definitely that the permeability of any given porous membrane for nonelectrolytic solutes depends on the molecular size of the latter, dropping off steeply above a certain molecular size characteristic for the membranes of a given porosity. Collander's results, taken today for granted, were 50 years ago a most important basic contribution. His findings on the electrolyte permeability of porous membranes were more complex, too complex to be reviewed here, but in the course of his work he corrected some long-standing erroneous ideas which had retarded progress for several decades.

As far as membranes in electrolytic systems were concerned, my first interest was to gain insight into the molecular mechanism of the electrochemical effects observed with porous membranes, particularly of their electromotive action. This blended later into a steadily growing desire to acquire all information that conceivably might be helpful in elucidating the mechanism of anomalous osmosis, a topic which tempted me as my next research project.

Anomalous osmosis, first described by Graham (1854), and assumed by him to be an electro-osmotic phenomenon, was one of the few electrical membrane effects that had attracted the interest of Bartell and his group (1914–1923). The explanations of this phenomenon suggested by them seemed unsatisfactory to me. Anomalous osmosis also was the topic of a long series of experimental studies by the biologist Jacques Loeb (1918–1920), famous for his work on parthenogenesis.

A thorough study of the papers of these and various other investigators convinced me that a satisfactory explanation of the mechanism of anomalous osmosis could result only from a deeper understanding of the electrochemistry of porous membranes. This topic had never been reviewed in a comprehensive, truly constructive manner. Even Freundlich in 1916 discussed at length only one investigation on porous membranes, a study of Bethe and Toropoff (1915) "On electrolytic processes at membranes" in which the polarization effects arising at porous membranes are treated in detail. The presentation of this subject was so overwhelmingly impressive that a brief introductory section was generally overlooked, which from the systematic point of view is really of much greater basic importance than the polarization studies. It was only much later that I began to realize, at least in part, the vast amount of insight present in it.

Bethe and Toropoff outlined clearly the most basic concept of the electrochemistry of membranes, namely, the correlation of the electrokinetic charge of the membranes and their electromotive properties, their electrolyte permeability, and their polarization behavior. Those charges (ions) which are fixed to the pore wall, and form the immovable part of electrical double layer at the membrane-matrix-solution interface, cannot participate in the movement of ions across the membrane either by thermal motion (diffusion) or under the influence of an electric current. Their counterions, however, which are dissociated off into the pore water, are freely movable and can move under the influence of a concentration or potential gradient. This concept is the ultimate basis of today's highly developed electrochemistry of porous membranes. The prolonged neglect of these and related insights of Bethe and Toropoff has undoubtedly retarded the development of the electrochemistry of porous membranes.

Relatively well known in the late 1920s was the then still current systematic work on the basic electrochemistry, particularly the electromotive behavior, of dense, "molecular sieve," collodion membranes of extreme ionic selectivity by Michaelis and collaborators (1925–1927). Michaelis expounded in a quantitative manner the relationship of the ratio of the "mean mobilities" (in reality the ratio of the transference numbers) of the anions and cations in a membrane to the electromotive action of the latter. His papers are still a delight to read; his review of his then still progressing work is a classic, showing a great mind at work. By his rigorous systematic work Michaelis laid the firm basis for all later work in this field in spite of the fact that it left many basic questions still open. It was my good fortune to have in later years many discussions with Professor Michaelis on membranes and other electrochemical topics. He was an exceedingly modest man of great scientific sagacity, originality, and amazing intuitive understanding.

For me, familiarity with Michaelis' work paid off soon. It furnished an essential part of the information which I utilized in 1929 to develop a theory of the mechanism of anomalous osmosis, which later was verified experimentally (1932). This work, to my great delight, found the approval not only of Professor Freundlich but also of such experts in electrochemistry as Nernst and Haber, and later also of Planck. This theory is based on the concept that porous membranes have pores of different diameter which, according to Michaelis, yield different pore potentials in concentration cells. Consequently, they must interact electrically with each other; local electric currents are set up and cause electro-osmosis, that is, anomalous osmosis. Expanding the view of membranes as micromosaic structures, I developed the quantitative theory of mosaic membranes composed of ideally anion and ideally cation selective parts (1932). It correlates in a rigorous manner the electrolyte permeability of such membranes under defined conditions to the electrical characteristics of their component parts. This is the first instance in which an exhaustive quantitative treatment of a complex electrochemical membrane system was shown to be feasible. Much later, after suitable "permselective" membranes had been developed in my laboratory, this theory was verified quantitatively; such membranes have recently attracted considerable interest.

After 1932, my feeling, and that of several older and much more experienced investigators I consulted, was that physicochemical-membrane research as carried out in the past had seemingly run its course. Some new theoretical idea and experimental approach seemed necessary to revitalize this field. The first step in this direction occurred when Theorell in 1935, and soon afterwards Meyer and Sievers (1936–1937) at greater length, proposed what is today known as the "fixed-charge theory of ionic membranes," which is the basis of today's electrochemical-membrane research. From then on, first slowly and later with increasing speed and breadth, research on electrolytic-membrane systems has been carried out both in academic and industrial laboratories; and soon nonelectrolyte-membrane systems also became a favored research topic. Today the literature on membranes and membrane effects is overwhelming. Therefore it is good to see that you have entered the ranks of those who try to organize and present in a book some large and important part of the current knowledge in this field.

Kind regards!

Sincerely yours,

KARL* [Sollner]

---

* *Comment.* The early experiences of Karl Sollner, the dean of American membranologists and equally esteemed on a global basis, should be an inspiration to all scientists. The sequel to Sollner's writings is regrettably too long to present here. It constitutes a masterful history and review of membrane science and is expected to be published as a separate work. (Karl Kammermeyer)

PREFACE

The purpose of this volume is to provide a unified treatise of all membrane separation processes. Much of the development of membrane technology is fairly recent, even though the history of membranes itself dates far back. Membranology is a multidisciplinary field. Various membrane processes have been developed so rapidly in several different disciplines of science and technology that the highly compartmentalized state of the art hardly allows full communication among the investigators. The literature covering membrane phenomena is also diversified and numerous. As a result, the terminologies are confusing and the theories need to be generalized in many cases. Thus, this book is written not only for those who are actively involved in membrane research but also for people who would like to know about membrane separations.

Biological membranes are important and hold a key in many questions of basic biology and medicine. However, since they exhibit distinct characteristics from artificial membranes, they should be treated separately. Indeed, there are many excellent books on this subject and we feel that it would be redundant to include it in this volume.

The first three chapters deal with basic definitions of membrane processes and present brief descriptions of various transport mechanisms. A generalized view is emphasized. In Chapters IV and XIII, we attempt to unify the existing theories in formulating engineering analysis. Chapters V through X discuss individual processes in detail. These are followed by two chapters covering what quantities are used and how they are measured in membrane processes. Chapter XIV explains how the membranes are prepared, and the last three chapters are devoted to specific applications.

Finally, we would like to express our appreciation to Miss Edna Wilson for typing the manuscript and to Dr. H. Rhim for his extensive assistance in preparing the manifold aspects of liquid-phase operations.

SUN-TAK HWANG
KARL KAMMERMEYER

Iowa City, Iowa
July 1974

HISTORY

In 1954 I had my first opportunity to present the results of many years of study in the gaseous diffusion field to a distinguished group at a Gordon Research Conference. I am greatly indebted to Letcher Jones for the opportunity to address the Conference on Separation and Purification. This event was a momentous occasion as it led to many friendships that have lasted through the years.

It also taught me a lesson. In my eagerness to put on a good show I attempted to cover too much in too short a time. Fortunately, I received good advice. One of the elder statesmen took me aside after the session and in a most friendly way said "Karl, I know you wanted to tell us all you did, but it would have helped a great deal if at the very beginning you would have told us what you were going to talk about."

So, here is the very basic explanation of what is involved in discussing membranes in separation. A membrane, usually a solid film, is placed in a vessel so that two compartments are established (see figure).

A feed stream of a mixture, gaseous or liquid, is introduced on the upstream side of the membrane. As it flows along the membrane some constituents permeate selectively, and an enriched permeate and a depleted reject stream are generated. It is now already evident that the process is

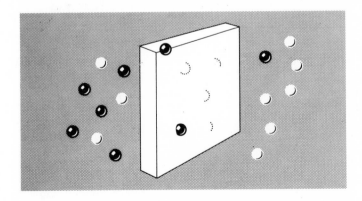

almost always one resulting in partial enrichment. Consequently, the usual installation will contain multiple stages.

With this admittedly very elementary introduction I have now followed my friend's advice, and it remains only to express Sun-Tak Hwang's and my appreciation to Karl Sollner for writing the Prologue. We find Dr. Sollner's philosophical approach fascinating and hope that the readers will enjoy his unique manner of presentation as much as we do.

KARL KAMMERMEYER

July 1974

CONTENTS

Chapter I

INTRODUCTION

This chapter is concerned with membranes that are used to separate materials in a great variety of applications. The growing interest in membrane processes utilized in industry, biomedical engineering, and space science is evidenced by the many publications that have appeared and continue to appear in the literature.

To treat the subject in a logical manner it is essential to cover the general scope of membrane processes and illustrate them with specific examples. On the whole, the engineering aspects of membrane technology have not received adequate attention, and therefore they will be emphasized here. To ensure satisfactory performance, the engineering and design of membrane processes are as important as the development of a new membrane of high performance characteristics.

It is interesting to speculate when the concept of using a membrane for separation was first realized. Although the phenomenon of osmosis was observed in 1748 by Abbe Nollet [1], its possible use as a separation process was not recognized until a much later date. Surely, when Fick developed his mathematical model of diffusion in 1855 (see Chapter 2, Section 1) he must have concluded that the varying rates of diffusion for different substances could be used as the basis of a separation process. As far as is known, the first reference to a separation procedure by means of a membrane is Graham's [2] use of a dialyzer in 1854, to separate a solution into its components. Subsequently, further evidence of the separation possibilities is contained in a statement made by Graham in his pioneer paper "On the Molecular Mobility of Gases" in 1863 [3]. In this treatise, which resulted in the formulation of Graham's law, the author mentions "the partial separation of mixed gases." Only a few years later, in 1866, Graham published another paper [4] dealing with the use of "colloid

1

septa" for the separation of gases. Thus, it would appear that Graham, as "master of the mint," deserves credit for originating the use of a membrane in spearation.

The natural membranes, which are of such fundamental importance in living organisms, belong to a class all by themselves. The structure of such membranes and the transport mechanisms by which they function are totally different from those of the "dead" membranes used in separation processes. Thus, it is beyond the scope of this chapter to deal with biological membranes in depth. A survey of essential information is, however, presented in Chapter X.

1 DEFINITION OF MEMBRANE

What is a membrane? Possibly the broadest definition is: a region of discontinuity interposed between two phases. This statement implies that membranes can be gaseous, liquid, or solid, or combinations of these phases. The term "region" in the definition is used to eliminate ordinary interfaces. Thus, the interfaces of two immiscible liquids, of a gas and a liquid, or of a gas and a solid would not ordinarily be considered as membrane structures.

An elusive, but nevertheless real gaseous membrane would be the front of a shock wave. So far, at least, we are not aware of any instance where this gaseous membrane has been used successfully for a separation process. Well-known examples of membranes are filmlike solids, porous or nonporous, and the recently developed liquid membranes of Li and co-workers [5], Barrer's authoritative book *Diffusion in and through Solids* [6] covers most or all of the solid membranes known to the present state of the art.

Membranes vary in their make-up from the relatively crude structure of a screen to extremely fine configurations only one molecular layer thick, as in the fatty acid spreading on liquid water. No synthetic membrane has as yet been created that can approach the performance of living membranes. The spark of life introduces factors that at present are beyond our ability to duplicate. A very informative presentation on the structure of cell membranes was published by Fox [7].

2 CLASSIFICATION OF MEMBRANES

Even a cursory examination of the variety of membranes that exist makes it evident that a single classification scheme is unlikely to permit a clear and concise presentation.

However, a rather informative picture is obtained by using multiple schemes, which can readily be interrelated. Such arrangements are represented by the following:

A. classification by nature of membrane
B. classification by structure of membrane
C. classification by application of membrane
D. classification by mechanism of membrane action

A. Classification by Nature of Membrane
    Natural membranes
    1. Living membranes
    2. Natural substances — modified or regenerated
    Synthetic membranes
    1. Inorganic
        metals
        ceramics
        glass
    2. Organic — Polymers
        films
        tubing
        hollow fibers

B. Classification by Structure of Membrane
    Porous membranes
    1. Microporous media
        compressed powders
        microporous glass (Corning)
        microporous ceramics
        AEC barrier
        microporous silver membrane (Selas)
        Millipore filters
        porous polymer structures
        cellulose acetate
        Nuclepore (General Electric)
    2. Macropores
        filters for gases and liquids, some Millipore filters
        ultrafilters
    Nonporous membranes
    1. Inorganic
        metal films
        glass
    2. Polymeric structures
        films
        tubes
        hollow fibers

laminated films

Morphological distinction

1. Crystalline
   inorganic (metals, ceramics)
   organic
2. Amorphous
   glass
   polymers

Liquid membranes

1. Monomolecular film
   fatty acid spreading on water
2. Liquid drop surrounded by stable liquid film, Li's work [5]

C. Classification by Application of Membrane

Here, a membrane, gaseous, liquid, or solid, is interspersed between phases, as, for instance, gas-membrane-gas.

Gaseous-phase systems

1. Molecular flow — size of pores and molecular velocity determine separation
2. Molecular flow with surface flow (surface or diffusive flow results from adsorptive properties of microporous media)
3. Sweep gas diffusion — use of carrier gas
4. Diffusive solubility flow of polymeric structures — solubility of diffusing gases or vapors is controlling parameter
5. Diffusive flow in solvated polymers — occurs with liquids or vapors that swell or modify membrane properties

Gas-liquid systems

1. Macropore structure
   removal of liquid entrainment in gas stream
   introduction of gas into liquid phase
2. Microporous structure
   ultrafine filters
3. Polymeric structures
   gas diffusion from or into liquid, that is, blood oxygenation (oxygenation and CO_2 removal)

Liquid-liquid systems

1. Gas transport from one phase to another liquid phase
   organ preservation devices, blood oxygenation
2. Osmotic processes
   reverse osmosis
   gel permeation
3. Liquid membranes

selective flow through liquid film

Gas-solid systems
1. Removal of particulate matter in gas by filter

Liquid-solid systems
1. Filtration of slurry with macroporous media
2. Biological waste treatment
3. Emulsion breaking

Solid-solid systems
1. Screening of solids on basis of particle size

D. Classification by Mechanism of Membrane Action

Adsorptive membranes
1. Microporous membranes
   porous Vycor
   activated charcoal, silica gel, and the like (as compressed powders)
2. Reactive membranes
   a chemically reactive material is contained in the membrane and reacts with one of the permeating constituents

Diffusive membranes
1. Polymeric membranes
   diffusive solubility flow
2. Metallic
   atomic-state diffusion
3. Glass
   molecular state or affinity phenomena

Ion-exchange membranes
1. Cation-exchange resins
2. Anion-exchange resins

Osmotic membranes
1. Regular osmotic membranes
2. Reverse osmotic membranes
3. Electro-osmotic membranes

Nonselective membranes (Inert behavior)
1. fritted glass
2. filters, screens

3 MEMBRANES IN SEPARATIONS

The role of a membrane in a separation process is to act as a selective barrier. It should permit preferred passage of a certain component out of the mixture. Macroscopic processes, like filtration, do not rely on the molecular properties of a membrane. However, microscopic processes, which constitute the majority of the membrane processes, are mostly due to molecular interactions between membranes and fluids. By virtue of the differences in their degree of affinity, the transmission rate through a membrane is expected to be different for each component. Thus, separation becomes possible.

For example, permeation through a polymeric membrane usually involves both diffusivity and solubility of the permeant species, and these properties are the result of molecular interactions. The sole exception would be the case of gaseous diffusion through a microporous membrane which proceeds strictly on the basis of free molecular or Knudsen diffusion. If surface diffusion is absent, or if it is so small that it can be ignored, bulk flow would not depend on the molecular characteristics of the membrane. As long as the pores are small enough to ensure Knudsen flow, it does not make any difference what kind of membrane is used. In actual practice, however, the adsorptive behavior of gases may not be negligible, depending on temperature and pressure of operation, and the phenomenon of surface flow will enter the picture.

In general, separation takes place on the surface of the membrane, or in the membrane, itself, where the molecular interactions compete with one another. Thus, it is necessary to understand what goes on in the vicinity and inside the membrane. The respective mechanism of permeation is the key.

In order to improve the separation efficiency, two factors should be examined thoroughly. One is selectivity, and the other is total flow for a given membrane. Both of these factors can be studied only after one has knowledge of the actual transport mechanism through the membrane. Taking the above example of the polymeric membrane, one is bound to seek higher solubility and diffusivity for a preferred component and lower solubilities and diffusivities for the other components. When such a membrane can be found it will give a high degree of separation.

References

1. Kirk-Othmer, *Encyclopedia of Chemical Technology,* 2nd ed., Vol. 14, Wiley-Interscience, New York, 1967, p. 345.
2. Thomas Graham, *Phil. Trans, Roy. Soc., (London)*, **144**, 177 (1854).
3. Thomas Graham, *Trans. Roy. Soc. (London)*, **153**, 385 (1863).
4. Thomas Graham, *Philos, Magazine and J. of Science Series, 4,* 32, (218), 401 (1866).
5. N. N. Li, R. B. Long, and E. J. Henley, *Ind. Eng. Chem.,* **57,** 18 (1965).
6. R. M. Barrer, *Diffusion in and through Solids,* Cambridge University Press, London, 1951.
7. C. Fred Fox, *Sci. Am.,* **226,** (2) 31 (1972).

General References

Barrer, R. M., *Diffusion in and through Solids*, Cambridge University Press, London, 1951.

Crank, J., and G. S. Park, Eds., *Diffusion in Polymers*, Academic, New York, 1968.

Gelman, Charles, "Membrane Technology, Part I. Historical Development and Applications," *Anal. Chem.*, **37**, 29A (1965); H. Z. Friedlander, and R. N. Rickles, Part II: "Theory and Development," *ibid.*, 27A.

Jost, W., *Diffusion in Solids, Liquids, Gases*, 4th ed., Academic, New York, 1965. *Note*. The main body of the book is essentially the same as the original version of 1952. An Appendix of 82 pages contains more recent findings and mostly literature references through 1959 and partially into 1960.

Kesting, R. E., *Synthetic Polymeric Membranes*, McGraw-Hill, New York, 1971.

Lacey, R. E., and S. Loeb, Eds., *Industrial Processing with Membranes*, Wiley-Interscience, New York, 1972.

Rickles, R. N., *Ind. Eng. Chem.*, **58**, (6) 18 (1966).

Rickles, R. N., and H. Z. Friedlander, *Chem. Eng.*, **73**, 111 (1966), *ibid.*, 121; *ibid.*, 163; *ibid.*, 217.

Rogers, C. E., "Permeability and Chemical Resistance," in *Engineering Design for Plastics*, Reinhold, New York, 1964.

Rogers, C. E., Ed., *Permselective Membranes*, Marcel Dekker, New York, 1971.

Spiegler, K. S., *Ind. Eng. Chem. Fund.*, **5**, (4) 529 (1966).

Tuwiner, S. B., *Diffusion and Membrane Technology*, ACS Monograph Series, Reinhold, New York, 1962.

Chapter II

FUNDAMENTALS

The process of separation involves several fundamental physical phenomena, as the constituents in the feed stream pass over the membrane, into it, through it, and finally leave the membrane on the other side.

Thus, the basic processes of permeation and diffusion, with the attendant surface phenomena, will be treated

1 PERMEATION AND DIFFUSION

Permeation is a phenomenon in which a certain species or component is passing through another substance, usually but not necessarily by means of diffusion. The term "diffusion" specifically refers to molecular diffusion, while permeation stands for a much more general phenomenon of mass transmission. In fact, permeation is a phenomenological definition. Thus the term encompasses a variety of transport mechanisms. A number of different kinds of driving forces can cause permeation. To name a few, the concentration gradient, pressure gradient, electric potential gradient, or even the temperature gradient may produce permeation. Depending on the specific permeation mechanisms the flow process may be called diffusion, osmosis, electrodialysis, electro-osmosis dialysis, reverse osmosis, ultrafiltration, electrophoresis, and so forth. These various phenomena will be discussed separately in Chapter III.

Nevertheless, it is necessary to define a quantity to express the overall degree of permeation in general, regardless of the actual transport mechanism. Even in the case where the actual permeation mechanism is known to be diffusion, the general definition of permeability may be more useful than the molecular diffusivity. To calculate diffusivity, the actual concentrations of diffusing

8

species in the membrane should be known. But these being impossible to measure experimentally, one therefore has to rely on the assumption that an equilibrium state exists at the membrane surface. This is, of course, undesirable, and frequently the assumption is subject to criticism.

A direct approach, that is, defining a phenomenological permeability, is not only practical but also a realistic way of expressing the overall degree of permeation through membranes. Permeability is defined in terms of concentrations or pressures, which exist outside of the membrane itself and thus are measurable experimentally.

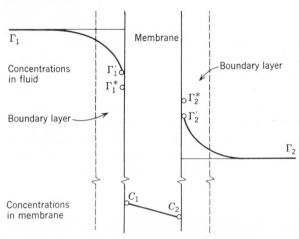

Fig. 2.1 Concentration profile across the membrane.

Referring to Fig. 2.1, the following equations can be written:

$$F = Q_\Gamma S \; \frac{\Gamma_1 - \Gamma_2}{l} \tag{2.1}$$

or

$$F = Q_p S \; \frac{P_1 - P_2}{l} \; . \tag{2.2}$$

These equations* may be used regardless of the actual permeation mechanism. As a matter of fact, the detailed mechanisms do not need to be known for the measurements and for the calculation of permeabilities by the above equations.

If the actual permeation mechanism is diffusion, Fick's first law can be used for a slow process. (See Eq. 3.13 for the general case.) For the same steady-state

*Symbols are defined in the Nomenclature List at the end of Chapter III.

permeation flux F,

$$F = DS \ \frac{C_1 - C_2}{l} \ . \tag{2.3}$$

Here, C represents the actual concentration within the membrane; however, Γ and P in Eqs. 2.1 and 2.2 represent the concentration and pressure, respectively, of bulk fluid outside of the membrane. Equation 2.3 is good in theory but not useful in practice when the concentrations in the membrane are unknown.

When permeation is due to other phenomena, we can similarly rewrite Eq. 2.3 in terms of new variables. However, this requires a model of the actual transport mechanism. As in the case of diffusion, we may be more specific with respect to permeation phenomena, but we must also accept the fact that these equations will be model dependent. For example, the permeation of gases through a microporous membrane is understood as the sum of Knudsen flow and surface flow. If we want to express the flow equation based on these phenomena, we have to agree upon a model of surface flow, which can describe the observed experimental results. There are many different models, sometimes conflicting with each other, therefore resulting in many different flow equations. Thus, a flow coefficient based on one model cannot be compared with a flow coefficient based on another model. However, if we choose to use the permeability equation, either (2.1) or (2.2), the problem does not exist.

There is one necessary precaution when we use the phenomenological permeability; that is, the phenomenological permeability is not a property of the membrane. Rather, it is a phenomenological quantity that depends on the experimental conditions under which the process is conducted. This overall permeability is a measure of permeation through not only the membrane but also the boundary layers on either side of the membrane, if there are any. Because the bulk concentrations or pressures are used in the equation, what goes on in the boundary regions is already included in the permeability. The boundary layer resistances can be analyzed and separated from the total permeability by a study of the membrane-thickness effect on permeability as discussed below.

A study of the thickness effect on permeability reveals explicit how the permeability and diffusivity are related. A hypothetical concentration profile and a schematic view across the membrane are shown in Fig. 2.1. It is assumed that both sides of the membrane are covered by thin layers of immobile fluid, which would give an extra resistance to permeation. The mass transfer through such a system consists of the following stepwise processes:

1. Diffusion through the boundary layer
2. Sorption into the membrane
3. Diffusion through the membrane

4. Desorption out of the membrane
5. Diffusion through the boundary layer

Each step represents a resistance to the gas transport of different magnitude. For practical purposes, however, the resistances of some steps are negligible in comparison with those of others. In the case of gas-phase permeation, processes 1 and 5 are not involved, and the resistances due to the steps 2 and 4 may even be negligible, However, for liquid-phase permeation, a large boundary resistance may result from steps 1, 2, 4, and 5. The presence of these boundary resistances reduces the available driving force for diffusion inside the membrane.

Using the definition of overall permeability given by Eq. 2.1, the steady-state flow equation is written as

$$F = Q'S \; \frac{\Gamma_1 - \Gamma_2}{l} \; . \tag{2.4}$$

If the diffusivity of the membrane is independent of concentration, the same steady-state flow rate for the inside of the membrane can also be expressed by

$$F = DS \; \frac{C_1 - C_2}{l} \; . \tag{2.5}$$

Here Fick's first law is used. For the sake of simplicity, the second term in the more rigorous Eq. 3.13 is neglected. Should diffusivity be concentration dependent, an average diffusivity can be used:

$$F = \overline{D}S \; \frac{C_1 - C_2}{l} \; , \tag{2.6}$$

where

$$\overline{D} = \int_{C_2}^{C_1} \frac{D}{C_1 - C_2} \, dC \tag{2.7}$$

Because it is very difficult to separate steps 2 and 4 from steps 1 and 5 experimentally, it is convenient to lump the resistances of steps 1 and 2 in one group, and steps 4 and 5 in another. Then one can write the flow equation for one side of the membrane:

$$F = S \; \frac{\Gamma_1 - \Gamma_1^*}{r_1} \; , \tag{2.8}$$

and for the other side:

$$F = S \; \frac{\Gamma_2^* - \Gamma_1}{r_2} \; , \tag{2.9}$$

where r_1 and r_2 are film resistances including resistances of sorption and desorption if they exist. The fictitious quantities, Γ_1^* and Γ_2^*, are the concentrations that would have produced the inside concentrations C_1 and C_2, respectively, under equilibrium conditions. If a linear isotherm (Henry's law) is applicable,

$$C_1 = S_m \Gamma_1^*, \tag{2.10}$$

$$C_2 = S_m \Gamma_2^*.$$

Combining Eqs. 2.8 to 2.11 with Eq. 2.5, and solving for F, yields

$$F = \frac{SDS_m(\Gamma_1 - \Gamma_2)}{DS_m(r_1 + r_2) + l}. \tag{2.12}$$

Comparing Eq. 2.12 with Eq. 2.4, the following is obvious:

$$Q' = \frac{DS_m l}{DS_m(r_1 + r_2) + l}. \tag{2.13}$$

This equation tells exactly how the observed permeability changes as the thickness of a membrane varies. Also, it shows that the film resistance could be significant when the diffusivity of the membrane is large, or when the thickness of the membrane is small. If there is no such film present, then the film resistance simply becomes zero, and the observed permeability reduces to the familiar form:

$$Q' = DS_m. \tag{2.14}$$

This equation has been used widely in many systems. However, it is clear from Eq. 2.13 that Eq. 2.14 holds only in a special case. Furthermore, Eq. 2.13 illustrates the fact that permeability is a phenomenological coefficient rather than a property of a given system, as given in Eq. 2.14. A change of an outside condition, such as film resistance or membrane thickness, alters the value of permeability. Therefore, in general, a comparison of two permeabilities for the same system but at different experimental conditions may not be meaningful.

The analysis of the thickness effect can be easily achieved by inverting Eq. 2.13:

$$\frac{1}{Q'} = \frac{1}{DS_m} + (r_1 + r_2)\frac{1}{l}. \tag{2.15}$$

When the inverse permeability is plotted against the inverse thickness, a straight line will result. From the intercept, the limiting value of permeability at infinite thickness will be obtained. The slope of the straight line gives the resistances of the boundary layer. A special example of the above analysis will be given in Chapter X for the dissolved oxygen permeation through a silicone rubber membrane.

2 NONEQUILIBRIUM THERMODYNAMICS

The phenomena of mass transmission across membranes are irreversible processes. Furthermore, a number of different driving forces can cause mass permeation through membranes, as mentioned in the previous section. Therefore, in order to discuss the general theory of membrane permeation, perhaps the best approach would be to start with the phenomenological theories of nonequilibrium thermodynamics.

There are three basic principles in nonequilibrium thermodynamics. The first law states that any driving forces can create any fluxes of nonequilibrium processes. The fluxes and forces are linearly related as follows:

$$J_i = \sum_j L_{ij}X_j. \tag{2.16}$$

The second law states that the phenomenological coefficients satisfy the Onsager reciprocal relationships:

$$L_{ij} = L_{ji}. \tag{2.17}$$

The third law specifies that the rate of lost work or entropy production multiplied by the temperature due to any irreversible processes can be expressed as the sum of products of conjugated fluxes and forces:

$$\phi = T \frac{d\,\Delta S}{dt} = \sum_i J_i\, X_i. \tag{2.18}$$

There are some exceptions to the above principles. The first one is the so-called Curie theorem: "For an isotropic system, fluxes and forces of different tensorial *character* (not rank) do not couple." The second one concerns the Onsager reciprocal relations. When an external field $|B$ is present, Eq. 2.17 is slightly modified to read

$$L_{ij}\,(|B) = L_{ji}\,(-\,|B), \tag{2.19}$$

where $|B$ may be an external magnetic field or the angular velocity of rotation for systems with Coriolis forces.

The linear law, Eq. 2.16, is just a generalization of the well-known laws of Fick, Fourier, Poiseuille, Ohm, or D'Arcy, which are linear relationships only between conjugated fluxes and forces. The generalized equation covers these existing laws, and all of the coupling phenomena. Such are the Dufour effect, Soret effect, electro-osmosis, electrophoresis, sedimentation potential, streaming potential, and the Donnan equilibrium. The Onsager reciprocal relationships are very valuable in studies of the coupled phenomena. They provide the symmetric results between a pair of coupled phenomena. For example, Saxén's relation in electrokinetic phenomena can be derived from this reciprocal equation [1].

The study of nonequilibrium thermodynamics is a phenomenological theory. It does not offer any explanation of how materials permeate through membranes. The actual mechanisms of transport cannot be unveiled by the result of a thermodynamic study. As in the case of classical equilibrium thermodynamics, nonequilibrium thermodynamics only tells the limits of certain phenomena. It describes a system by macroscopic variables. It is especially useful for the study of systems in which couplings take place between two or more processes.

In membrane permeation, nonequilibrium thermodynamics is of no value if there is only one kind of driving force that causes one type of flux. However, this phenomenological study is very helpful when there are two or more driving forces generating two or more fluxes and coupled phenomena between them.

As an example, let us consider the electrokinetic phenomena. A porous membrane separates an isothermal system of multicomponents carrying electrical charges into two compartments. From conservation of mass, energy, and charge, and utilizing the Gibbs equation, the entropy production rate is obtained from the following equation [2]:

$$\phi = T \frac{d \, \Delta S}{dt} = J \, \Delta P + I \, \Delta E. \tag{2.20}$$

This equation corresponds to Eq. 2.16. We can write the linear phenomenological equations for fluxes and forces from Eq. 2.17:

$$I = L_{11} \, \Delta E + L_{12} \, \Delta P, \tag{2.21}$$

$$J = L_{21} \, \Delta E + L_{22} \, \Delta P. \tag{2.22}$$

Equation 2.21 tells us that both the electrical potential and the pressure gradient can cause the electric current. Likewise, Eq. 2.22 tells us that both the electric potential and the pressure gradient are responsible for the bulk volume flow. From the Onsager reciprocal relationship, Eq. 2.18,

$$L_{12} = L_{21}.\qquad(2.23)$$

The above equations are basic laws of nonequilibrium thermodynamics for our system. Now, let us see how these equations describe the various electrokinetic phenomena.

First, the streaming potential or mechanoelectric effect is an electric potential buildup across the membrane due to a pressure difference when there is no electric current. From Eq. 2.21:

$$\left(\frac{\Delta E}{\Delta P}\right)_{I=0} = -\frac{L_{21}}{L_{11}}.\qquad(2.24)$$

Second, the electro-osmosis or electroendosmosis is defined as the bulk volume flow due to the electric current when the pressure drop across the membrane is zero. From Eqs. 2.21 and 2.22:

$$\left(\frac{J}{I}\right)_{\Delta P=0} = \frac{L_{21}}{L_{11}}.\qquad(2.25)$$

Third, the electro-osmotic pressure is the pressure buildup across the membrane due to an electrical potential difference when there is no mass flow. From Eq. 2.22:

$$\left(\frac{\Delta P}{\Delta E}\right)_{J=0} = -\frac{L_{21}}{L_{22}}.\qquad(2.26)$$

Fourth, the streaming current is an electric current caused by a bulk flow of material across the membrane when there is no electrical potential drop. From. Eqs. 2.21 and 22:

$$\left(\frac{I}{J}\right)_{\Delta E=0} = \frac{L_{12}}{L_{22}}.\qquad(2.27)$$

Using the Onsager reciprocal relationship, Eq. 2.23, we can combine Eqs. 2.24 to 2.27 into the following two equations:

$$\left(\frac{\Delta E}{\Delta P}\right)_{I=0} = -\left(\frac{J}{I}\right)_{\Delta P=0},\qquad(2.28)$$

$$\left(\frac{\Delta P}{\Delta E}\right)_{J=0} = -\left(\frac{I}{J}\right)_{\Delta E=0}.\qquad(2.29)$$

The above equations are known as Saxén's relations [1], and these have been verified experimentally [3].

References

1. U. Saxén, *Ann. Phys. Chem.*, **47**, 46 (1892).
2. S. R. DeGroot, *Thermodynamics of Irreversible Processes*, North-Holland, Amsterdam, 1961, p. 186.
3. H. Freundlich, *Colloid and Capillary Chemistry*, Dutton, New York, 1922, p. 246.

Chapter III

MECHANISMS OF MEMBRANE TRANSPORT

The mechanisms involved in the variety of possible membrane processes fall into categories of basic transport phenomena.

Thus, the scope of this chapter is a presentation of definitions for the specific processes and the delineation of appropriate mass-transfer models. This should clarify the involved processes and provide pertinent distinctions, for instance, between reverse osmosis and ultrafiltration.

1 DIFFUSION

Diffusion is a universal phenomenon by which matter is transported from one point to another under a concentration gradient. Whenever there is a concentration difference of a species, between two points, the random motions of molecules yield a net transfer of that species from the region of higher concentration to that of lower concentration.

The mathematical theory of diffusion is based on the fundamental principle of nonequilibrium thermodynamics, that is, diffusional flux is proportional to the concentration gradient. In his original papers [1, 2] Fick stated that he could apply Fourier's law for heat conduction and that of Ohm for the electric conduction to diffusional processes. This unified view of transport phenomena of various physical entities was later generalized by Onsager [3] in his formulation of nonequilibrium thermodynamics. The differential form of Fick's first Law of diffusion is

$$\frac{F}{S} = - D \frac{\partial C}{\partial X} .$$
(3.1)

This equation should be recognized as a special case of the general linear relationship Eq. 2.16. For the purpose of our discussions of membrane diffusion, we can safely avoid the complicated vector notations, since the transport will be usually unidirectional, and membranes are relatively thin.

In the case of unsteady-state diffusion, Fick's second law applies:

$$\frac{\partial C}{\partial t} = \frac{\partial}{\partial x} \left(D \frac{\partial C}{\partial x} \right).$$
(3.2)

This equation describes the accumulation of material at a given point as a function of time. The above two laws are the most fundamental equations of diffusion. Now, let us examine how these laws apply to our membrane system. Despite the fact that numerous books and articles on transport phenomena have appeared in recent years, most people still do not recognize and differentiate the multiplicity of material fluxes. Most authors take it for granted that Eqs. 3.1 and 3.2 can be used to interpret their permeation data through membranes without suspecting that this may not be justified. Even though the errors might be small, it is in principle wrong to use the above equations directly for the membrane system. Therefore, we shall attempt here to clarify the situation and to present a correct method.

As stated before, diffusion is mass transfer due to molecular motions. Bulk motion cannot be called diffusion. Only "diffusional" motions cause diffusion. What, then, is the *diffusional* motion? We can define precisely what the diffusional motions are in terms of molecular velocities, but first let us limit our discussion to only one specific kind of unit: molar quantities. In general, there can be a number of other units and combinations thereof [3]. For instance, in defining concentration, one can have the mass concentration g/cc; the molar concentration, (g)(mole)/cc; or the mass fraction, g/g. Similarly, there can be many different kinds of fluxes.

Using the molar concentration, we may define a local molar average velocity v^* as

$$v^* = \frac{\sum\limits_i C_i v_i}{\sum\limits_i C_i} \quad , \tag{3.3}$$

where v_i is the velocity of the ith species with respect to a stationary coordinate system. This velocity is not the velocity of an individual molecule; it refers to the local average velocity of the ith species.

Evidently, then

$$N_i = C_i v_i \quad . \tag{3.4}$$

This is the molar flux of the ith species with respect to stationary coordinates, but it may not be the diffusional flux in general. If v^* is nonvanishing, then Cv^* is the local flow rate per unit cross-sectional area. Thus, we can define the "diffusional" flux for the ith species as

$$J_i^* = C_i \left(v_i - v^* \right) \quad , \tag{3.5}$$

which indicates the diffusional flux with respect to the local molar average velocity v^*. Using the definition of N_i, Eq. 3.5 becomes

$$J_i^* = N_i - C_i v^* \quad . \tag{3.6}$$

From this expression, it is obvious that the molar diffusional flux is the difference between the molar flux with respect to stationary coordinates and the bulk flow of the ith species due to the local molar flux of the mixture.

It is important to note that the diffusional flux in Fick's first law should be identified as the molar diffusional flux as defined above:

$$\frac{F_i}{S} = J_i^* \quad . \tag{3.7}$$

However, in many cases it can be approximated with N_i because v^* is negligibly small. In those cases, from Eqs. 3.1, 3.6, and 3.7 we obtain

$$N_i = J_i^* = - D \frac{\partial C_i}{\partial x} \; . \tag{3.8}$$

This is the equation that most people use in membrane permeation. They interpret the diffusional flux as being the permeation flux. Even though this may be a good approximation for many systems, we now clearly know that Eq. 3.8 is not valid in general. The error involved in this approximation is the second term of Eq. 3.6, that is, $C_i v^*$. When the permeation flux is very large, this may not be negligible.

To illustrate this point, let us consider a single-component system. There is only one species A permeating through the membrane. However, the membrane will act as a second component in our diffusion system. Using the subscript M for the membrane, Eq. 3.6 becomes

$$J_A^* = N_A - C_A v^* = N_A - \frac{C_A}{C} (N_A + N_M); \tag{3.9}$$

since the membrane does not move,

$$N_M = 0 \; . \tag{3.10}$$

Then, Eq. 3.9 is reduced to

$$J_A^* = \left(1 - \frac{C_A}{C}\right) N_A \tag{3.11}$$

or

$$N_A = \frac{C}{C_M} J_A^* = - \frac{C}{C_M} D \frac{\partial C_A}{\partial x} \; . \tag{3.12}$$

Re-expressing Eq. 3.12 yields

$$N_A = - D \frac{\partial C_A}{\partial x} - \frac{D}{2C_M} \frac{\partial C_A^2}{\partial x} \; . \tag{3.13}$$

The permeation flux of A observed in the laboratory is N_A, not J_A^*. Thus in the general case any permeation data obtained experimentally should be correlated by using Eq. 3.13 and not Eq. 3.8.

The difference between Eq. 3.13 and Eq. 3.8 is the second term in Eq. 3.13, which may be quite small in many systems. This term is second order with respect to C_A. Thus, we may expect, in general, a nonlinear relationship between the *observed* permeation flux and the overall concentration gradient for membrane processes. An important point is that the system still obeys so-called Fickian diffusion.

Similarly, Fick's second law should be re-examined for membrane systems. A material balance for component A gives

$$\frac{\partial C_A}{\partial t} + \frac{\partial N_A}{\partial x} = R_A, \tag{3.14}$$

where R_A is the molar rate of A produced per unit volume of the system due to chemical reaction. Substituting Eq. 3.13 into 3.14 for a system with *no* chemical reaction yields

$$\frac{\partial C_A}{\partial t} = \frac{\partial}{\partial x}\left(D\,\frac{\partial C_A}{\partial x}\right) + \frac{\partial}{\partial x}\left(D\,\frac{C_A}{C_M}\,\frac{\partial C_A}{\partial x}\right). \tag{3.15}$$

Comparing this equation with Eq. 3.2, we note that the new term, which is nonlinear with respect to C_A, appears in our modified second law of diffusion. It is still a Fickian type of diffusion. These modifications were necessary because the permeation flux N_i, which is the quantity measured in the laboratory, is not the diffusion flux $J_i{}^*$. The material flux going through a membrane should never be confused with the true diffusional flux.

The significance of the preceding treatment is as follows:

For *membrane diffusion* processes of all types, the correct transport equations are Eqs. 3.13 and 3.15. They differ from Fick's basic diffusion equations 3.1 and 3.2 in that second terms are added that are due to the presence of the membrane. In many instances, particularly in gaseous diffusion processes, the effect of the second term may be negligible.

2 HYDRODYNAMIC FLOW

When membranes are porous and the fluid flow is laminar, a simple hydro-dynamic theory applies. In the case of gas-phase flow, the pores must be large enough compared with the mean free path of gas molecules to ensure viscous flow. For many membranes, the structures of pores are very complicated and frequently not known. Various models have been used with different degrees of success. Numerous pore-structure factors are introduced, such as average pore diameter, porosity, equivalent hydraulic radius, average length of capillary, particle diameter, pore size distribution, tortuosity, and specific surface area. We shall begin with the phenomenological discussions of hydrodynamic flow.

Darcy's Law

This basic law governing the flow of fluid through porous media was originally developed by Darcy [4]. It states that the flow rate is directly proportional to the pressure gradient causing flow. The linear relationship is once again an example of the general phenomenological equations of fluxes and forces in nonequilibrium thermodynamics (see Eq. 2.16), so that

$$\frac{V}{St} = \frac{K}{\mu} \frac{\Delta P}{l} . \tag{3.16}$$

Here the flow coefficient is divided by the viscosity of the fluid, and K is called the permeability. This definition of permeability differs from our earlier definition given by Eqs. 2.1 and 2.2. However, this practice is widespread in petroleum technology and in soil science.

The viscosity of a fluid is a measure of internal friction between fluid laminae flowing at different velocities. If a fluid undergoes laminar flow, this friction yields shear forces. When a flowing fluid contacts a solid surface, it adheres to the surface, resulting in a zero fluid velocity. As a consequence of the viscosity and the adhering property of the fluid, the solid surface experiences a drag force. The viscous resistance is a counter force to this drag.

Darcy's law expresses that the flow resistance is due to viscous drag, and the permeability K contains all the properties of the porous medium. This definition of permeability is intended to separate the fluid property, that is, the viscosity, from the pore-structure properties in the overall flow coefficient. The unit of Darcy's permeability is a *darcy*, which is a flow of 1 cc/(sec)(cm^2) with a pressure gradient of 1 atm/(cm) for a fluid with 1 cP viscosity. The explicit expression of permeability as a function of pure-structure parameters will depend on the specific pure models.

Capillary Model (Hagen-Poiseuille Equation)

When a porous membrane consists of straight cylindrical capillaries of equal size, the Hagen-Poiseuille equation should apply directly to describe the flow rate, and

$$\frac{V}{t} = \frac{nS\pi r^4}{8\mu} \frac{\Delta P}{l} . \tag{3.17}$$

In terms of molar units:

$$F = \frac{V}{t} \frac{P}{RT} = \frac{nS\pi r^4 P}{8\mu RT} \frac{\Delta P}{l} . \tag{3.18}$$

Comparing Eq. 3.16 and Eq. 3.17, we can obtain the Darcy permeability for the capillary model. Noting the porosity for the capillary membrane to be $\epsilon = n\pi r^2$, the permeability becomes

$$K = \frac{r^2 \epsilon}{8} ,$$ (3.19)

which contains only the properties of pores.

Kozeny-Carman Equation

For a porous membrane consisting of a capillary bundle with noncircular cross section, Kozeny [5] developed a hydrodynamic equation. He assumed that the path of fluid flow would be tortuous. Using the concept of the hydraulic radius, Kozeny derived the following equation:

$$\frac{V}{St} = \frac{\epsilon^3}{k'(1 - \epsilon)^2 S_0^2 \, \mu} \frac{\Delta P}{l} ,$$ (3.20)

where k' is a dimensionless constant, dependent only on the pore structure. A detailed discussion of this constant is given by Carman [6, 7].

Comparing Eq. 3.16 and Eq. 3.20, the Darcy permeability becomes

$$K = \frac{\epsilon^3}{k'(1 - \epsilon)^2 S_0^2} .$$ (3.21)

Numerous modifications of the original Kozeny theory have been developed [6-11], and all of these present discussions of the various structural theories of permeability for different pore models.

For a laminar flow of macroscopic scale, the Kozeny equation is widely used from filtration problems to flow through packed beds. Thus, if a membrane is made of compressed powder or other filter materials and the flow mechanism can be assumed as laminar, we may apply the Kozeny-Carman equation to calculate the flow rate.

3 DIALYSIS

Dialysis is basically a diffusion process. While diffusion refers to the phenomenon itself, dialysis ascribes the separation of substances in solution by means of their unequal diffusion rates through membranes. Thus, dialysis is achieved by imposing a concentration gradient across the membrane. Historically, dialysis is the first membrane process of separation. Graham [12] was successful in separating a colloid solution using a device called a dialyzer. It was a simple design, consisting of a membrane covering the open end of a cylinder immersed in a container of water. The cylinder contained the solution to be dialyzed. The dialyzing cell was kept in place until the desired separation was achieved.

Since then, many modifications have been introduced to improve the efficiency of dialyzers. However, the basic feature of the unit remained virtually the same. The arrangement of membranes and their supporting system have been developed greatly. Mechanical agitation was introduced to reduce the resistances of mass transfer on either or both sides of the membrane.

Today, the range of available dialyzers is very wide. From small, laboratory-scale dialysis cells to large, commercial dialyzers, many different types of assemblies can be obtained. Dialysis cells (Chapter XII) for laboratories are often made of rigid acrylic plastic with polished surfaces for easy sealing and visibility from all angles. An equilibrium dialysis cell consists of two inter-changeable compartments, which are separated by a semipermeable membrane. A continuous-flow dialysis cell makes it possible for a small volume of substance to be dialyzed into a large volume of buffer solution. The large-scale industrial dialyzers were developed along three lines: the tank type, the tube type, and the filter-press type. Some of these commercial dialyzers have very complicated structures.

The governing equation in dialysis is the diffusion equation:

$$W = D_0 S \frac{C_1 - C_2}{l} \quad . \tag{3.22}$$

If the resistances of the stagnant boundary layers on both sides of the membrane contribute significantly to the total resistance to transfer, the overall dialysis coefficient can be separated into

$$\frac{l}{D_0} = \frac{l}{D_m} + \frac{l}{D_1} + \frac{l_2}{D_2} \quad . \tag{3.23}$$

where D_m is the membrane diffusivity, D_1, D_2 are diffusivities of boundary layers, and l_1, l_2 are the thicknesses of boundary layers. Depending on the relative magnitudes of l_1, l_2, and D_1, D_2, D_m, Eq. 3.23 will indicate which term should be the rate-determining factor in dialysis.

4 OSMOSIS

Historically, the term osmosis was originally used to represent the transport of solutes as well as solvents across membranes. Two different words were employed to distinguish the two different transport phenomena by Dutrochet [13] in 1823. The transfer of solvent was called endosmosis: we now call it just osmosis; and the passage of some solute together with solvent was termed exosmosis: today we call it dialysis or diffusion.

Osmosis is also a diffusion process. However, the term osmosis is confined to a process in which there is a flow of solvent only. As an example, a process is called osmosis if only the solvent permeates through the membrane from the less concentrated side to the more concentrated solution. If any of the solutes are transferred, then it is called dialysis. When the high-concentration side is sealed to prevent the transfer of solvent, a hydrostatic pressure will build up on that side of the membrane, and this pressure is known as the osmotic pressure. It is convenient to use this osmotic pressure as a driving force in describing any osmosis process.

In 1885, van't Hoff showed that the osmotic pressure π_A in dilute solutions is related to the concentration of solute C_A by

$$\pi_A = RTC_A \ , \tag{3.24}$$

However, in many cases, even for a dilute solution, a small correction is needed to compensate for the slight deviation from ideality of the solution, so that

$$\pi_A = aRTC_A \ , \tag{3.24a}$$

where the osmotic coefficient a is usually estimated from vapor-pressure data or from the freezing-point depression of the solution involved. For solutions of higher concentrations of solute, the following general equation can be used to evaluate the osmotic pressure:

$$\pi_A = \frac{RT}{v_B} \int_0^{C_B} \left(\frac{\partial \ln a_B}{\partial C_B} \right)_{P,T} dC_A \ , \tag{3.24b}$$

where v_B and a_B are the partial molar volume and the activity of the solvent, respectively. When the simple equation (3.24a) is obeyed, osmosis is said to be normal. If the observed osmosis is either greater or less than that predicted by van't Hoff's law, Eq. 3.24, it is called anomalous osmosis.

5 ANOMALOUS OSMOSIS

The osmotic phenomena for systems of charged membranes and electrolytes as solutes may be quite different from the normal osmosis discussed above. When the electrolyte solute transfers from one side of the charged membrane to another, the solvent also transfers in the opposite direction. The diffusion of solute causes an uneven electric field across the membrane, thus resulting in an added driving force for ions. If this yields an enhanced solvent transfer, it is called positive anomalous osmosis. On the contrary, if the transfer of solvent is less than predicted by Eq. 3.24, it is called negative anomalous osmosis.

Both positive and negative anomlaous osmosis have been observed for many ion-exchange membranes [14-17]. Grim and Sollner [17] have developed a method by which they could isolate the anomalous part of total osmosis. The transport of solvent across charged membranes consists of both a normal and an anomalous osmotic component. When membranes are electrically neutral, there is no anomalous osmosis. Thus, if the membrane can be charged and discharged reversibly, the normal component of osmosis can be measured with the membrane in the electroneutral state. Then the difference between the total osmosis and the normal osmosis will be the net portion of the anomalous osmosis. By changing the pH of the various electrolyte solutions, different degrees of anomalous osmosis were observed.

6 REVERSE OSMOSIS

When a hydrostatic pressure exceeding the osmotic pressure is applied to a system that is separated by a semipermeable membrane, the solvent is forced to flow from the high-concentration side to the low-concentration side. Since the flow direction is the reverse of ordinary osmosis, this process is called reverse osmosis. Ordinarily, reverse osmosis requires tremendously high pressure to yield any appreciable amount of solvent transfer. The membrane used in the early studies of Reid and co-workers [18, 19] were conventional cellulosic types. Considerable difficulties were encountered with membrane degradation, and although some progress was made to improve the life of membranes and to some extent flow rates through the membranes, the overall improvement was not great enough to better the economic competitiveness of the process.

The vastly improved cellulose acetate membrane developed by Loeb and Sourirajan [20, 21] resulted in a surge of the reverse-osmosis process for desalination of brime. The Loeb-Sourirajan-type porous cellulose acetate membranes are made by incorporating a suitable water-soluble additive in the film casting solution and leaching out the additive with water after casting. This technique created a very porous spongelike membrane structure covered with a thin surface layer, which has the preferential permeability of water to solute. The flow mechanism proposed by Sourirajan [20] to explain the observed separation is based on the concept of the so-called critical pore diameter. He assumed a microporous structure for the surface layer of membrane. The concentrated water due to a negative adsorption of the solute at the membrane interface then flows through the micropores under a pressure difference. If the pore diameter is large, the flow rate will be high, but solute separation will be low since the bulk solution will also go through the pores. When the pores are small, the separation will be great, but the total flow rate will be small. Thus, there is an optimum pore size that is taken as the critical pore diameter. This is about twice the thickness of the interfacial sorbed water layer.

Lonsdale et al. [22] and Banks and Sharples [23] are reluctant to accept the concept of a porous surface layer, and the question of the actual mechanism in reverse osmosis is likely to remain a controversial subject for some time to come.

Regardless of the flow mechanism or separation mechanism, one can write the transport equations for solvent and solute based on the phenomenological equations. For the solute, the diffusion equation applies:

$$N_A = D_A k \frac{C_A' - C_A''}{l} \quad , \qquad (3.25)$$

where k is Henry's constant and C_A', C_A'' are concentrations of solute in upstream and downstream, respectively. For the solvent, a permeation equation holds:

$$N_B = \frac{Q}{l} \left[\Delta P - \sum_{i \neq B} \pi_i \right] , \qquad (3.26)$$

where π_i is the osmotic pressure due to solute i, and ΔP is the applied hydrostatic pressure difference.

An extensive treatment of reverse osmosis is given by Sourirajan in his book *Reverse Osmosis* [24].

7 PIEZODIALYSIS

At times the term *piezodialysis* (pressure dialysis) is used to describe a phenomenon where some solutes permeate a membrane preferentially by virtue of a pressure difference applied across the membrane. In this case, the pressure is equivalent to the electrical potential difference in electrodialysis. Both these processes rely primarily on the external driving force to maintain the steady state of separation. The piezodialysis should not be confused with reverse osmosis or ultrafiltration. The difference is in the moving component. In piezodialysis, solutes permeate through the membrane, whereas in reverse osmosis, solvent is one component moving across the membrane.

In recent developments of the water-desalination process, piezodialysis has been reported as a successful new process. As opposed to reverse osmosis, a concentrated salt solution is permeated through the membrane, leaving a dilute product stream. The types of membrane used in piezodialysis are mostly mosaic membranes. These membranes possess heterogeneous composite structures, such as cationic and anionic sequences to form charge-mosaic membranes. Synthesis of block copolymers promises to be a future method of preparing such membranes.

8 ULTRAFILTRATION

The term *ultrafiltration* was first used by Bechhold [25] in 1907. Ultra-filtration is a separation process in which large molecules or colloidal particles are filtered from the solution by means of suitable membranes. In the literature, however, ultrafiltration is frequently used as a synonym for reverse osmosis. In fact, there are several interchangeable names used to represent the process in question: ultrafiltration, hyperfiltration, molecular filtration, microfiltration, and reverse osmosis. The usage of different terms depends largely on the particular field of study and authors. The difference between ultrafiltration and reverse osmosis is obscure and arbitrary to some extent. In some cases, they overlap each other. Nevertheless, it is possible and convenient to distinguish these two processes. Michaels [26] proposed to use "reverse osmosis" for membrane separations involving solutes whose molecular dimensions are within one order of magnitude of those of the solvent, and to use "ultrafiltration" to describe spearations involving solutes of molecular dimensions greater than 10 solvent molecular diameters and below the limit of resolution of the optical microscope (0.5 μ). There are some common characteristics for these two processes. For both processes, the hydrostatic-pressure forces the solvent to permeate a membrane, keeping solute on one side of the membrane. Two different terms are used just because the sizes of particles to be separated are different. Both processes require external application of energy in the form of hydraulic pressure to accomplish separation of components, whereas dialysis or a regular diffusional process does not require such force, since the separation takes place spontaneously.

As to the transport equation for ultrafiltration, essentially the same equation, Eq. 3.26, as for the reverse osmosis will apply when there is an osmotic pressure difference. If the osmotic pressure does not exist or cannot be estimated, the hydrodynamic equation, Eq. 3.16, should be used to calculate the flux of solvent across the membrane.

9   ELECTRODIALYSIS

Electrodialysis is a process in which solute ions move across membranes by application of an electrical field. Thus electrodialysis requires an external energy to maintain the separation process in a form of electrical potential. The application of electrical energy will drive electrolytes from a dilute to a more concentrated solution. Although electrodialysis was started as a modification of ordinary dialysis by adding a couple of electrodes [27], the two processes are distinctly different in many ways. Ordinary dialysis depends on the concentration gradient; therefore the flow of solute is always from the more concentrated to less concentrated solution. However, since electrodialysis utilizes electrical energy as a driving force, the direction of transport can be either way, depending on choice. In ordinary dialysis, the concentration gradient may diminish gradually as a result of the mass transport. In electrodialysis, the external electrical potential can be easily maintained until the desired degree of separation is achieved. The types of membranes used in the two processes are also different. Usually, ion-selective membranes are employed in electrodialysis, although ordinary dialysis does not use ion-exchange membranes. As a result, the two processes differ in membrane requirements, equipment design, and objectives.

Since electrodialysis involves two separate driving forces, that is, concentration difference and electrical potential difference, the net transport of an ion in question can be expressed as the sum of two flow rates:

$$\frac{V}{St} = \frac{b_1}{R'} \Delta E \pm \frac{b_2 D \, \Delta C}{l} \, , \tag{3.27}$$

where b_1 and b_2 are proportionality constants, ΔE is the electrical potential difference, and R' is the electric resistance of the cell. The first term in Eq. 3.27 represents the ion flux driven by the electrical potential, which is directly proportional to the electric current across the membrane. The second term is the flow rate due to simple diffusion under a concentration gradient. If the applied electrical potential gradient is in the same direction as that of the concentration gradient, the plus sign should be used. But usually the electrical potential is applied opposite to the concentration gradient in order to produce a more concentrated solution in the practice of electrodialysis processes. If this is the case, the minus sign should be taken, because the movements of ions by the two forces are in opposite directions.

10 ELECTRO—OSMOSIS (ELECTROENDOSMOSIS)

Electro-osmosis is the transport of solvent due to an electric current when there is no applied hydrostatic pressure across the membrane. This is the reverse phenomenon of the streaming potential as mentioned in Chapter II. Electro-osmosis differs from electrodialysis in its moving components. Both processes occur under an electrical field, but the passage of solvent is called electro-osmosis whereas the transfer of solute ions is named electrodialysis. Therefore, in many cases, when an electrical voltage is applied across the membrane, it is likely that both processes may take place simultaneously.

The volumetric flow rate per unit area is given by

$$\frac{V}{St} = \frac{E'ZI}{4\pi\mu\lambda} \, . \tag{3.28}$$

11 DONNAN EFFECT

When the membrane is ion selective and the solution contains several different free ions, an unequal distribution of ions will result across the membrane.

Consequently, osmotic and electrical potential differences, called "Donnan potential" or "membrane potential," will be created. This membrane potential was first measured by Donnan [28], and the results agreed with the simple theory for ideal solutions. There are usually two cases in which the Donnan effect may occur. One is with ion-exchange membranes placed in a solution of strong electrolytes. The other is a system of colloidal solution. Normally, colloidal electrolytes are too large to permeate through a membrane, thus the membrane will act like an ion-selective membrane by passing only small ordinary ions.

The simplest, but most elucidating illustration of Donnan equilibrium is given by Taylor and Glasstone [29], and here it will be repeated to explain how Donnan equilibrium is attained. An aqueous solution of NaCl is separated by a membrane into two compartments as shown in Fig. 3.1. The membrane is

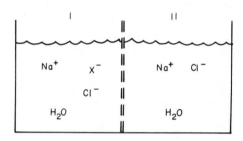

Fig. 3.1 Diagram of Donnan equilibrium.

permeable to both ions Na^+ and Cl^-. Now add another salt of sodium NaX, which is also completely dissociated into Na^+ and X^-, to one side of the membrane only. Then, the common ions Na^+ and Cl^- will permeate through the membrane to the other side until an equilitrium is obtained, although the noncommon ion X^-, being impermeable, will remain in the original compartment. The condition of the thermodynamic equilibrium requires that the activities of any diffusible component should be equal on both sides of the membrane:

$$a^I_{H_2O} = a^{II}_{H_2O} \quad , \tag{3.29}$$

$$a^I_{NaCl} = a^{II}_{NaCl} = a^I_{Na^+} \, a^I_{Cl^-} = a^{II}_{Na^+} \, a^{II}_{Cl^-} . \tag{3.30}$$

Assuming the behavior of ideal solutions, we can replace the activities with concentrations:

$$c^{I}_{Na^{+}} \, c^{II}_{Cl^{-}} = c^{II}_{Na^{+}} \, c^{II}_{Cl^{-}} \quad . \tag{3.31}$$

The maintenance of electroneutrality is also required so that

$$c^{I}_{Na^{+}} = c^{I}_{Cl^{-}} + c^{I}_{X^{-}} \quad , \tag{3.32}$$

$$c^{II}_{Na^{+}} = c^{II}_{Cl^{-}} \quad . \tag{3.33}$$

Combining the above equations, the Donnan ratio can be obtained:

$$\left(\frac{c^{II}_{NaCl}}{c^{I}_{NaCl}} \right)^{2} = 1 + \frac{c^{I}_{NaX}}{c^{I}_{NaCl}} \tag{3.34}$$

From this equation, it is obvious that an unequal distribution of the diffusible salt is achieved at the equilibrium state. The ratio of the NaCl concentrations in the two compartments is always greater than 1 because the concentrations cannot be negative.

12 KNUDSEN FLOW

The viscous flow governed by the Poiseuille formula is limited in the pressures and temperatures such that the mean free path of fluid molecules is very small compared to the diameter of the pores. If the mean free path becomes comparable or larger than the pore size, the fundamental concept of viscous flow, that is, the continuum concept, breaks down. In the continuum regime, the collision frequency between gas molecules exceeds greatly that of gas to wall. However, when the mean free path is much larger than the pore diameter, the collisions between gas molecules become much fewer than the collisions between gas molecules and the wall. This situation is known as a rarefied gas, and the flow in this limit is usually referred to as "free-molecule diffusion" or "Knudsen flow."

The original work of Knudsen in 1909 [30] contains his theory of free-molecule diffusion as well as experimental data of several gases. A typical example is given in Fig. 3.2 for the flow of CO_2 through a glass capillary (Capillary No. 4). In the high-pressure range the flow obeys the Poiseuille equation very closely. As the pressure becomes lower, the data points deviate from the straight line. After going through a minimum point, further lowering

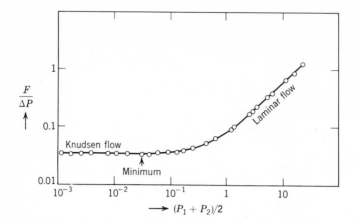

Fig. 3.2 Knudsen's experimental data for carbon dioxide flow in Capillary No. [30].

of pressure causes the flow to increase slightly until it approaches asymptotically a limiting value. In the extremely low pressure range, the flow becomes independent of pressure, which is the true Knudsen regime. In the intermediate pressure region, no satisfactory theory can explain the flow data. Based on the kinetic theory of gases, Knudsen derived the following equation for free-molecule flow through a long circular capillary:

$$F = \frac{8\pi r^3}{3(2\pi RMT)^{1/2}} \frac{P_1 - P_2}{l} \quad . \tag{3.35}$$

where r is the radius of the capillary and M is the molecular weight of the gas. Basically, the same equation applies to all microporous membranes with a little modification of a geometric factor G.

$$F = \frac{GS}{(2\pi RMT)^{1/2}} \frac{P_1 - P_2}{l} \quad . \tag{3.36}$$

Many authors have attempted to express the geometric factor G in terms of more explicit quantities, such as tortuosity and porosity, based on different pore models.

For short capillaries, which can be considered as the model of actual porous materials, Clausing [31] developed the methods of calculating the flow as a function of the ratio of capillary length over capillary radius. At the limit of zero length, or a hole in an infinitely thin membrane, the flow equation is reduced to

$$F = \frac{S}{(2\pi RMT)^{\frac{1}{2}}} \; (P_1 - P_2). \tag{3.37}$$

This limiting flow is known as molecular effusion or Knudsen effusion. The above equation is simply the difference between two impinging molecular beams from both sides of the membrane through a cross section of S.

Pollard and Present [32] also performed similar computations, but they extended their theory to the transition region, where the mean free path is comparable to the pore diameter. Their theory explains the minimum occurrence in the flow rate for a long capillary and its absence for a short capillary and for porous media. This fact has been well established experimentally for a long time. In the intermediate pressure range, the so-called "slip" flow is added customarily to the Poiseuille formula, so that essentially the Poiseuille equation can still be used with a slight modification. However, it should be pointed out that this approach is not rigorous.

There are a number of qualitative differences between Poiseuille flow and Knudsen flow. The Poiseuille formula, Eq. 3.18, contains the fourth power of the radius whereas the Knudsen equation, Eq. 3.35, contains the cube of the radius. The Poiseuille flow is inversely proportional to the viscosity of the fluid; however, the Knudsen equation does not involve the viscosity because there is no concept of viscosity in the rarefied gas (or Knudsen regime). The viscous flow increases linearly as the pressure drop increases: this point is identical with Knudsen flow; however, the Poiseuille equation is also dependent upon the average pressure whereas the Knudsen flow is not. When a mixture of gases flows through a tube according to the Poiseuille equation, each component cannot behave independently because of the molecular chaos. Therefore, the flow will be as if one kind of gas is going through the capillary with the average viscosity. There is never a separation. However, in the Knudsen regime, the component gases will flow through the capillary independently of each other. Thus, a separation takes place due to the difference in their molecular weight. This is the principle involved in the isotope separation of ^{235}U by means of the gaseous diffusion process.

In porous media, when the flow is in the free-molecule regime, the surface diffusion is also present. The relative magnitude of surface flow with respect to the total flow is determined by many factors. A detailed discussion will follow in a later chapter.

13 THERMAL EFFECTS

So far our discussions have been limited to the phenomena taking place in isothermal systems. Imposing a temperature gradient across membranes, we can also expect a mass-transfer process based on the theory of nonequilibrium thermodynamics, Eq. 2.16. When the temperature gradient causes diffusion, this process is known as thermal diffusion or the "Soret effect." If the temperature gradient is responsible for the flow of solvent, it is called thermo-osmosis. Yet, if the flow is in the Knudsen regime, it is called thermal effusion.

Denbigh [33] was the first to demonstrate experimentally that either CO_2 or H_2 passes through a nonporous rubber membrane under a temperature gradient. The theoretical interpretation of this coupling phenomenon is well presented by Katchalsky [34]. In the liquid phase, Rastogi and co-workers [35] reported measurements of the thermo-osmotic permeability of a Du Pont 600 cellophane membrane using water. However, Carr and Sollner [36] found no transport of water through a collodion membrane. A speculative discussion was presented by Spanner [37] on the significance of thermo-osmosis for biological membranes.

Thermal effusion is a much more established phenomenon, both theoretically and experimentally. Since it is rarefied gas flow, the effusion equation can be obtained from Eq. 3.37 as

$$F = \frac{S}{(2\pi RM)^{\frac{1}{2}}} \left(\frac{P_1}{T_1^{\frac{1}{2}}} - \frac{P_2}{T_2^{\frac{1}{2}}} \right). \tag{3.38}$$

At equilibrium, when there is no flow, the pressures at both sides of the membrane are related by

$$\frac{P_1}{P_2} = \left(\frac{T_1}{T_2} \right)^{\frac{1}{2}}. \tag{3.39}$$

As in the case of Knudsen flow for an isothermal system, a separation is also possible for two different gases if their molecular weights are different.

The above thermal phenomena do not play a significant role in any membrane-separation process, since the magnitudes of the coupled phenomena are usually quite small compared to the conjugated phenomena (principal phenomena). However, they may have some applications in an unusual environment.

14 FACILITATED DIFFUSION

All the phenomena hitherto described have been pure mass-transport processes under a concentration or pressure gradient. When the simple diffusion process is coupled with chemical reactions, the net transport rate may be greatly affected. Usually, it is facilitated. Thus, the name *facilitated diffusion* is given to a system in which the rate of transport is increased from the expected value based on the simple diffusion theory alone. However, facilitated diffusion cannot change the final equilibrium that may be reached by simple diffusion.

It only speeds up the rate of attainment of the equilibrium. Another characteristic feature of this facilitated diffusion is the fact that the rate of permeation increases proportionally with increase of concentration gradient only to a limit, beyond which it varies little with further increase of the driving force. This limiting permeation rate suggests several possible mechanisms. A "carrier" system is one, and an augmenting chemical reaction is the other.

The carrier model assumes a carrier substance inside the membrane that shuttles back and forth between the two interfaces carrying the permeant molecules. A good example of this model would be the oxygen transport via hemoglobin molecules studied by Scholander [38]. He observed more than eightfold increase in oxygen transfer rate when hemoglobin molecules are present. This facilitated diffusion was explained as being based on the concept of a hemoglobin carrier model. The oxygen molecules are transmitted from one hemoglobin molecule to the next in a chain. As the pressure of air was changed, the rate of nitrogen diffusion increased proportionally, but the oxygen-transport rate remained constant over a wide pressure range.

The second model assumes chemical reactions that are taking place inside the membrane, and the reaction products actively participate in the diffusion process. A simplified version of this model will be given here for a binary system. A membrane containing a suitable solution divides the system into two parts. One side of the membrane is filled with species A and the other with species B initially. Both A and B will diffuse through the membrane according to their own concentration gradients. Inside the membrane, A and B react to form AB, which will also diffuse within the membrane due to its own concentration gradient:

$$A + B \underset{k_2}{\overset{k_1}{\rightleftharpoons}} AB \; . \tag{3.40}$$

The schematic picture is shown in Fig. 3.3. If the equilibrium conditions for the above reaction are such that the concentration gradient of the product AB is in the same direction as the concentration gradient of A, the diffusion of AB will take place in the same direction as the diffusion of A. When the product AB reaches the other side of the membrane interface, it will decompose into A and B, thus releasing extra A out of the membrane into the B phase. Since the total rate of A transport through the membrane consists of the rate of A diffusion and the rate of AB diffusion, the net transport of A is facilitated. It is also easy to explain the limiting rate of facilitated diffusion that does not depend on the increase of the concentration gradient A based on this model. If the chemical reaction is the rate-determining step, the overall transport rate of A will be determined by the reaction rate and the capacity of the reaction in

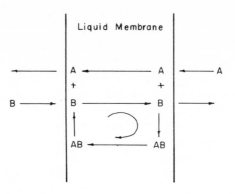

Fig. 3.3 Diagrammatic scheme of facilitated diffusion.

the membrane. Therefore, the increase of the concentration A produces little effect on the overall rate of A transport.

A detailed theoretical and experimental investigation of nitric oxide diffusion through a liquid film of ferrous chloride solution was published by Ward [39]. The basic governing equations are three transport equations for A, B, and AB:

$$D_A \frac{d^2 C_A}{dx^2} = k_1 \, C_A \, C_B - k_2 \, C_{AB} \, , \qquad (3.41)$$

$$D_B \frac{d^2 C_B}{dx^2} = k_1 \, C_A \, C_B - k_2 \, C_{AB} \, , \qquad (3.42)$$

$$D_{AB} \frac{d^2 C_{AB}}{dx^2} = -k_1 \, C_A \, C_B + k_2 \, C_{AB} \, . \qquad (3.43)$$

With appropriate boundary conditions, the simultaneous differential equations are solved for these cases. His experimental data confirm the theoretical analysis.

The most recent publication dealing with carrier-facilitated transport, by Smith et al. [40], deals with the physical and mathematical characteristics of the process. The rather extensive analysis resulted in mathematical formulations that permit interpretation of available data, even with wide variations in operating parameters.

15 ACTIVE TRANSPORT

All the phenomena discussed above are for nonliving membranes. Mass transport takes place in the direction from high concentration to low concentration, or from high electrochemical potential to low electrochemical potential. For living membranes, such as cell membranes, the transport phenomena are generally quite complicated. In some cases, a substance may be transported through a cell membrane against a concentration gradient or electrochemical potential gradient. This does not mean that the second law of thermodynamics is violated. In addition to the diffusion process, there must be other mechanisms that cause this phenomenon by supplying energy to the system. Thus, the cell is doing work to move the substance through the cell membrane against a concentration gradient. Since the cell membrane is "actively" transporting the matter, the term "active transport" is used for such a process. Some authors use the phrase "passive transport" for all the other permeation processes that are taking place in the same direction as the concentration gradient in order to contradistinguish the two types of transport processes. This is shown in Fig. 3.4; the arrows indicate the directions of

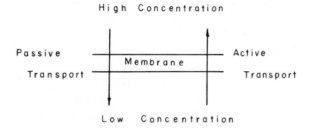

Fig 3.4 Diagrammatic scheme for active transport.

material transport. Since active transport requires extra energy released by metabolic reactions, it is often called "metabolically linked transport." A whole chapter is devoted to this subject in the excellent book *Cell Physiology* by Giese [41]. A theoretical treatment of the active transport based on the nonequilibrium thermodynamics will be found in a chapter of Katchalsky's book [34].

Numerous examples have been cited in the literature, and many different mechanisms have been suggested to explain the active transport. However, only a generalized (but simplified) scheme will be given here to show how the active transport is possible for cell membranes. A model scheme is depicted in Fig. 3.5, which is quite similar to Fig. 3.3 for the facilitated diffusion. But the main difference is in their directions of mass flow with respect to the concentration gradient. Another different aspect of the active transport model lies in the fact that the second component B does not have to be involved in the overall permeation processes, instead, it plays an indispensable role in the metabolic cycle of active transport within the cell membrane. Suppose that there is a certain enzyme E_1 near the surface of low-concentration A. This enzyme promotes the chemical reaction between A and B to yield a compound AB within the cell membrane. According to its concentration gradient, the compound AB will diffuse through the membrane to the other side, where there

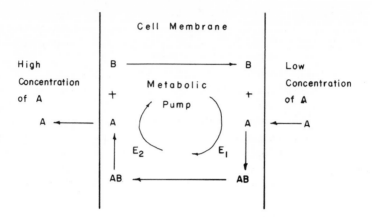

Fig. 3.5 Diagram for mechanism of active transport.

is another enzyme E_2. The enzyme E_2 does just the opposite of what E_1 does, that is, it expedites the decomposition of AB into A and B. Then the component A will diffuse out of the membrane to the high-concentration side, because there is a buildup of A locally inside the membrane. The carrier molecule B is too large to pass through the cell membrane, thus being contained within the membrane. Therefore, these metabolic reactions together with appropriate enzymes make an engine that generates power to pump the component A against its concentration gradient.

The accumulation of potassium ions in cells is a well-known example of active transport. In many plant and animal cells, the accumulation of potassium has been found up to one thousand times higher than the ambient medium. Another example is the active transport of dyes through the cells of the kidney tubule. A carefully prepared kidney tubule is placed in a dilute solution of phenol red. After awhile, the dye is accumulated in the cavity showing deep pink in color. But placing this on ice, the dye diffuses out and makes a homogeneous color. Warming up enhances the accumulation of the dye again, which strongly suggests that the metabolic supply of energy is valid in active transport.

Nomenclature

Symbol	Meaning
a	Activity
$\mathbb{B}$	External force field
C	Concentration

D	Diffusivity
$\bar{D}$	Average diffusivity
D_0	Overall diffusivity
D_m	Membrane diffusivity
E	Electrical potential
E'	Dielectric constant
F	Total flow rate
G	Geometric factor
I	Electric current
J	Generalized flux
J^*	Diffusional flux
K	Darcy permeability
L	Phenomenological coefficient
M	Molecular weight
N	Total mass flux
P	Pressure
Q'	Permeability, concentration units
Q	Permeability, pressure units
R	Gas constant
R'	Electrical resistance
R_A	Rate of A formation
S	Area
ΔS	Entropy change
S_m	Solubility in membrane
S_0	Surface per unit particle volume
T	Absolute temperature
V	Volume flow
W	Rate of solute transfer
X	Generalized force
b	Proportionality constant(s)
k	Henry's law constant
l	Dimensionless constant(s) in Eq. 3.21
n	Number of capillaries per unit area
r	Radius of capillary
r_1, r_2	Film resistances
t	Time
x	Direction coordinate
v_B	Velocity
v^*	Molar average velocity
a	Osmotic coefficient
Γ	Concentration in bulk phase
Γ^*	Equilibrium concentration

ϵ	Porosity
ζ	Zeta potential
λ	Specific conductivity
μ	Viscosity
π	$3.141\overline{6}$
π_i	Osmotic pressure, $i = A,B,\ldots$
ϕ	Rate of lost work

Subscripts

1,2	Position indicater; also phase designation
A,B,i,j	Components
M,m	Membrane property

References

1. A. Fick, *Ann. Physik Chem.*, **40**, 59 (1855).
2. A. Fick, *Phil. Mag.*, **10**, (4) 30 (1855).
3. R. B. Bird, W. E. Stewart, and E. N. Lightfoot, *Transport Phenomena*, Wiley, New York, 1960, p. 495.
4. H. P. G. Darcy, *Les Fontainer Publiques de la Ville de Dijon*, Victor Dalmont, Paris, 1856.
5. J. Kozeny, *Sitzber. Akad. Wiss. Wien. Math-naturw. Kl., Abt. IIa*, **136**, 271 (1927).
6. P. C. Carman, *Trans. Inst. Chem. Engr. (London)*, **15**, 150 (1937).
7. P. C. Carman, *Trans. Inst. Chem. Engr. (London)*, **16**, 168 (1938).
8. C. S. Brooks and W. R. Purcell, *Trans. AIME*, **195**, 289 (1952).
9. N. T. Burdine, L. S. Gournay, and P. P. Reichertz, *Trans. AIME*, **189**, 195 (1950).
10. H. P. Grace, *Chem. Eng. Progr.*, **49**, 303 (1953).
11. J. M. Coulson, *Trans. Inst. Chem. Engr. (London)*, **27**, 237 (1949).
12. T. Graham, *Phil. Trans. Roy. Soc. (London)*, **144**, 177 (1854).
13. H. Dutrochet, *Nouvelles Recherches sur l'Endosmose et l'Exosmose*, J. B. Bailliere, Paris, 1828.
14. K. Sollner, *Z. Elektrochem.*, **36**, 234 (1930).
15. K. Sollner and A. Grollman, *Z. Elektrochem.*, **83**, 274 (1932).
16. R. Schlögl, *Z. Physik Chem. (Frankfurt)*, **3**, 73 (1955).
17. E. Grim and K. Sollner, *J. Gen. Physiol.*, **40**, 887 (1957); *ibid.*, **44**, 381 (1960).
18. C. E. Reid and E. J. Breton, *J. Appl. Polymer Sci.*, **1**, 133 (1959).
19. C. J. Breton, Jr. and C. E. Reid, *A.I.Ch.E. Chem. Eng. Symp. Ser.*, **24**, 171 (1959).
20. S. Sourirajan, *Ind. Eng. Chem. Fund.*, **2**, 51 (1963).
21. S. Loeb and S. Sourirajan, *Advan. Chem. Ser.*, **38**, 117 (1963).

22. H. K. Lonsdale, U. Merten, and R. L. Riley, *J. Appl. Polymer Sci.*, **9**, 1341 (1965).
23. W. Banks and A. Sharples, *J. Appl. Chem.*, **16**, 153 (1966).
24. S. Sourirajan, *Reverse Osmosis*, Academic, New York, 1970.
25. H. Bechhold, *Z. Physik Chem.*, **60**, 257 (1907).
26. A. S. Michaels, "Ultrafiltration," in *Progress is Separation and Purification*, E. S. Perry, Ed., Wiley, New York, 1968, p. 297.
27. H. N. Morse and J. A. Pierce, *Z. Physik Chem.*, **45**, 589 (1903).
28. F. Donnan, *Z. Elektrochem.*, **17**, 572 (1911).
29. H. S. Taylor and S. Glasstone, *A Treatise on Physical Chemistry*, Van Nostrand, New York, 1951, p. 524.
30. M. Knudsen, *Ann. Physik*, **28**, 75 (1909).
31. P. Clausing, *Z. Physik*, **66**, 471 (1930); *Ann. Physik*, **12**, 961 (1932); *Physica*, **9**, 65 (1929).
32. W. G. Pollard and R. D. Present, *Phys. Rev.*, **73**, 762 (1948).
33. K. G. Denbigh, *Nature*, **163**, 60 (1949).
34. A. Katchalsky and P. F. Curran, *Nonequilibrium Thermodynamics in Biophysics*, Harvard University Press, Cambridge, Mass., 1967.
35. R. P. Rastogi, R. L. Blokhra, and R. K. Aggarwala, *Trans. Faraday Soc.*, **60**, 1386 (1964).
36. C. W. Carr and K. Sollner, *J. Electrochem. Soc.*, **109**, 616 (1962).
37. D. C. Spanner, *Symp. Soc. Exptl. Biol.*, **8**, 76 (1954).
38. P. F. Scholander, *Science*, **131**, 585 (1960).
39. W. J. Ward III, *A.I.Ch.E.J.*, **16**, 405 (1970).
40. K. A. Smith, J. H. Meldon, and C. K. Colton, *A.I.Ch.E.J.*, **19**, (1) 102 (1973).
41. A. C. Giese, *Cell Physiology*, Saunders, Philadelphia, 1968.

Chapter IV

EQUILIBRIUM RELATIONSHIPS

In the previous chapters, the fundamental concepts and the mechanisms of membrane transport were discussed. Furthermore, it was pointed out how separation becomes possible under a given driving force across the membrane. However, in the actual operation of any membrane process, there is an additional aspect of importance, that is, how the available driving forces are distributed along the membrane. In other words, it is imperative to know the interrelationship between the concentrations of permeating components on both sides of the membrane. For the sake of convenience, the word *equilibrium* will be used to denote this situation.

There exists a striking similarity between membrane processes and other well-known stage operations, such as distillation, extraction, and absorption. In these mass-transfer processes the term *equilibrium relationships* is commonly used to express the interrelationship between the concentrations in the two contacting phases. Although these equilibrium relations are not necessarily the true thermodynamic equilibria, they represent the dynamic steady-state conditions.

Two different equilibrium relations will be used throughout this book: *point equilibrium* and *stage equilibrium.*

1 POINT EQUILIBRIUM

The point equilibrium relates compositions at any point X in the reject stream and Y in the permeate stream as indicated in the diagram, Fig. 4.1. If these conditions were to be plotted against distances along the membrane, they would describe two concentration profiles.

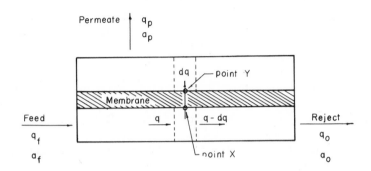

Fig. 4.1 Membrane module.

As discussed previously, the driving force for the selective transport of a species across a membrane can generally be represented by its chemical activity with appropriate standard and reference states. To be of practical use, the activity will then be replaced by a proper set of driving forces such as concentration, pressure, and electrical potential. In terms of such abstract activity a_i of a species i, the mass conservation over a differential membrane area dS yields the following permeation equation:

$$-dq_i = dS \ K_i \ (a_i^I - a_i^{II}) \ . \qquad (4.1)$$

Here, q_i is the flow rate of the species i in the reject stream and superscripts I and II refer to the reject and permeate streams. The cross-coupling (see Chapter II) is assumed to be negligible so that one single transfer coefficient K_i suffices to express the flux across the membrane. For a multicomponent system, the above equation can then be set up for each component. However, these permeation equations alone are not sufficient to give the point-equilibrium relationship in many complicated membrane processes, where there are some additional transport phenomena such as concentration polarization, longitudinal mixing, and the pressure loss along the membrane. When the equations are set up for these secondary processes, the differential material balances, represented by Eq. 4.1, can be integrated to obtain the point equilibrium.

These secondary transport phenomena are treated in detail in subsequent chapters. However, it is desirable to have them identified in principle at this point. Therefore, brief summaries of the phenomena and their respective effects upon the degree of separation are presented in the following subsections.

Secondary Transport Phenomena

The most important phenomena that influence equilibria in general, and point equilibrium in particular, are the occurrence of concentration polarization, longitudinal mixing, and pressure loss in the membrane channel.

Concentration Polarization

This refers to the buildup or depletion of some chemical species within a thin boundary layer adjacent to the membrane phase. The selective transport is primarily responsible for such a phenomenon. The polarization generally hampers the transport because of a reduction in the available driving force. In gaseous diffusion, the hampering effect is almost absent since the diffusion coefficient is very large. The polarization effect is, however, very severe in most liquid-phase processes because of the extremely small diffusion coefficient. In an osmotic process, for example, the solute to be rejected is accumulated near the membrane surface in the reject stream. Such polarization reduces the osmotic flux and enhances the solute leakage. In electrodialysis, the concentration at the membrane surface is depleted in the dialysate compartment, but accumulated in the transfer compartment. Thus, it enhances the diffusional leakage of the electrolyte solute from the transfer compartment back into the dialysate compartment, Also, the depletion of the solute greatly increases the electrical resistance, and the electric current density as well as the electric transference of the electrolyte solute are accordingly reduced. The adverse effects can be reduced by employing a high longitudinal velocity in a very thin membrane channel, or some mechanical turbulence promoters.

Longitudinal Mixing

As the reject and permeate streams are enriched in or stripped of some of their constituents, their concentrations are distributed along the membrane channel. Such distribution induces selective transport in the longitudinal direction through molecular diffusion. When some turbulence promoters are used in the channel in order to minimize the adverse effect of concentration polarization, the eddy diffusion will be mainly responsible for the selective longitudinal transport. Longitudinal dispersion always tends to produce a more uniform concentration along the channel and, therefore, in many cases will have an adverse effect on separation. In subsequent chapters, two extreme cases are discussed: complete mixing and plug flow. The longitudinal mixing is negligible in plug flow, whereas in complete mixing the concentration is uniform along the membrane channel.

Pressure Loss Along the Membrane Channel

As already mentioned, the adverse effect of concentration polarization and longitudinal mixing can be reduced by using a high stream velocity in a thin channel. When this is done the pressure loss along the channel becomes significant because of the frictional drag. Thus, when the membrane separation is achieved through the use of external pressure as a driving force, the pressure loss greatly reduces the degree of separation.

Effects on Equilibrium

The qualitative effect of such disturbances in the transport mechanism is, in general, a reduction in the overall driving forces through the membrane, resulting in concomitant changes in the point equilibrium. The spread between permeate and reject concentrations as pictured in Fig. 4.2 for a gaseous-diffusion example can become greater or less at different positions of the membrane. Just which situation prevails will depend on the occurrence of any of the secondary effects.

On an overall basis, however, the net result will be a reduction in the average concentration spread and thus, in general, a lesser degree of separation. The significant fact is therefore that the point equilibrium is affected by the primary transfer mechanism through the membrane as well as the secondary effects.

In addition, changes in operating parameters will influence the equilibrium. These parameters are factors that are under the direct control of the designer. Primarily, they are such items as the ratio of the pressures on both sides of the membrane and the ratio of permeate stream to feed stream. These matters are treated in appropriate details in Chapter XIII.

Example of Gaseous Diffusion

In this process, the driving force across the membrane becomes a partial pressure difference and can be expressed conveniently in terms of mole fractions, x and y, of the reject and permeate phases. Figure 4.2 shows experimental data for point equilibria obtained by Breuer [1] in the separation of a mixture of helium and nitrogen. When these data are plotted in the conventional style of the x-y diagram, the graph appears as in Fig. 4.3. The data of Fig. 4.2 also give information on the local driving forces as concentration gradients, which vary along the membrane channel.

There are two such gradients that must be distinguished. The gradient in the transverse direction is responsible for flow perpendicular to the membrane, that is, through the membrane. It can be read from the diagrams, either Fig. 4.2

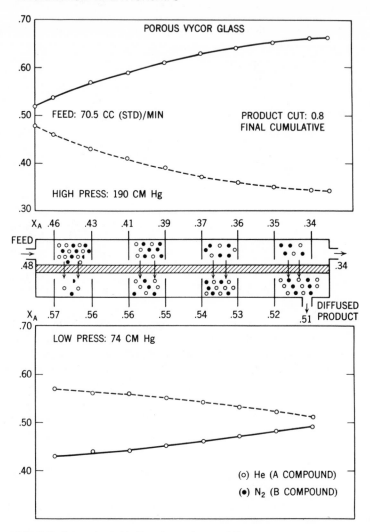

Fig. 4.2 Separation of helium from nitrogen [1].

or Fig. 4.3, as the respective composition differences at each specific location. Another gradient is the driving force that will result in longitudinal dispersion along the membrane channel. It can be obtained only if a diagram such as in Fig. 4.2 is available. The gradient itself is represented by the slope of the concentration profile at any point. Inspection of Fig. 4.2 reveals that the slower premeating component, that is, nitrogen, will be subject to backdiffusion in the longitudinal direction. When this occurs, the concentration relationship

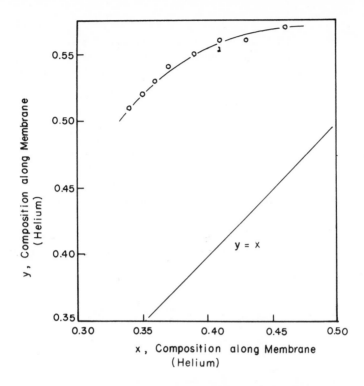

Fig. 4.3 The x-y diagram in separation of helium from nitrogen.

will, of course, be altered. Breuer's [1] contribution was an analysis of this phenomenon and development of a reasonably successful model, which is presented in Chapter XIII.

2 STAGE EQUILIBRIUM

When a desired degree of separation cannot be achieved with a single membrane module, or when the pressure loss along the membrane channel becomes too large, a cascade system of such modules is necessary. In this case, the concept of the "stage equilibrium" is very useful. The equilibrium relates the activities of the overall streams of a module in the system, that is, a_p versus a_0, a_p versus a_f, or a_0 versus a_f. The relationship can, of course, be obtained by integrating the differential material balances, Eq. 4.1, over the entire membrane surface. It should be noted that the relationship is independent of how the streams are recycled between stages in the system.

Consider a cascade system of gaseous-diffusion modules where the adverse effect of concentration polarization is absent. The total flow rate is so small and also the membrane channel is so short that concentrations of the permeate and reject phases are uniform within a membrane module by the longitudinal mixing. The pressure loss along the channel is also negligible. In this case, the integration of Eq. 4.1 over the entire membrane surface S_T for a binary mixture of species A and B yields

$$q_p y_p = K_A S_T (P^I x_0 - P^{II} y_p), \tag{4.2}$$

$$q_p (1 - y_p) = K_B S_T [P^I (1 - x_0) - P^{II} (1 - y_p)] . \tag{4.3}$$

Here, P^I and P^{II} are the total pressures of the reject (high-pressure) and the permeate (low-pressure) phases, respectively, and mole fractions x and y refer to species A. Note that the activities are replaced by the corresponding partial pressures. Combining these equations, the following stage equilibrium relationship can be obtained:

$$\frac{y_p}{1-y_p} = a^* \frac{x_0 - \text{Pr } y_p}{1-x_0 - \text{Pr}(1-y_p)} , \tag{4.4}$$

where the so-called ideal separation factor a^* and the pressure ratio Pr are defined as follows:

$$a^* = \frac{K_A}{K_B} ; \quad \text{Pr} = \frac{P^{II}}{P^I} . \tag{4.5}$$

Thus, the relationship that relates the permeate mole fraction y_p to the reject mole fraction x_0 depends on a^* and Pr, but not on the flow rates of the streams. When the stage equilibrium relates y_p to the feed mole fraction x_f, or x_0 to x_f, the relationship, however, depends on the cut $\theta = q_p/q_f$ through the overall material balance:

$$x_f = \theta y_p + (1 - \theta) x_0 . \tag{4.6}$$

If the longitudinal mixing is not complete, even the stage equilibrium that relates y_p to x_0 strongly depends on the cut θ as shall be seen in Chapter XIII. In this respect, the membrane process contrasts with the stage equilibrium of the conventional stagewise fractionation column. The effects of the pressure ratio Pr and the cut θ on the stage equilibrium are illustrated in Fig. 4.4 and 4.5 in the separation of carbon dioxide (A) from oxygen (B) with a silicone rubber membrane ($a^* = 5.983$). Here, both the reject and the permeate phases are assumed to be in complete mixing. Suppose the gaseous mixture of $x = 0.2$

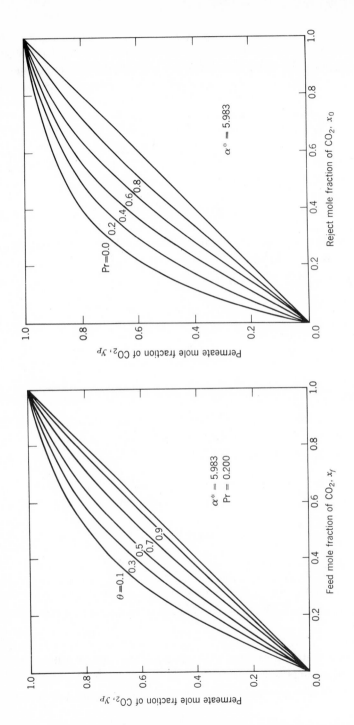

Fig. 4.4 Effect of Pr on the $x_0 - y_p$ stage equilibrium.

Fig. 4.5 Effect of θ on the $x_f - y_p$ stage equilibrium.

to be enriched up to a concentration of carbon dioxide more than $x = 0.9$ by using a simple cascade system of modules as shown in Fig. 4.6. The cascade system is operated with a constant cut $\theta = 0.3$. In this case, the number of

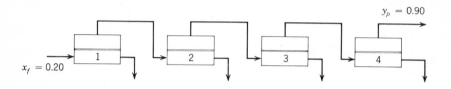

Fig. 4.6 Simple cascade system.

modules required for the separation can be easily found to be four, by directly using the y_p versus x_f stage equilibrium as shown in Fig. 4.7.

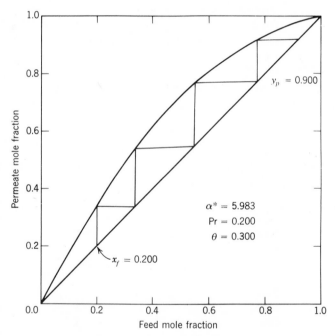

Fig. 4.7 Evaluation of the number of modules in a simple cascade system from the $x_f - y_p$ stage equilibrium.

This somewhat sketchy treatment of stage equilibrium and how it is affected by operating parameters is intended only to introduce the basic concepts. The rather involved effects of operating parameters and also of secondary transport phenomena are treated in depth in Chapter XIII.

Nomenclature

Symbol	Meaning
A,B	Refer to chemical species
K_i	Transfer coefficient of species i through a membrane
P	Absolute pressure
Pr	Pressure ratio (P^{II}/P^{I})
S	Membrane area
S_T	Total membrane area
X, Y	Position indicators
a	Chemical activity
q	Total flow rate
q_i	Flow rate of species i
x	Mole fraction of species A in the reject phase
y	Mole fraction of species A in the permeate phase
a^*	Ideal spearation factor (K_A/K_B)

Superscripts

I	Refers to the reject phases
II	Refers to the permeate phases

Subscripts

A,B	Refer to specific chemical species
f	Refers to the feed stream
i	Refers to an arbitrary chemical species
o	Refers to the reject stream
p	Refers to the permeate stream

Reference

1. M. E. Breuer and K. Kammermeyer, *Sepn. Sci,* **2**, 3 319 (1967).

Chapter V

GASEOUS DIFFUSION

The phenomena that may enter into a membrane permeation process can make the overall mechanisms rather complex. Thus, the attempt to clarify the situation becomes difficult enough so that a clear-cut exposition must consider many facets. The following approach may not be completely satisfactory, but it is the best we know.

Even though the process of separation implies that one deals with a mixture of two or more components, it is of help first to consider the behavior of single components. As a matter of fact, there are many situations where the permeation of mixtures can be interpreted simply on the basis of individual component behavior. Specific examples of such cases are the usual gas transfer through plastic films used in the packaging of foods. Here, the flows of oxygen, nitrogen, water vapor, and carbon dioxide through the film in either direction proceed independently of each other.

In principle one must differentiate between flow through porous and nonporous mdeia. When the criterion is that of separative flow, then the term porous is limited to *microporosity,* where the pore size must be such that molecular diffusion is predominant, that is, molecular collisions with the pore wall greatly exceed any collisions between gas molecules. This means that as a general guide line, the pore sizes must be in the range of 50 to 300 Å.

With nonporous membranes the behavior is primarily controlled by solubility and diffusivity relationships. Thus, the separation mechanism is completely different from that encountered in microporous media.

An overall view of membrane classification was presented in Chapter I and for separative considerations the terms "adsorptive" and "diffusive" membranes were employed. As a microporous medium represents in effect a rather active capillary system, it will exert adsorptive forces on the gaseous components and create conditions of flow that can differ greatly from pure gas-phase flow in the open pore structure. Intuitively, the flow phenomenon in a diffusive membrane, which is in essence a pore-free material, will be that of a so-called "diffusive solubility flow" type.

An additional factor is the behavior of the gas phase in the gas-membrane system. While it is convenient to talk in terms of gas (noncondensable) and vapor (condensable) behavior at ambient temperatures and pressures, such a simple distinction cannot, in principle, be extended to membrane systems. The separative processes may entail low and high temperature, as well as low and high pressure, and a variety of combinations of temperature and pressure. Therefore, helium, for instance, which one is likely to consider as a noncondensable gas, will behave as a condensable vapor at appropriately low temperatures. Similarly, a condensable vapor such as methyl bromide will behave as a gas at a high enough temperature or at an adequately low pressure.

A generalized schematic representation of the processes encountered with microporous membranes in gaseous separation is shown in Table 5.1. Here, molecular flow, slip flow, and laminar flow occur in the gas phase, and adsorbed or surface flow is thought to occur on the surface of the solid membrane. The degrees to which the different flows enter are determined by the pore size of the membrane and the properties of the gases.

It is important to realize that the surface flow, when present, is in addition to the normal molecular and bulk flow, so that it results in increased total flow through the membrane.

With nonporous membranes, the chief considerations become solubility and temperature. For the most part the nonporous materials are likely to be polymers. The behavior of metal foils and of glass membranes requires separate treatment.

Table 5.1   Variables in Flow through Microporous Membranes

Adsorptive Membranes	*Microporous Structure*
(1) Low Pressure (vacuum), high temperature (200-500°C)	Gas flow follows molecular diffusion; no adsorption phenomena take place
(2) Low pressure (vacuum), medium temperature (30-100°C)	Adsorption takes effect; molecular diffusion plus adsorbed flow
(3) Medium pressure (atmospheric), medium temperature (30-100°C)	Increased adsorptive effects, molecular flow still possible
(4) Medium pressure (atmospheric), low temperature (0-20°C)	Adsorptive effects predominate; slip flow possible
(5) High pressure (up to 40 atm), low temperature (-30-0°C)	Absorptive effects control, laminar flow may occur

When using polymers, the generalized scheme presented in Table 5.2 will also be affected by such factors as glass-transition temperature of the polymer itself, the presence or absence of crystallinity, solvation effects resulting in polymer swelling, and pressure effects particularly with the more condensible types of substances. Obviously, when temperature connotations are used, the "high" temperatures will be relatively low compared to possible high temperatures that can be employed with nonpolymeric porous membranes.

Both of the generalized schemes must be interpreted with caution. The pressure and temperature levels that are listed will vary greatly with the membrane and gaseous materials. Therefore, for instance, for methyl bromide through polyethylene a high pressure would be about 700 mm Hg, but for helium through Teflon a high pressure would be about 60 psi.

Also, the allowable temperature levels will depend greatly on the membrane materials. Thus, with polymeric membranes one is likely to work below about 100°C and more often below 40°C. But, with microporous membranes of inorganic composition, temperatures of 200 to 300°C would be quite acceptable.

A concise and well-arranged tabulation of the effects of variables upon gas and vapor permeability was created by Friedlander and Rickles [1] as shown in Table 5.3.

The actual interplay of these factors will become more evident in the subsequent discussions of flow phenomena.

Table 5.2 Variables in Flow through Nonporous Membranes

Nonporous Membranes	*Diffusive Membranes*
(1) Low pressure (vacuum), high temperature (up to 100°C)	No pressure effect on permeability; usually higher permeability; solubility essentially determined by temperature
(2) Low pressure (vacuum), medium temperature (20-60°C)	No pressure effect; normally lower permeability at lower temperature; solubility will increase with lower temperature
(3) Medium pressure (atmospheric), medium temperature (20-60°C)	No pressure effect; mainly same as in (2)
(4) Medium Pressure (atmospheric), low temperature (-10-10°C)	Some pressure effect, will increase permeability; solubility may increase to give further increase in permeability
(5) High pressure (up to 100 psi), low temperature (-10-10°C)	Appreciable pressure effect; will increase permeability; solubility same as (4)

1 FLOW THROUGH MICROPOROUS MEMBRANES

The flow through microporous media represents a variation of the flow through fine capillaries. While Graham [2] was the original investigator of microporous-media flow, it remained for Knudsen [3] to study the flow of gases through single capillaries. A microporous membrane, of course, is just a conglomerate of capillaries, large and small, straight and tortuous, and thus the interpretation of flow through single capillaries is of fundamental importance to the mechanism prevailing in a porous matrix.

Knudsen's studies were of a pioneering nature, and it has thus become customary to use the term *Knudsen flow* to represent the phenomenon of free molecular diffusion.

In order to effect separation of a gaseous mixture by means of a microporous membrane, strictly on the basis of free molecular diffusion, the following conditions must be met:

•Pore diameter must be appreciably less than the mean free path of the diffusing constituents.

•Temperature must be high enough to avoid appreciable surface flow, unless such flow would enhance separation (this situation will be illustrated in a

Table 5.3 The Effect of the Principal Variables upon Gas Permeability[a]

	Effect
Temperature	(1) Arrhenius-type effect—for permanent and other small-size gases
	(2) Activation energy is function of temperature for CO_2 and H_2O above glass temperature
	(3) Arrhenius-type effect does not hold for organics
Pressure, concentration	(1) For permanent gases and CO_2 independent of pressure
	(2) For most other gases, permeability is a function of pressure
Penetrant nature	(1) Permeability is a decreasing function of molecular diameter
	(2) Mutual compatibility and ease of condensation increases permeability
Plasticizing effect	(1) Ability to plasticize polymer increases permeability
Polymer structure	(1) Crystalline region acts as an impermeable phase so rates are proportional to volume fraction of amorphous phase
	(2) Thermal and mechanical working history is important
	(3) Solvent conditioning is important
	(4) Cross-linking reduces diffusion rates
Grafting	(1) Grafting greatly affects permeability
Irradiation	(1) Irradiation causes cross-linking, which reduces diffusion
Film thickness	(1) Usually has no effect but may be important for very thin films or for films with appreciable concentration gradients

[a] Reference 1.

later section).
·Pressure must be low enough so as not to reduce the mean free path to approach the pore diameter, or to cause adsorption.

When these conditions are met, then gaseous components will flow through the pores at different velocities, resulting in a corresponding change in composition. In accordance with kinetic theory the component of lower molecular weight will exhibit the greater speed and become enriched in the permeated stream.

Graham [2] originally established the molecular weight dependency of free molecular flow, so that

$$\frac{n_1}{n_2} = \left(\frac{M_2}{M_1}\right)^{\frac{1}{2}} = a^* \, , \tag{5.1}$$

where $n_{1,2}$ are the moles of lighter and heavier component diffusing in unit time and $M_{1,2}$ are the respective molecular weights, the attainable degree of separation thus being, expressed as an ideal separation factor a^*. Equation 5.1 is the well-known expression used to interpret measurements made in molecular effusion.

Surface Flow of Pure Components

The existence and occurrence of so-called surface flow has been adequately documented. Numerous investigators have published data that show conclusively that such a phenomenon does indeed exist. The expressions used by various authors are surface flow, surface diffusion adsorbed flow, condensed flow, and others. The preferred term to be used here is *surface flow*. A fully satisfactory explanation of its mechanism has not yet been provided, but the several model concepts do a reasonably good job to represent the available findings (see Table 5.4). Most representative of the progress that has been made in recent years are papers by Oishi [4], Barrer [5], Metzner [6], Hwang [7], and their respective co-workers.

A rather comprehensive review of the status of surface flow of pure components, and of the effects of pressure and temperature upon such flow, was contained in Volume 1 of *Progress in Separation and Purification* [8]; it would be redundant to repeat this coverage. However, it is worthwhile to present some material that helps to elucidate the concept of surface flow, and, of course new findings that have been reported in the recent literature.

Models for mechanisms are summarized in Table 5.4. The underlying concept inherent in all models is that gas phase flow and surface flow proceed in parallel.

Table 5.4 Models of Mechanisms

Investigator	Basic Mechanism	Ref.
Oishi	Site hopping; statistical analysis	4
Metzner	Force parameter effects	6
Barrer	Activated site hopping	5
Flood	Hydrodynamic flow with additive surface-flow term; pore size large enough to get laminar flow	9
Gilliland	Spreading pressure and force balance	10
Hwang	Site hopping with local equilibrium gas and absorbed phase	7

Surface Flow—Pressure Effect

Several reports in the literature indicate that the permeability of condensable vapors in the high-surface-coverage region reaches a maximum when it is plotted against pressure, as shown in Fig. 5.1. No quantitative explanation has been proposed, but a qualitative interpretation can be fashioned by dividing the pressure range into three regions and applying an appropriate equation to each flow mechanism.

In the lower-pressure region, the ordinary gas-phase and surface-flow equations are applied, giving a continuous increase of permeability with pressure. The intermediate-pressure region yields some pores plugged with the condensed liquid; therefore the permeability drops rapidly with the increase in pressure. The ordinary viscous-flow equation is applied to this region, giving a rapid increase of the flow resistance due to the buildup of condensate. Finally, in the higher-pressure region, the entire pore matrix is filled with the condensed liquid. The flow mechanism is again the viscous liquid flow, but the permeability does not change much with the pressure. It should be noted that capillary force plays no role in the total driving force across the porous medium, since both sides of the medium are at the same temperature and can be considered to have the same average pore radius.

Most of the available information on surface-flow behavior with changing pressure has been obtained using easily condensable gases at pressure levels below their saturation pressure. Obviously, it is of considerable interest to have information on behavior at high pressures to see if operation at elevated pressures would present advantages in separation.

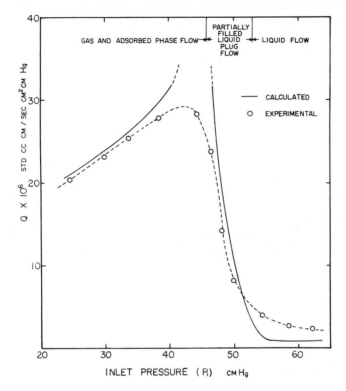

Fig. 5.1   Effect of pressure on permeability of *n*-butane in microporous Vycor glass at 0°C [8, 11].

Recently, Stahl [12] investigated high pressures with a microporous Vycor glass membrane and found that there is little if any effect on flow when the gases are above their critical temperature. Figure 5.2 shows the data for seven compounds up to about 1000 psi.

The ordinate in this figure represents values of $Q(MT)^{1/2}$, and, therefore, if no surface flow were present the data for all compounds at any temperature should fall on one line located at 2.7×10^{-4}. Thus, the results show that surface flow is present, but that increasing pressure has only little effect above the critical point.

An even more significant behavior is illustrated in Fig. 5.3, where Stahl's data for the permeability Q of carbon dioxide are presented at two temperatures below the critical point, that is, 0 and 25°C, and also at 50°C, a condition above the critical point. The behavior at rather high pressures parallels that shown in Fig. 5.1, and although this is not unexpected, it is gratifying to

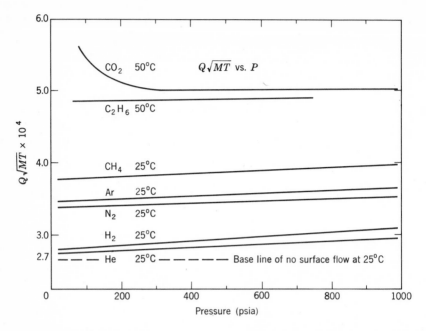

Fig. 5.2 Effect of high pressure upon flow of gases through microporous Vycor glass [12].

know that one can expect the same phenomenon to occur with other gases and other porous membranes that meet the proper pore size requirements.

Surface Flow—Temperature Effect

The effect of temperature upon surface flow in a microporous membrane can be quite pronounced, even with the so-called inert gases. The surface flow will be particularly high at low temperatures, will decrease progressively with increasing temperature, reach a minimum and then begin to increase again with further increase in temperature. This behavior is strikingly illustrated by Hwang's [7] data for microporous Vycor glass, presented in Fig. 5.4.

Thus at temperatures higher than the highest one investigated, a maximum point should occur, because at very high temperatures the surface flow should disappear entirely, and the total flow should approach asymptotically for free molecular-flow level at a value of about 4×10^{-4}. If surface flow were not present all of the gases should fall on this single base line.

It has not as yet been possible to carry experiments to the higher temperatures because a truly microporous membrane has not been available. The porous Vycor membrane unfortunately will not stand temperatures higher than about $600°K$ without change in structure resulting in shrinkage to the Vycor-brand glass.

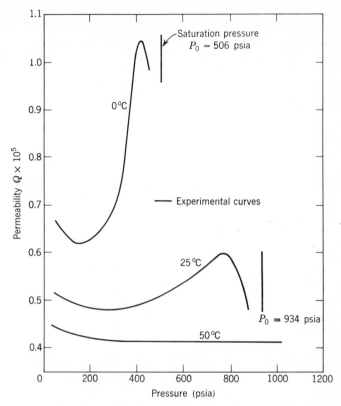

Fig. 5.3 Permeability of CO_2 through microporous Vycor glass as a function of pressure [12].

In terms of magnitude the differences in $Q(MT)^{1/2}$ values are not large, and it was considered desirable to extend the helium curve to lower temperatures. This was done by Hwang [13] at liquid-hydrogen temperature: the result is shown in Fig. 5.5. The close agreement with the working equation presented below was very gratifying, but the important implication is that a definitely appreciable difference in the flow coefficient could be demonstrated.

The so-called working equation was developed from a rather complicated fundamental equation given in Ref. 7 by making simplifying assumptions. Assuming the site-hopping model as the mechanism of the surface migration, the surface diffusivity can be calculated, as the adsorption isotherm provides the relationship between the surface concentration and the gas-phase pressure, The final equation is as follows:

$$Q(MT)^{1/2} = A + BT \exp\left(\frac{\Delta}{T}\right) , \qquad (5.2)$$

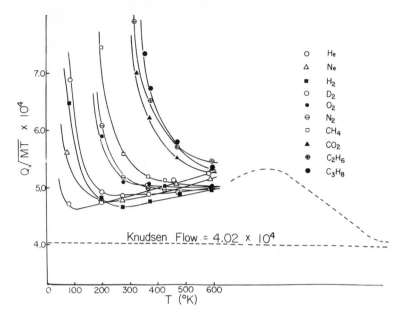

Fig. 5.4 Effect of temperature on adjusted permeability of gases and vapors in microporous Vycor glass [7].

where Q is the total permeability, M is molecular weight, T is absolute temperature, A is a constant corresponding to the gas-phase flow, and B and Δ are constants that determine the surface flow. The second term represents the contribution of surface transport. This equation then permits the determination of the surface diffusion in relation to the total flow, and also predicts the permeability of a new gas for a given microporous medium. It can be used in the study of separation factors for a binary mixture as a function of temperature. Figure 5.4 shows how well the equation fits the data for various gases over an experimentally wide temperature range. This analysis indicates that the surface diffusion of helium in microporous Vycor glass may be 13 to 25% of the total flow in the temperature range of 77 to 600°K. Therefore, every study in this field should be re-examined, as all previous investigations have assumed that helium will not give any surface flow whatever.

Surface Flow—Effect on Separation

The effect of surface flow on separation of a gaseous mixture by means of a microporous membrane is, of course, related to variation of temperature or pressure, or of both of these parameters.

The generalized situation is well illustrated by Fig. 5.6, where standard permeabilities obtained by Tock [8, 14] are plotted against temperature for the flow of a number of gases through microporous Vycor glass.

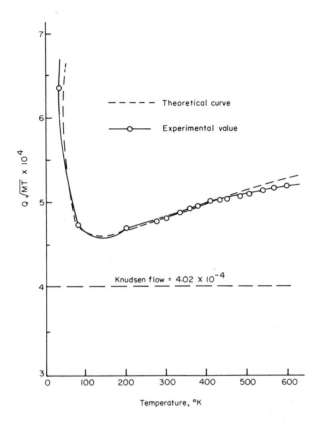

Fig. 5.5 Helium flow through microporous Vycor galss as a function of temperature. Reprinted from Ref. 13, by courtesy of Marcel Dekker, Inc

What makes this graph particularly important is that every intersection of any two curves represents a nonseparative composition, that is, an azeotrope. Strictly speaking, the term *azeotrope* implies a boiling mixture. The term has, however, been used to signify a nonseparative condition in its broad sense. The data presented on this graph are the first comprehensive evidence of the existance of nonseparation mixtures in the gas phase, although a few instances in previous work [7] indicated that such situations may exist. Therefore, although gaseous diffusion can usually be counted upon to be useful in separating distillation azeotropes (constant boiling mixtures, CBMs), the gaseous-diffusion process will be plagued with its own "gaseous-diffusion azeotropes," GDAs, in analogy to the CBMs of distillation. However, even though the composition of distillation azeotropes can be varied only by changing the distillation pressure (excluding the use of extractive distillation), a GDA can be manipulated independently by a change in either pressure or temperature.

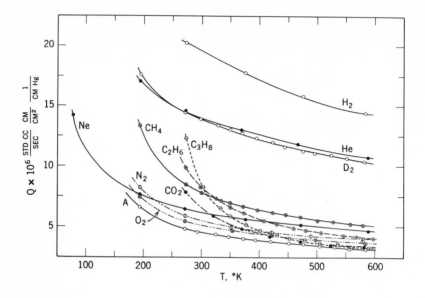

Fig. 5.6 Transfer coefficient chart for standard permeabilities of gases and vapors in microporous Vycor glass, as a function of temperature [8, 14].

An example of the separation of a distillation azeotrope as presented by Hagerbaumer [15] is given in Fig. 5.7. This example represents a variation of both temperature and pressure. Pressure variation does not appear on the graph, but similar data at other pressure levels show that more separation was accomplished at 2 atm pressure on the high-pressure side of the membrane.

The significant fact is that the membrane-separation process results in the enrichment of the compound that has the higher molecular weight, that is, benzene. The data of Fig. 5.7 also demonstrate the effect of superheating of vapors on separation. Better separation was obtained the closer the operating temperature approached the saturation temperature. As the temperature approached the critical temperature, less separation was obtained. Further superheating, that is, bringing the vapors into a more "gaslike" state, would result in behavior in accordance with Graham's law, so that the compound of lower molecular weight would become enriched in passing through the membrane, because under such conditions surface flow would become negligible, and free molecular flow should control.

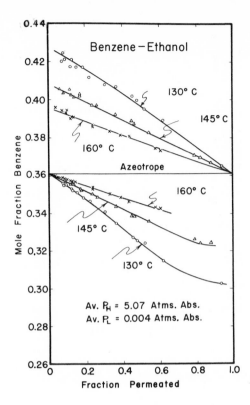

Fig. 5.7 Superatmospheric pressure separation of benzene-ethanol with microporous Vycor glass [15]. Courtesy of the American Institute Chemical Engineers.

Carbon dioxide and propane have essentially the same molecular weight, 44.010 and 44.094, respectively. Thus, it would be rather difficult to separate mixtures of the two gases on the basis of only free molecular flow through a membrane. However, due to the occurrence of surface flow it is possible to separate mixtures with relative ease. Wyrick [16] performed the separations and established the effect of pressure as shown in Fig. 5.8 as well as of temperature variation (not shown in Fig. 5.8).

A later investigation by Tock [14] covered the temperature effect, and his data are presented for the CO_2-C_3H_8 system in Fig. 5.9. Tock used the separation factor a as a criterion: it is instructive to note that this factor approaches the ideal value of 1.0 with increasing temperature. As plotted, the

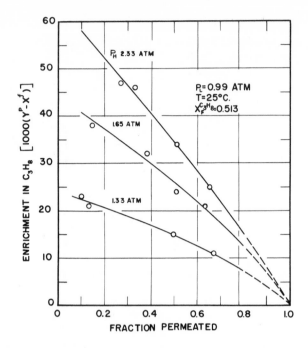

Fig. 5.8 Effect of pressure upon separation of CO_2-propane mixtures with microporous Vycor glass. Reprinted by permission from Ref. 16. Copyright 1958 by the American Chemical Society.

factors are less than 1.0, because propane exhibits more surface flow than carbon dioxide.

The conclusions to be drawn from the illustrated examples are as follows:

1. Surface flow in microporous membranes exists as a definite phenomenon and can be manipulated to effect gaseous separations.

2. Its occurrence can and does impede separations that are intended to be carried out on the basis of free molecular flow only. This situation is discussed later by the example of uranium isotope separation in gaseous diffusion plants. If surface flow is to be avoided, one must use as high a temperature and as low a pressure as possible.

3. The presence of surface flow can lead to the formation of nonseparable gas mixtures. However, selection of proper operating conditions, that is, temperature and pressure, will generally permit a set of operating parameters that will allow separation of such mixtures.

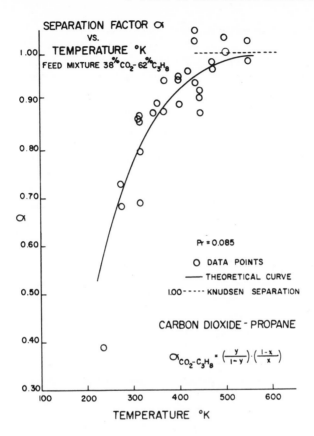

Fig. 5.9 Temperature effect upon separation of CO_2-propane mixtures with microporous Vycor glass [14].

2 FLOW IN NONPOROUS MEMBRANES

There exists an unresolved question as to whether any membrane, so-called porous or nonporous, will act to some extent as a microporous medium. Even the membranes that we call "nonporous" may have a number of minute pores whose diameter will be in the 5-to-10-Å range. However, rightly or wrongly, such structures are considered nonporous, and their behavior would seem to justify this viewpoint.

Nonporous membranes fall into two primary classifications, that is, inorganic and organic structures. The inorganic membranes comprising ceramic and metallic materials will be discussed in a separate chapter. Thus, the main objective of the following discussion will be the elucidation of gas flow in organic membranes, which means the behavior of *polymer-gas systems.* A rather instructive and up-to-date review of the transport in permselective membranes was published in 1971 by Rogers and Sternberg [17].

Almost any polymeric material that can be prepared in the form of a thin-walled shape is likely to be suitable as a separative membrane. The flow mechanism through polymer membranes is generally accepted as consisting of sorption (adsorption or absorption) of the permeating medium, gas or liquid, on the polymer surface, solution into the polymer, diffusive flow through the polymer matrix, and desorption from the other side of the membrane. Where some disagreement exists is in the concept of flow within the polymer. However, it has been well established that the solubility of the permeating species can play an important role, and thus the term "diffusive solubility flow" is perhaps most descriptive.

Gas and Vapor Permeation—Single Component

The status of diffusive solubility flow in polymers can be characterized by stating that flow rates, that is, permeabilities, for gases and vapors can be taken as those of the pure components, when determined under the proper conditions of pressure and temperature.

Methods for obtaining permeability data will be discussed in Chapter XII. In principle, there are two methods in general use, one utilizing the permeation of a gas from a definite high pressure on one side of a membrane to a lower pressure, usually vacuum, on the other side of the membrane: the so-called *variable-pressure method,* as originally developed by Barrer [18]. The other method, called the *variable-volume method,* measures the amount of gas that penetrates at a constant pressure on the low-pressure side of the membrane [19].

The most informative set of data published on flow in polymers is that of Szwarc and co-workers [20] on the behavior of methyl bromide in polyethylene. As pictured in Fig. 5.10, the permeability curves illustrate the complete range of behavior in polymers as affected by both temperature and pressure.

The plot is of the conventional Arrhenius type, that is, semilogarithmic, with the logarithm of permeability plotted against reciprocal absolute temperature. The curve at 100 mm absolute pressure is typical of the behavior of gases in polymers. All gases and vapors above their critical temperature exhibit the straight-line relationship when there is no pressure effect, and the usual trend is to increased permeability with increasing temperature (moving to the left on the abscissa scale). There are some exceptions, however, such as the flow of carbon dioxide through silicone rubber as shown in Fig. 5.11 [21].

Another phenomenon that can be expected is that of breaks in the line as given in Fig. 5.12 and 5.13. The data in Fig. 5.12 were obtained by Brubaker [22]; it is evident that curves 3, 4, and 9 represent anomalous behavior. The

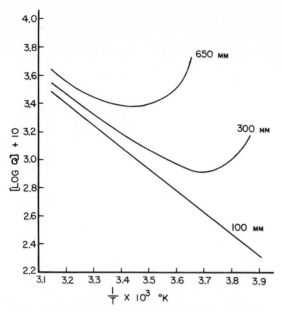

Fig. 5.10 Effect of temperature and pressure upon permeability of methyl bromide in polyethylene. Reprinted by permission from Ref. 20. Copyright 1957 by the American Chemical Society.

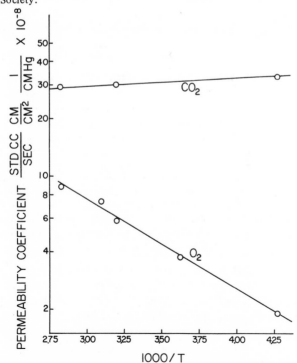

Fig. 5.11 Permeability versus temperature in silicone-rubber membrane [21].

69

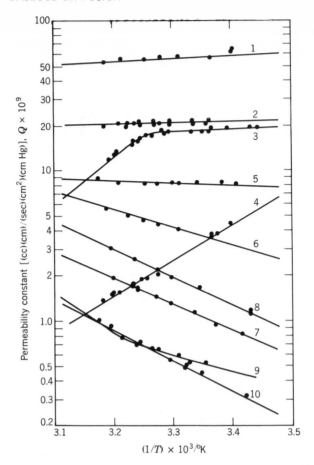

Fig. 5.12 Gas permeability range of plastic films. 1: H_2, vinyl film, commercial calendered, no directional pattern; 2: CO_2, vinyl film, commercial calendered, no directional pattern; 3: H_2, vinyl film, experimental press polished, no directional pattern; 4: CO_2, vinyl film, experimental press polished, no directional pattern; 5: H_2, vinyl film, commercial calendered, no directional pattern. 6: CO_2, vinyl film, commercial calendered, no directional pattern; 7: H_2, polyethylene film, commercial, extruded, directional pattern; 8: CO_2, polyethylene film, commercial, extruded, directional pattern; 9: H_2, vinyl film, experimental, extruded, no strain pattern; 10: CO_2, vinyl film, experimental, extruded, no strain pattern. Reprinted by permission from Ref. 22. Copyright 1952 by the American Chemical Society.

break in curves 3 and 9 are similar to the observations of Meares [23] pictured in Fig. 5.13. These breaks reflect changes in polymer structure at the respective temperature levels. Stannet and co-workers more recently reported data on the permeability of nobel gases [24], and definite breaks in the temperature lines

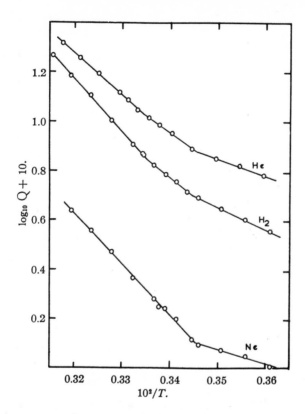

Fig. 5.13 Permeabilities for helium, hydrogen and neon in polyvinyl acetate. Reprinted by permission from Ref. 23. Copyright 1954 by the American Chemical Society.

were evident at the glass-transition temperature of poly(methyl acrylate).

The literature on permeabilities of gases and vapors in polymers is so extensive that treatment of individual publications would call for a separate volume.

The simultaneous effects of pressure and temperature are clearly evident from Fig. 5.10 at the 300- and 650-mm pressure levels. The normal tendency to lower permeabilities at lower temperatures is overcome by the effect of increasing pressure upon the solubility of the gaseous component in the polymer. Li and Long [25, 26] investigated pressure effects up to about 70 atm (~1000 psi) and attributed the increased permeability at higher pressure to a plasticizing effect exerted by dissolved gases and vapors.

One would expect that some relatively inert gases may show a similar behavior at sufficiently high pressures. This has indeed been substantiated by Li and Long with polyolefin films [26] and by Stern's [27] investigations at high pressures. The behavior of the system methane-Teflon is illustrated in Fig. 5.14.

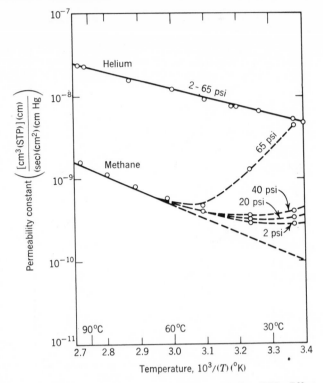

Fig. 5.14 Permeabilities of helium and methane in Teflon FEP. Effect of transition temperature. Reprinted by permission from Ref. 27. Copyright 1965 by the American Chemical Society.

However, Stern cautions that the methane behavior may represent a transient condition resulting from progressive changes in the polymer structure at temperatures below 60°C.

Helium does not yet show any pressure dependence at the 65-psi level. Later studies, also by Stern et al. [28] give more details on the behavior of helium flow in polyethylene. Here too, there is essentially no effect of pressure, up to 65 atm, upon the permeability over the temperature range of 0 to 50°C. Somewhat surprising, analogous data for nitrogen under the same conditions, also indicate the absence of a pressure effect, and similar data of Li and Long [26] are somewhat inconclusive.

However, data for the system carbon dioxide-polyethylene as shown in Fig. 5.15 indicate conclusively that a pressure effect is present [28]. A pressure effect also was observed for some of the lower-molecular-weight hydrocarbons and N_2O.

In summary, the pressure effect on flow of gases, and vapors in polymers is greatly dependent on the critical temperature; the studies of Li and co-workers

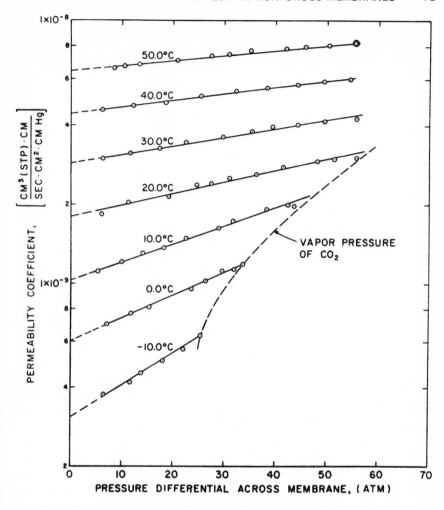

Fig. 5.15 CO_2-polyethylene system. Dependence of permeability on pressure difference across membrane. Reprinted from Ref. 28, by courtesy of Marcel Dekker, Inc.

[25, 26, 29, 30] and those of Stern and co-workers represent the most up-to-date and comprehensive coverage of the subject [27, 28, 31, 32].

Effect of Membrane Thickness

The significance of membrane thickness can be summed up by stating that increased thickness reduces flow of gases and thus requires larger units for a given production. Therefore a major effort in the preparation of polymeric membranes will always be the production of as thin a membrane as possible, consistent with minimal requirements of strength and handling qualities. The normal effect of thickness is a linear dependency of flow, so that doubling the thickness will halve the permeation rate.

In a very general sense ½ mil thickness (0.0005 in.) is a practical lower limit, and 1 mil is a more practical value. Some specialty membranes have been prepared as thin as 0.1 mil, and some even less, by vapor deposition on a substrate, but these have so far not been found practical, except for some specialized experimentation. When membranes are used in tubular shape or in the form of hollow fibers, one finds that practically attainable wall thicknesses are from 5 to 10 mils on up. The difficulty here is that spinning techniques place a feasible lower limit on thickness.

Many investigators have reported that there is no unusual effect of thickness, at least up to about 10 mils. Some experiments have been reported with thicker membranes, as, for instance, Brewer's [33] work up to 30 mils and Hwang's [34] up to 50 mils, both with silicone rubber. In neither case was there any evidence that these rather thick membranes resulted in anything but proportional reduction in flow. Until some contradicting evidence becomes available, it is safe to assume that the thickness effect is linear.

Effect of Polymer Structure

Numerous publications have treated the relationship of polymer structure and permeability. Examples of possible effects are illustrated in Fig. 5.12 and 5.13. A detailed discussion of the many possible ramifications would be very lengthy and is not deemed justified.

Suffice it to refer to a rather good presentation in Kesting's book on *Synthetic Polymeric Membranes* [35] and an investigation by Roberts [36], which had the specific objective of relating polymer structure and gas permeation.

The results, which have been observed many times, show conclusively that crystallinity reduces permeabilty. Furthermore, Roberts' [36] work in particular gives strong indication that a stress condition, such as orientation, as well as membranes in a stretching frame, will show decreased permeability. As expected, and as mentioned previously, polymers that may become plasticized by solution of the permeant and, of course, swollen membranes will show increased permeabilities.

A number of special situations have been studied that deal with the influence of structural factors upon gas and vapor flow in polymer films, for example, Michaels' and Hausslein's [37] investigation of the effect of elastic factors upon sorption and transport properties of polyethylene, the work of Rogers [38] on membranes with a gradient of inhomogeneity, the study of Ash, Barrer, and Palmer [39] of diffusion in multiple laminates, and a similar sugject by the Szwarc school of investigators on permeability valves [40]. The problems encountered in using a silicon-rubber membrane for a model of a blood oxygenator were studied in detail by Buckles [41]. This work includes variations in filler content of polymers. A rather specific study by Koh [42] related the permeation of carbon dioxide and water vapor to density and crystallinity in polyethylene, polypropylene, Nylon 6, and cellulose acetate.

That considerable interest continues to exist in the role played by polymer structure was evidenced by a number of papers presented at the American Chemical Society Meeting, August-September 1972 in New York. Stallings et al. [43] reported the permeability behavior of neon, argon, and krypton in blends of polystyrene and poly(phenylene oxide), and observed nonadditivity of diffusion coefficients. The sorption and diffusion of 2,3,4-TriMe-pentane, benzene, chloroform, and dichloromethane in amorphous and crystalline polybutadienes was studied by Brown et al. [44]. An interesting finding was that solvent annealing of a 94% transpolymer led to a decrease in the diffusion coefficient, while shock cooling raised the coefficient. Rates of diffusion of toluene in high-density linear polyethylene films of varying degrees of orientation were reported by Kwei [45], who found that quench-rolled films showed a linear dependence of the diffusion coefficient on concentration, claimed to be the first report of such behavior. According to Paul and Kemp [46] the effect of adsorptive fillers on the diffusion time lag can be very large and cause errors in sorption and diffusion determinations. But they will have only a minor effect on steady-rate permeation.

The complexity of the interacting effects of structural and environmental parameters excludes the possibility of presenting data in a uniformly correlative manner. However, it is essential to recognize the permeability behaviors at least in qualitative terms. A comprehensive collection of permeability data was prepared by Choi [47]. As a correlating parameter he used a ratio of permeabilities, so that the permeability of a given gas or vapor was referred to a base material such as oxygen or nitrogen. Choi's study also portrays the multitudinous effects of variables, such as pressure, temperature, polymer density, and degree of crystallinity. Water-vapor transmission data are treated in a similar manner.

Diffusive Flow—Effects on Separation

There is a fair amount of information on the use of polymeric membranes in gas or vapor separation on a commercial or even semitechnical scale: see Chapter XV. On the other hand, one can find numerous proposals for industrial uses with computed performance results that are realistic and plausible. These examples illustrate the combined effects of pressure, temperature, and polymer selection for the membrane as described under previous headings.

The most comprehensive design analyses of several gaseous systems have been carried out by Stern [8, 27]. The following examples are treated in detail: (a) recovery of helium from natural gas, (b) separation of oxygen from air, and (c) carbon dioxide control in breathing atmospheres.

The helium recovery by means of a Teflon FEP barrier appears to be the most promising of the processes. The technological requirements are formidable, as the barrier septum has to operate at a pressure drop of some 900 psi. The optimum operating conditions developed from the analysis are presented in Table 5.5.

Table 5.5   Recovery of Helium from Natural Gas by Permeation through Teflon FEP^a Membranes^b

Optimum Operating Conditions for Large Plants
Feed gas: 0.45 He, 17.06 H_2, 76.43 CH_4, 6.06 hydrocarbons
Feed gas compressed to 1000 psig
0.001 in. thick membrane
Three permeation stages
80°C operating temperature 7.3-psi permeator backpressure
60% recovery of helium in natural gas
Interstage gas compressed to 950 psig
72% helium in product gas

^{a} Teflon FEP: a copolymer of tetrafluoroethylene and hexafluoropropylene.
^{b} See Refs. 8 and 27.

In 1966, Stern concluded that: "The described process is the most efficient permeation technique now available, but is still not quite competitive with the more advanced cryogenic processes for the large-scale recovery of helium. However, the development of more permeable membranes, along with more sophisticated permeation cycles, could greatly increase the efficiency of the membrane process" [27]. Since then, there have been several notable developments, and the present intensive activity in hollow fiber technology will undoubtedly have a highly beneficial effect and a growing number of semiscale or even full-scale gaseous-diffusion installations should come into existence. Several examples are presented in Chapter XV.

Oxygen separation [27] from air was analyzed with the use of an ethyl cellulose film (1 mil thick). In the single-stage process, Stern computes enrichment of oxygen to 32.6%, and states that a 91.1% O_2 concentration should be attainable with five stages. On the basis of economics, however, the process could not be competitive on a large scale with conventional technology.

The carbon dioxide removal in space cabins [27] was computed on the basis of a three-man mission, using silicone rubber as the selective membrane [48]. Stern estimates that the permeation equipment will weigh a total of about 400 lb, including auxiliaries.

The intriguing problem of removing the carbon dioxide from the air of a space cabin led Major and Tock [49] to develop a cross-flow gaseous-diffusion cell with multilayers of silicone-rubber films as the barrier material. Previous studies had shown that silicone rubber was well suited for CO_2 permeation. In order to permit a compact cell design, it was necessary to devise the anchoring of the silicone rubber on a porous paper substrate without permitting appreciable penetration of the polymer into the porous substrate. The permeability values of the gases, that is, CO_2, O_2, and N_2, were as expected and so were the separation results. The objective of building a large area into a given volume was accomplished to a satisfactory degree, and cell efficiencies of 96% or better were attained routinely. Some details of construction are presented in Chapter XII.

It appears that the only comprehensive information on actual laboratory scale separations is that of Brubaker [50], where experimental data are reported for the following gas-polymer systems:

Gas	Membrane
He-O_2	Polyethylene (Visking)
CO_2-H_2-O_2-N_2	Polyethylene (Visking)
NH_3-H_2-N_2	Polyethylene (Visking)
SO_2-O_2-N_2	Polyethylene (Visking)
CO_2-H_2-O_2-N_2	Cellulose acetate butyrate (Tennessee Eastman)
CO_2-H_2-O_2-N_2	Trithene (Visking)
NH_3-H_2-N_2	Trithene (Visking)

An interesting observation in the separation of multicomponent mixture is that the components of intermediate permeabilities exhibit a maximum point in the enrichment curve, as illustrated in Fig. 5.16 for a ternary mixture. This is similar to the behavior of multicomponent mixtures in distillation. In this example, the high solubility of ammonia results in an appreciable enrichemnt of this component, even though its molecular weight is so much greater than that of hydrogen.

One recent study by McCandles [51] presents permeability data for 14 films and single-stage separations of CO-H_2 mixtures through four types, with pressure differentials across the 1-mil-thick films up to 500 psi, and temperature levels from 22 to 117°C. The Weller-Steiner Case I and II models permitted reasonably good presentation of the data. Best separations were attained with films of caprolactam, Dacron, Parylene C, and Polyimide. Computer calculations are presented for a two-stage unit.

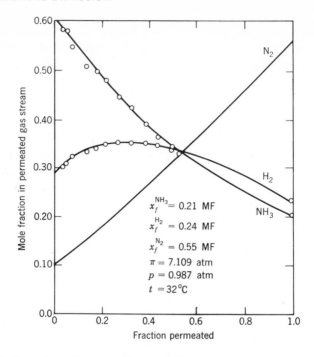

Fig. 5.16 Separation of ternary mixture ammonia-hydrogen-nitrogen with polyethylene membrane. Reprinted by permission from Ref. 50. Copyright 1954 by the American Chemical Society.

3 WATER-VAPOR TRANSMISSION AND EFFECT OF WATER VAPOR ON GAS PERMEATION

Water-vapor permeation through polymers has received a considerable amount of attention, primarily because of the use of polymeric films as moisture barriers in packaging and general moisture protection.

Possible interaction of water vapor (or liquid water) with the polymer system will influence the mechanism of permeation. Obviously, in hydrophilic polymers, solution of water in the polymer will occur, resulting in a softening or plasticizing effect, and it may cause considerable swelling. Thus, the diffusion process inside the membrane, as expressed by the solubility coefficient and the diffusion constant, will be highly concentration dependent, which means that permeation rates will be greatly affected by relative humidity, that is, vapor pressures. On the other hand, what might be considered hydrophobic polymers, such as the polyolefins, are likely to exhibit permeabilities that are independent of vapor pressure. This situation is illustrated in Fig. 5.17, and what makes these data particularly interesting is that Stannett and Yasuda [52] have checked out the situation at 100% relative humidity by using liquid water contact, thus

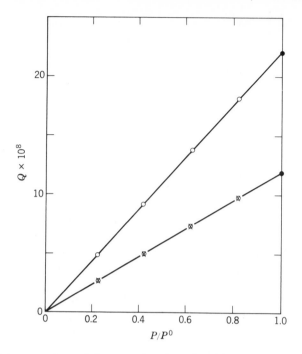

Fig. 5.17 Water-vapor transmission as a function of relative vapor pressure. ○ = vapor polyethylene (0.922 density); ⊕ = vapor polypropylene (0.907 density); and ● = liquid water [52].

proving that liquid-contact permeation should give results as at saturated vapor conditions.

Water-Vapor Transmission (WVT)

Only a few investigations seem to be available on the effects of extrinsic factors upon water-vapor transmission (WVT): Sciarra and Patel [53], however, recently demonstrated that water-vapor permeability of ethyl cellulose and polyamid resin films can be controlled by plasticizer additions. This behavior parallels that of other polymers in regard to gas or vapor transmission, as a general trend is that increasing plasticizer content will loosen the structure. A word of caution should be added: although loss of plasticizer is not likely to be high, some of it will be evaporated by the permeating component and show up as an increase in permeability.

Most measurements of WVT are made by a weight-determination method. There are, of course, a number of variations that can be used, but they are all exemplified by the basic methods prescribed by the American Society for Testing and Materials, that is, ASTM Standard E96-66 [54]. The best way to describe this method is to quote from the ASTM directions:

"The material to be tested is fastened over the mouth of a dish, which contains either a desiccant or water. The assembly is placed in an atmosphere of constant temperature and humidity, and the weight gain or loss of the assembly is used to calculate the rate of water vapor movement through the sheet material under the conditions prescribed. . ." (by a set of temperature and relative humidity conditions).

Most of the WVT measurements in industrial use are carried out by a gravimetric method, and the usual permeability units shown in Fig. 5.17 are used more in research investigations.

Thus, when one looks up WVT values they are likely to be given as water-vapor permeabilities in (g) (cm thickness)/(24 hr)(m^2)(mm Hg) and variations thereof. Often, the thickness factor is not specified or the pressure differential is not given, and then such figures are of questionable validity. In addition it is important that the respective test conditions are known, as the effect of humidity in particular is very pronounced. In general, then, care has to be excercised if one wants to convert one set of units into another set.

Appropriate details of the method, often called the cup method, will be presented in Chapter XII.

A variation of the cup method was recently proposed by Osburn et al. [55], where the plastic material is used as a tube, or formed into a sealed envelope, and WVT is measured by weighing. Liquid can be placed inside the plastic to determine the outflow of water vapor (or solvent vapor), or a desiccant can be placed inside and vapor flow determined to the inside from a surrounding atmosphere of variable humidity. A more detailed description is also given in Chapter XII.

Of course, several methods have been used where the permeating water vapor is measured by a sepcific detector. Thus, an apparatus called the Infrared Diffusiometer uses an IR sensing beam to detect water-vapor buildup in a test cell. It was described by Wood [56] who states, however, that this apparatus is not a primary standard in the usual sense, and that it is basically a comparator. Salami [57] believes that the IR detector method is more accurate and rapid than the other methods. Details of this equipment are discussed in Chapter XII. An apparatus using helium as a sweep gas and a thermal conductivity cell as detector was investigated by Varsanyi [58]. In principle, it corresponds to the use of a gas chromatograph for measuring the transmitted water vapor.

Effect of Water Vapor on Gas Permeation

The existing situation can be summarized by saying that procedures for obtaining the permeability of *moist gases* is still in a state of flux. An ASTM committee has been engaged with this problem for some time and has conducted a number of round-robin tests, but, so far, it has not been possible to devise a fully acceptable procedure.

The main difficulty is the lack of agreement of results obtained by different investigators. Some variation is probably due to problems in humidity and temperature control, conditioning of the polymer film, swelling or nonswelling of the polymer, and the like, all of which require delicate control for consistency of operation. But, there are many instances where investigators' measurements vary by multiples of magnitude that could hardly be ascribed to lack of control over testing parameters.

For instance, Simril and Hershberger [59] report a 50-fold increase in CO_2 permeability through cellophane, when the relative humidity (RH) goes from 0 to 100%, while Pilar's [60] data would give some 1000-fold increase, both at $25°C$. Similar discrepancies were reported by other authors for CO_2 and other gases. This situation prompted Kantesaria [61] to conduct a specific study of the permeability of a number of gases in films of polyvinyl alcohol, Nylon-6, and silicone rubber. An essential feature of the study was to investigate two representative methods. One of the methods is similar to that used by Pilar [60], designated as Method A and the other method is an adaptation of the usual permeability method, called Method B.

In *Method A*, the film is preconditioned so that it has a uniform moisture content throughout, before a measurement is started. The partial pressure of water vapor is constant throughout the system when the test gas is introduced, so that permeation takes place at essentially constant absolute humidity and film moisture content.

In *Method B*, the test gas is humidified to the desired humidity, and is then allowed to permeate through the film. The downstream humidity depending on the relative permeation of water vapor, a moisture gradient is thus established in the film.

The results obtained with the two methods are shown in Fig. 5.18 and 5.19, for the permeation through polyvinyl alcohol and a tenfold difference in the methods is evident from the data. In each case, the permeability of CO_2 is plotted as a relative value Q_R, which represents the ratio of CO_2 permeability with water vapor to the permeability of the dry CO_2.

In general, the permeability of a gas increases as the moisture content of the film increases. Since Method B results in a film in which parts are drier than in Method A, it is to be expected that the permeability using Method B would be less. For films with a strong dependence of permeability on moisture content, this behavior was observed. For other films, the difference between the two methods is within experimental error.

Evidently, then, one has to consider that more than one method may have to be used, depending on the specific instance of the film application. Furthermore, one is tempted to speculate that Method B might give the same results as Method A if the test were carried out long enough so that the film would reach the appropriate moisture equilibrium.

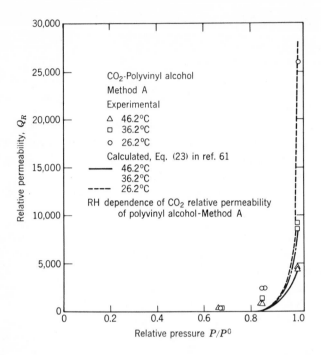

Fig. 5.18 Effect of water vapor on permeability of CO_2 by Method A [61].

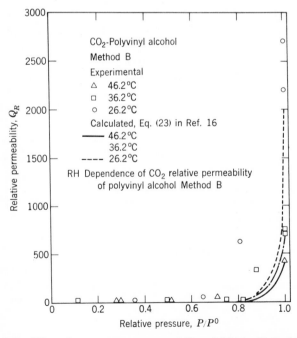

Fig. 5.19 Effect of water vapor on permeability of CO_2 by Method B [61].

4 TUBULAR MEMBRANES AND HOLLOW FIBERS

As indicated previously, the advent of fine-bore tubing and hollow fibers has had a tremendous impact on the matter of diffusion-cell design and construction. Before such materials became available, the greatest handicap to equipment design was the problem of crowding large areas into a minimum of space. Another equally vexing situation was the arrangement for support of membranes, which, as they are usually wanted as thin as possible, always presented difficult handling, installation, and maintenance problems. In addition, the inherent fragility of thin membranes gave rise to an undesirably high rate of ruptures in operation.

With present hollow-fiber technology it has become possible to house a large diffusion area in a relatively small space. One example of this is the Permasep equipment of the Du Pont Company for water purification, where as many as 1.7 million hollow fibers can be accommodated in a tube of about 10 cm outside diameter and 2.5 m length. Also, numerous proposals and prototype models of blood oxygenators and artificial kidney units can be found in the literature. Thus, it is to be expected that this type of arrangement will go far to make gaseous-diffusion operation competitive with existing separation operations.

The potential usefulness of fiber bundles was investigated by Blaisdell [62] by constructing a number of diffusion cells and using several varieties of hollow fibers and capillary-type tubing. For the sake of simplicity, most trials were made with air and its components. A not unexpected finding was that silicone-rubber tubing expanded in the studied pressure range. The expansion effect seemed to be dependent on the physical size of the tubing. For relatively large silicone-rubber tubing the change in size with pressure was hardly noticeable, but for one particularly thin tube the change in physical size was remarkable. This tubing was obtained from the Medical Engineering Corporation, Racine, Wisconsin.

By using the expandability of silicone-rubber tubing the diffusion rate of a gas in a given length of tubing is improved in two ways: first, when pressure is applied to the inside of a tube the diameter increases thus causing the walls to become thinner, and second, when the diameter increases the transfer area becomes larger. For most silicone rubber tubing this increase in permeability is not substantial, but for the special expansion discovered in the particular type, this increase gave rise to a multiple of the normal diffusive gas flow.

Figure 5.20 is a diagram of the rather simple type of structure used to build tubular cells. In some instances a rather long tubing bundle was coiled into the shell.

For most silicone-rubber tubing a maximum of 10 to 20% increase in the permeated flow rate would be obtained from the expansion of the tube when the pressure on the high side ranged up to 150 cm Hg over-pressure. The exceptional tubing, with an o.d. of 0.381 mm and on i.d. of 0.254 mm, showed

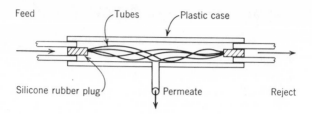

Fig. 5.20 Experimental tubular cell for expansible tubing [62]. Courtesy of the American Institute of Chemical Engineers.

an increased permeated flow rate of almost 20 times that obtained under unexpanded conditions. Results with oxygen permeation are presented in Fig. 5.21. Here, the ordinate is given as an "effective" permeability constant Q'_{O_2},

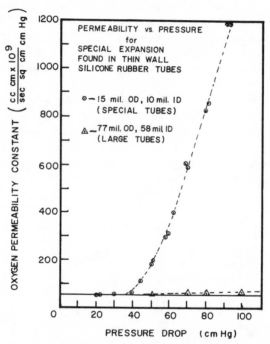

Fig. 5.21 Apparent permeability of silicone-rubber tubings [62]. Courtesy of the American Institute of Chemical Engineers.

which means that the values are calculated assuming that the tubes do not change in diameter. A base value of silicone rubber is 60×10^{-9} (std. cc)(cm)/-(sec)(cm^2)(cm Hg), as shown by the solid base line. The relatively thick-walled tubing gave only a slight increase with increasing pressure-pictured by the triangular points. Experiments with other gases proved that the increased flow rate was not due to pinholes. Actual separation results obtained for air as the feed stream are reported in the original paper [62].

It is evident that the tubular designs will receive considerable attention and find many new applications. Calculations for surface-area requirements will be treated in Chapter XIII; pertinent computer programs are contained in Appendices B and C.

5 MEMBRANES IN ISOTOPE SEPARATION

When isotopic molecular weights are appreciably different, as with hydrogen and deuterium, a membrane (barrier) will act as in normal gaseous diffusion either by microporous separation or by diffusive solubility flow with nonporous polymerics.

On the other hand, separation of isotopes such as the uranium hexafluorides becomes much more difficult because of rather small differences in molecular weight. Thus, the molecular-weight ratio of the UF_6 isotopes (^{238}U and ^{235}U, respectively) is about 1.008, and this is, of course, a measure of the respective ease of separation.

There are two basic considerations that must be examined in deciding which type of barrier should be used. When the Atomic Energy Commission's first gaseous-diffusion plant was established, the available membrane technology dictated the choice of a microporous metallic barrier [63]. At that time polymeric materials were not available that could stand up to the operating conditions of the cascade and the unavoidable presence of some HF. Also, in general, flow rates through nonporous polymeric membranes are lower by a factor of about 100 than rates through an equivalent thickness of an effective microporous medium.

The inherent adsorptive properties of a microporous structure will give rise to adsorption and consequent adsorbed or surface flow. When compounds that are to be separated exhibit preferential adsorption, it is possible to utilize this behavior to enhance the degree of separation. However, isotopes with small molecular-weight differences, specifically those of UF_6, are likely to adsorb to the same degree. Thus, the resulting surface flow will work against the molecular flow separation in the gas phase. Consequently, it will be of advantage in such systems to operate at reduced pressures and elevated temperatures. Both of these factors will reduce surface flow and thus improve the effectiveness of the separation process.

When polymeric membranes are used, the effect of selective solubility will control the degree of separation. Here, also, it is likely that little or no selectivity will exist between the solubilities of isotopes possessing only small differences in molecular weights. As the separative action of nonporous polymeric membranes is entirely dependent on differences in solubilities, it seems that such membranes will not, in general, be suitable for the separation of such isotopes.

Isotope separation by means of barrier diffusion is a highly specialized subject, and as such it has been treated extensively in the literature. In particular, there are two excellent reference texts: *The Theory of Isotope Separation* [64] and *Nuclear Chemical Engineering* [65]. A pertinent treatise of separation of isotopes with very small separation factors is presented in Chapter XVI.

Some years ago a comprehensive literature survey was carried out [on separation of stable isotopes] in Eastern Europe. This has been published as a P.B. report [66], and the available literature did not contain any essential information on barrier separations. Western European efforts in barrier separation seem to be limited to the French work, which, while lacking in detail, indicates that an inorganic microporous barrier will be used in an enrichment cascade [67, 68]. Although there is no definite information available, it is generally accepted that Soviet and Chinese production units for fissionable isotopes are based on United States technology.

6 MASS-DIFFUSION PROCESS

In 1940 Maier [69] authored a U.S. Bureau of Mines bulletin entitled "Mechanical Concentration of Gases." An intensive study was carried out on a diffusional separation process by using a rather porous septum and a so-called sweeping gas or vapor on one side of the septum. Maier resurrected the term "atmolysis," originally used by Graham, for the process. Subsequent publications by Benedict [70] and Benedict and Boas [71] make use of the term "mass diffusion." Cichelli et al. [72] label it "sweep diffusion," and Schwertz [73] calls it "free double diffusion."

Figure 5.22 indicates diagrammatically how the process operates. Benedict's description [70] of the process is very appropriate, as follows:

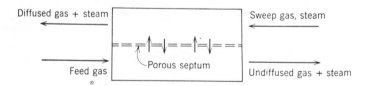

Fig. 5.22 Schematic of sweep gas-diffusion process.

"The individual stage consists of a vessel divided into two compartments by a porous barrier or screen made up of relatively coarse holes. The barrier serves only to retard mass mixing of the gases in the two compartments, and unlike the gaseous diffusion process, does not promote molecular effusion. Steam, or some other cheap, readily condensible vapor, is charged on one side of this barrier, and flows countercurrent to the gases to be separated, which are charged on the other side of the barrier. In one method of operating this process, the gases in each compartment are at substantially the same pressure. Under these conditions steam will diffuse through the barrier into the gases to be separated, and the gases will diffuse into the steam, with a total diffusive flow rate equal to that of the steam. The component with the higher diffusion coefficient, D_1, into steam will tend to be concentrated in the diffused gas relative to the component with the lower diffusion coefficient into steam, D_2. If the pressures are balanced exactly and the mole fraction of steam on each side of the barrier is high, the separation factor, a, is $a = D_1/D_2$.

In another method of operating this process, the steam pressure is greater than the gas pressure and there is a net flow of steam into the gas, which greatly reduces the diffusion rate of each component. However, the rate of diffusion of the component of lower diffusion coefficient is retarded to a greater extent than the rate of diffusion of the other component, so that the separation factor is increased significantly."

The role of the separating medium is actually not that of a diffusive membrane. Therefore, strictly speaking, this process is in a fringe area of membrane separation. However, it has received enough attention that it should, at least, be presented in highly abbreviated form. The publications by Benedict and Boas [71] and Cichelli [72] carry extensive mathematical interpretations, as well as experimental separation data.

Nomenclature

Symbol	Meaning
A,B	Constants in equations
M	Molecular weight
P	Pressure
P°	Saturation pressure
$\overline{P}$	Average pressure
P_H	High pressure
P_L	Low pressure
P_r	Pressure ratio P_L/P_H
Q	Permeability
Q_R	Relative permeability
Q'	Apparent permeability

Symbol	Meaning
T	Absolute temperature
n	Number of molecules diffusing
x	Composition unpermeated
y	Composition, permeated
a^*	Ideal separation factor
Δ	Constant in equation
θ	Fraction permeated

Subscripts	
1, 2	Denotes components 1, 2

Superscripts	
f	Feed
p	Product (Permeate)
o	Reject

References

1. H. Z. Friedlander and R. N. Rickles, *Anal. Chem.*, **37**, 27A (1965).
2. Thomas Graham, *Trans. Roy. Soc. (London)*, **153**, 385 (1863).
3. M. Knudsen, *Ann. Physik*, **28**, 75 (1909).
4. K. Higashi, H. Ito, and J. Oishi, *J. Atom. Energy Soc., Japan*, **5**, 846 (1963).
5. R. M. Barrer, *Proc. Brit. Ceram. Soc.*, **5**, 21 (1965); *A.I.Ch.E. – Inst. Chem. Engrs (London), Symp. Ser.*, **1**, 112 (1965).
6. J. A. Weaver and A. B. Metzner *A.I.Ch.E.J.*, **12**, 655 (1966).
7. S-T. Hwang and K. Kammermeyer, *Sepn. Sci.*, **1**, 629 (1966); *Can. J. Chem. Eng.*, **44**, 82 (1966).
8. K. Kammermeyer, "Gas and Vapor Separations by Means of Membranes." in *Progress in Separation and Purification*, E. S. Perry, Ed., Vol. 1, Interscience, New York, 1968, p. 335.
9. R. F. Bartholomew and E. A. Flood, *Can. J. Chem.*, **43**, 1968 (1965).
10. E. R. Gilliland, R. F. Baddour, and J. L. Russell, *A.I.Ch.E.J.*, **4**, 90 (1958).
11. J. F. Haman, Ph.D. thesis, Univeristy of Iowa, 1965.
12. D. E. Stahl, Ph.D. thesis, University of Iowa, 1971.
13. S-T. Hwang and K. Kammermeyer, *Sepn. Sci.*, **2**, 555 (1967).
14. R. W. Tock and K. Kammermeyer, *A.I.Ch.E. J.*, **15**, 715 (1969); R. W. Tock, Ph.D. thesis, University of Iowa 1967.
15. D. H. Hagerbaumer and K. Kammermeyer, *Chem. Eng. Prog. Symp. No. 10, Collected Res. Papers for 1954*, **50**, 25 (1954).
16. K. Kammermeyer and D. D. Wyrick, *Ind. Eng. Chem.*, **50**, 1309 (1958).
17. C. E. Rogers and S. Sternberg, *J. Macromol. Sci.–Phys. B*, **5**, 189 (1971).
18. R. M. Barrer, *Diffusion in and through Solids*, Cambridge University Press, London, 1951.

19. H. E. Huckins and K. Kammermeyer, *Chem. Eng. Progr.*, **49**, 180, 294, 517 (1953).
20. I. Sobolev, J. A. Meyer, V. Stannett, and M. Szwarc, *Ind. Eng. Chem.*, **49**, 441 (1957).
21. R. W. Tock, M.S. thesis, University of Iowa, 1964.
22. D. W. Brubaker and K. Kammermeyer, *Ind. Eng. Chem.*, **44**, 1465 (1952).
23. P. Meares, *J. Am. Chem. Soc.*, **76**, 3415 (1954).
24. W. H. Burgess, H. B. Hopfenberg, and V. T. Stannet, *J. Macromol. Sci.—Phys. B*, **5**, 23 (1971).
25. N. N. Li and R. B. Long, *Progress in Separation and Purification*, Vol. 3, E. S. Perry and C. J. Van Oss, Eds., Wiley, New York, 1970.
26. N. N. Li and E. J. Henley, *A.I.Ch.E. J.*, **10**, 666 (1964); N. N. Li and R. B. Long, *ibid.*, **15**, 73 (1969).
27. S. A. Stern, T. F. Sinclair, P. J. Gareis, N. P. Vahldieck, and P. H. Mohr, *Ind. Eng. Chem.*, **57**, 49 (1965); S. A. Stern, "Industrial Applications of Membrane Processes: The Separation of Gas Mixtures," *Proc. of Symposium*, May 19-20, 1966, Southern Research Institute, Birmingham, Alabama.
28. S. A. Stern, S. -M. Fang, and R. M. Jobbins, *J. Macromol. Sci.—Phys. B*, **5**, 41 (1971).
29. N. N. Li, R. B. Long, and E. J. Henley, *Ind. Eng. Chem.*, **57**, 18 (1965).
30. N. N. Li, *I E C Prod. Res. Dev.*, **8**, 281 (1969).
31. S. A. Stern, J. T. Mullhaupt, and P. J. Gareis, *A.I.Ch.E. J.*, **15**, 64 (1969).
32. S. A. Stern, S. -M. Fang, and H. L. Frisch, *J. Polymer Sci. A2*, **10**, 201 (1972).
33. M. E. Breuer and K. Kammermeyer, *Sepn. Sci.*, **2**, 319 (1967).
34. S. -T. Hwang, T. E. S. Tang, and K. Kammermeyer, *J. Macromol. Sci.—Phys. B*, **5**, 1 (1971).
35. R. E. Kesting, *Synthetic Polymeric Membranes*, McGraw-Hill, New York, 1971.
36. R. W. Roberts and K. Kammermeyer, *J. Appl. Polymer Sci.*, **7**, 2175 (1963).
37. A. S. Michaels and R. W. Hausslein, *J. Polymer Sci. C*, **10**, 61 (1965).
38. C. E. Rogers, *J. Polymer Sci. C*, **10**, 93 (1965).
39. R. Ash, R. M. Barrer, and D. G. Palmer, *Brit. J. Appl. Phys.*, **16**, 873 (1965).
40. C. E. Rogers, V. Stannett, and M. Szwarc, *Ind. Eng. Chem.*, **49**, 1933 (1957).
41. R. G. Buckles, Ph.D. thesis, Massachusetts Institute Technology, 1966.
42. K. W. Koh, Ph.D. thesis, University of Iowa, 1970.
43. R. L. Stallings, H. B. Hopfenberg, and V. Stannett, American Chemical Society, 164th Mtg, New York, August-September, 1972, p. 131.
44. W. R. Brown, R. B. Jenkins, and G. S. Park, American Chemical Society, 164th Mtg, New York, August-September 1972, p. 139.
45. T. K. Kwei and T. T. Wang, American Chemical Society, 164th Mgt, New York, August-September 1972, p. 146.
46. D. R. Paul and D. R. Kemp, American Chemical Society, 164th Mtg, New

York, August-September 1972, p. 156.
47. C. K. Choi, M.S. thesis, University of Iowa, 1972; Dept'l Rept., Chemical & Materials Eng'g, Univeristy of Iowa, 1974.
48. K. Kammermeyer, *Ind. Eng. Chem.*, **49**, 1685 (1957); U.S. Patent 2,966,235, (1960).
49. C. J. Major and R. W. Tock, in *Atmosphere in Space Cabins and Closed Environments*, K. Kammermeyer, Ed., Appleton-Century-Crofts, New York, 1966, p. 120
50. D. W. Brubaker, K. Kammermeyer, and J. O. Osburn, *Ind. Eng. Chem.*, **46**, 733 (1954).
51. F. P. McCandles, *I&EC Proc. Des. and Dev.*, **11**, 470 (1972).
52. V. Stannett and H. Yasuda, "Permeability." in *Crystalline Olefin Polymers*, Part II, R. A. F. Raff and K. W. Doak, Eds., Interscience, New York, 1964, Chapter 4.
53. J. J. Sciarra and S. P. Patel, *J. Soc. Cosmet. Chem.*, **23**, 605 (1972).
54. ASTM Standard Methods, E96-66, D1434-66 (Reapproved 1972) *1972 Annual Book of ASTM Standards*, Part 27, American Society for Testing and Materials, Philidelphia, Pa., 1972.
55. J. O. Osburn, K. Kammermeyer, and R. Laine, *J. Appl. Polymer Sci.*, **15**, 739 (1971).
56. R. Wood, *Modern Converter*, September-October (1970).
57. M. Salami, ACS Meeting, New York, August-September, 1972.
58. I. Varsanyi, Central Food Res. Inst., Budapest, Hungary; Departmental Rept. Chem. Eng., Univ. of Iowa, 1967.
59. V. L. Simril and A. Hershberger, *Modern Plastics*, **27**, 95 (1950).
60. F. L. Pilar, *J. Polymer Sci.*, **45**, 205 (1960).
61. P. P. Kantesaria, Ph.D. thesis, University of Iowa, 1970.
62. C. J. Blaisdell and K. Kammermeyer, *A.I.Ch.E. J.*, **18**, 1015 (1972).
63. J. Wilson, *Des Moines Register*, February 25 (1952).
64. K. Cohen, *The Theory of Isotope Separation*, McGraw-Hill, New York, 1951.
65. M. Benedict and T. H. Pigford, *Nuclear Chemical Engineering*, McGraw-Hill, New York, 1957.
66. K. Kammermeyer, *Soviet Research and Development in Separation of Stable Isotopes*, P.B. 61-31660, U.S. Department of Commerce, Office of Technical Services, Washington, D.C. (1961).
67. "Europe Looks at U-235," *C E News*, **36**, 68 (1958).
68. *Proc. 2nd Intern. Conf., Peaceful Uses At. Energy*, Geneva, 380 (1958).
69. C. G. Maier, *U.S. Bureau of Mines Bulletin*, **431**, Gov. Printing Office, Washington, D.C. (1940).
70. M. Benedict, *Trans. Am. Inst. Chem. Engr.*, **43**, 41 (1947).
71. M. Benedict and A. Boas, *Chem. Eng. Progr.*, **47**, (2), 51 (1951); *ibid.*, (3), 111 (1951).
72. M. T. Cichelli, W. D. Weatherford, Jr., and John R. Bowman, *Chem. Eng. Progr.*, **47**, (2) 63 (1951); *ibid.*, (3), 123 (1951).
73. F. A. Schwertz, *Ind. Eng. Chem.*, **45**, 1592 (1953).

Chapter **VI**

NONPOROUS INORGANIC MEMBRANES

In some respects the nonporous inorganic membranes are of great interest and potential value for industrial processes. This is because there are some such membranes that permit virtually complete separation of certain constituents in a single-stage operation. The examples of hydrogen separation with palladium foils, of helium separation with fused quatrz, and possibly of oxygen separation with silver are the only ones known today that approach a go-no-go separation process. Technologically, however, these processes have to be conducted at rather high temperatures, and this requirement introduces great difficulties in design and operation of equipment.

There is a fair number of reports on gas flow through metals and glasses, and the older literature is well covered in the earlier books of Barrer [1] and Jost [2] ; the latter has brought the coverage up to 1959.

A detailed résumé of the available information is hardly justified. However, a summary-type presentation, especially of more recent data, is of definite interest and also informative.

1 TRANSPORT MECHANISM

For inorganic membranes, the permeability is also considered to be a product of the solubility and diffusivity. However, depending on the mechanism of dissolution, the permeation equation will assume different forms. For systems of gas permeation through silica, zeolite, and inorganic salts, the permeation equation is the same as that for organic polymer membranes, that is,

$$F = QS \frac{P_1 - P_2}{l}$$ (6.1)*

where the permeability is given by

$$Q = DS_m \quad .$$ (6.2)†

The flow equation describing the linear dependence on the pressure, Eq. 6.1, however, was found not to be valid for many systems of gaseous permeation through metals. It has been a well-known fact that the gas molecules (especially diatomic molecules such as H_2, O_2, N_2, and CO) become dissociated when they are adsorbed on a metal surface. Thus, the adsorption rate of these gases on metals is proportional to the square root of the gas pressure instead of the pressure itself. Now, if we assume that the gas dissolved in the metal is in its atomic state, that is, dissociated state, rather than its molecular state, we can easily deduce that the permeation rate should be proportional to the square root of the gas pressure:

$$F = Q_M S \frac{P_1^{1/2} - P_2^{1/2}}{l} \quad ,$$ (6.3)

where Q_M is also a product of D and S_m in appropriate units. This equation was first developed by Richardson et al. [3] and subsequently verified by a number of investigators [1, 4-6]. Except at very low pressure, the Richardson equation is well obeyed for gas permeation through metals. At low pressures, Smithells and Ransley [6] accounted for the deviation from the square-root law by the use of an adsorption model.

The temperature dependence of the permeability through metals and glasses is of typical Arrhenius type as shown in Fig. 6.1, 6.2, 6.4 and 6.5. The activation energies range from 5000 to 45,000 cal/g atom.

2 PERMEABILITY IN METALS

Although the investigations of earlier workers were mostly of a qualitative nature, one must marvel at the astuteness of the investigators in observing flow of gases into and through metals. The incomparable Graham [7] already observed in 1866 that hydrogen would penetrate palladium, and Deville and Troost [8] reported as early as 1863 that hydrogen would diffuse through platinum.

*Also Eq. 2.2.
†Also Eq. 2.14.

For the present purpose it seems appropriate to list compilations of more recent data. There are two investigators, in particular, who have reported extensively on metals. They are F. J. Norton of the General Electric Company, and W. Eichenauer of the Technical University of Darmstadt, Germany: it is of interest to note the anomalous situation, that neither author ever refers to the other's work in their numerous publications.

The most important data on metals are summarized in Fig. 6.1 and 6.2. Data in Fig. 6.1 are for hydrogen permeation in a variety of metals, in Fig. 6.2 for several metal-gas combinations. Both graphs are plotted for a fixed

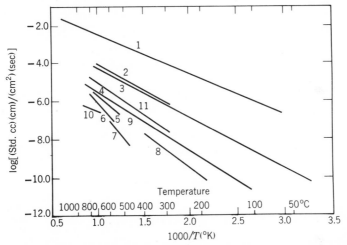

Fig. 6.1 Permeation flux of hydrogen through metal foils; $P_1 = 76$ cm Hg, $P_2 = 0$. 1: Pd [13] ; 2: Ni [13] ; 3: Fe (low C) [13] ; 4: Fe/27Cr [13] ; 5: Pt [13] ; 6: Mo [13] ; 7: Al [13] ; 8: Cu [13] ; 9: Ag [14] ; 10: Au [15] ; 11 321 s. steel ($l \geqslant 0.2475$ cm) [10].

pressure drop of 76 cm across the metal membrane. Thus, if one wishes to obtain permeabilities in the commonly accepted units for metals, the values read on the ordinate should be divided by $76^{1/2}$ to give

$$\frac{(\text{std. cc})(\text{cm thickness})}{(\text{sec})(\text{cm}^2 \text{ area})(\text{cm Hg pressure drop})^{1/2}}$$

A rather comprehensive study of quite recent origin is that of Phillips and Dodge [9, 10] on permeation of hydrogen through stainless steel. Of particular interest in that they investigated various thicknesses of metal and found that surface phenomena had an effect on permeability. Their data are plotted in Fig. 6.3 by the analysis scheme of Hwang [11], which permits the interpretation of interfacial resistances. The plot shows conclusively that the permeability behaves as the usual phenomenological coefficient, and thus increases with increasing thickness. The slope of the straight line in Fig. 6.3 represents the

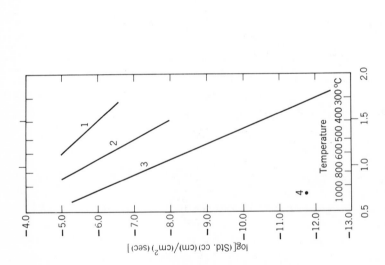

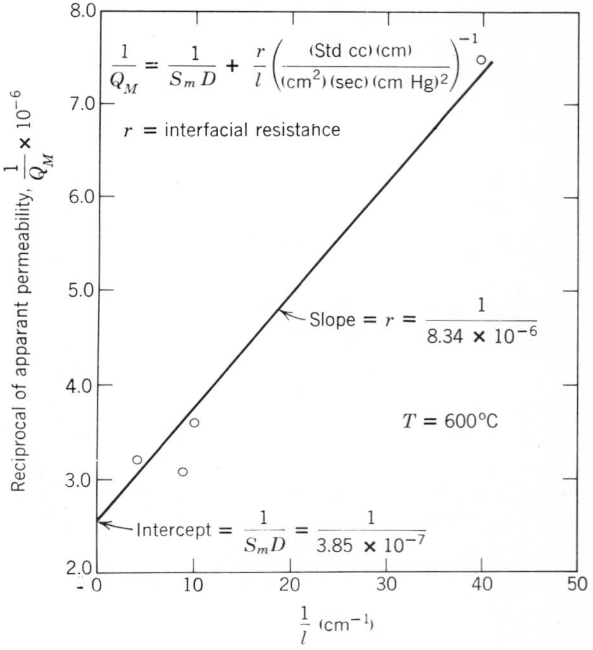

Fig. 6.3 Analysis of permeation data for interfacial resistance; hydrogen through stainless steel [9].

Fig. 6.2 Permeation flux of various gases through metal foils: $P_1 = 76$ cm Hg, $P_2 = O$. 1: Ni-D$_2$ [16]; 2: Ag-O$_2$ [15]; 3: Fe-N$_2$ [17]; 4: Pt-O$_2$ [18].

interfacial resistance for all thicknesses. The intercept on the ordinate, which is the reciprocal of Q gives the limiting, that is, maximum, permeability at infinite thickness, where interfacial resistances become negligible. Thus, the value of 3.85×10^{-7} agrees well in magnitude with the permeability read at $600°C$ from Fig. 6.1 on curve 11, when that value is divided by $(76)^{1/2}$ to convert to the same units. Also, it should be noted that the data in Fig. 6.3 were obtained earlier than those plotted in Fig. 6.1.

Information on the use of metals as separative membranes is presented in Chapter XV.

A very up-to-date collection of papers on hydrogen diffusion in metals was published in the August 1972 issue [12] of *Berichte der Bunsengesellschaft*. There are five major treatises and fifteen abstracts that cover all aspects, experimental and theoretical, of diffusional behavior of hydrogen and its isotopes in numerous metals and alloy combinations.

3 PERMEABILITY IN GLASSES

Besides the metals, the only other inorganics that have some promise as separative membranes are the ceramics, specifically high-silica materials. The fact that gases would penetrate glass at a fair rate has been known since 1900 when Villard [19] made the statement that "Fused silica, heated to redness, is permeable to hydrogen just like platinum, but to a lesser degree." Also, Watson [20] reported in 1910 his experiments on atomic-weight determination, where he purified helium by diffusion through silica glass. Additional observations were published by a variety of investigators, but little use of this property was made until Norton reinvestigated the phenomenon and published extensive data [13, 17, 18, 21].

The best available data on helium permeation through various glasses are correlated in Fig. 6.4, and a few data available on flow of other gases through siliceous materials are presented in Fig. 6.5. Two of the most recent studies are those of Masaryk [22] and of Walters [23]. Here, Masaryk was concerned with the effect of hydroxyl ion content of fused silica, as it affected the permeation of helium and hydrogen. The study by Walters developed data for the permeation of helium and deuterium through a borosilicate glass. Magnitudewise the new data agree with those given in Fig. 6.4. However, Walters found a hysteresislike behavior when taking readings at assending and descending temperatures, respectively. A possible interpretation of the observed differences is that the permeabilities may be affected by the thermal history of the glass.

So, at the necessary operating temperatures, that is, above at least $400°C$ for silica and above $600°C$ for metals, the permeabilities of such membranes are of the same order of magnitude as plastic films. However, the selectivity of

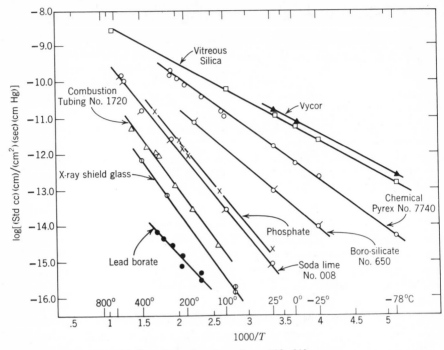

Fig. 6.4 Permeation of helium through various glasses [13, 21].

some of the metal-gas and glass-gas combinations reported upon is decidedly better than for any other known membrane-gas system. Thus, hydrogen through palladium, and helium through silica are almost exclusion-type processes as far as other gases are concerned.

The severe operating conditions called for by the use of inorganic membranes then become the deciding factor whether the technology is able to cope with the requirements. Presently available information indicates that the only instance of industrial application is hydrogen purification with palladium alloy foils, which is covered in some detail in Chapter XV.

Nomenclature

Symbol	Meaning
D	Diffusivity
F	Flux
P	Pressure
Q	Permeability

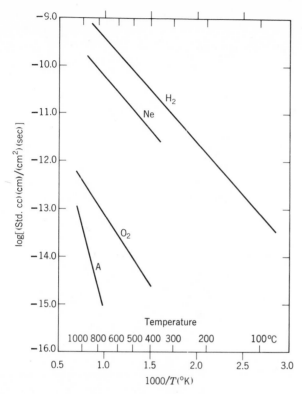

Fig. 6.5 Permeation of gases through vitreous silica [13]

Q_M	Permeability through metals
S	Area
S_m	Solubility in membrane
l	Membrane thickness

Subscript

| 1,2 | Position |

References

1. R. M. Barrer, *Diffusion in and through Solids,* Cambridge University Press, London, 1951.
2. W. Jost, *Diffusion in Solids, Liquids, Gases,* Academic, New York, 1965.
3. O. Richardson, J. Nicol, and T. Parnell, *Phil. Mag.,* 8, (6) 1 (1904).
4. V. Lombard and C. Eichner, *Bull. Soc. Chim. France,* 53, 1176 (1933).

5. W. Ham and J. D. Sauter, *Phys. Rev.*, **47**, 337 (1935).
6. C. J. Smithells and C. E. Ransley, *Proc. Roy. Soc. A*, **150**, 172 (1935).
7. T. Graham, *Phil. Mag.*, **32**, (4) (1866).
8. H. Deville and L. Troost, *Compt. Rend.*, **56**, 977 (1863).
9. J. R. Phillips and B. F. Dodge, *A.I.Ch.E. J.*, **8**, 93 (1963).
10. J. R. Phillips nad R. F. Dodge, *A.I.Ch.E. J.*, **14**, 392 (1968).
11. S. -T. Hwang. T. E. S. Tang, and K. Kammermeyer, *J. Macromol. Sci.-Phys. B*, **5**, 1 (1971).
12. *Ber. Bunsenges.*, **76**, 783 (1972).
13. F. J. Norton, in *1961 Transactions of the 8th Vacuum Symposium and 2nd International Congress*, Pergamon Oxford, 1962, p. 8.
14. W. Eichenauer, H. Kunzig, and A. Pebler, *Z. Metallkde*, **49**, 220 (1958).
15. W. Eichenauer and D. Liebscher, *Z. Naturforsch.*, **17a**, 355 (1962).
16. W. Eichenauer, *Memoirs Scientifique Rev. Met.*, **57**, 943 (1960).
17. F. J. Norton, *General Electric Rev.*, September 1952.
18. F. J. Norton, *J. Appl. Phys.*, **29**, 1122 (1958).
19. P. Villard, *Compt. Rend.*, **130**, 1752 (1900).
20. W. Watson, *J. Chem. Soc., (London)*, **97**, 810 (1910).
21. F. J. Norton, *J. Amer. Ceram. Soc.*, **36**, 90 (1953).
22. J. S. Masaryk, Univ. Calif., AEC Contract No. W-7405-eng-48; PB 182175, 1968.
23. L. C. Walters, *J. Amer. Ceram. Soc.*, **53**, 288 (1970).

Chapter VII

PERVAPORATION

The process where a liquid mixture is in direct contact with one side of a membrane, and where the permeated product is removed from the other side is frequently called *liquid permeation*. This term was used by Binning [1] in his pioneering studies with polymer membranes. Because of the hybrid nature of the process, that is, the presence of both liquid and vapor phases, the term *pervaporation* was introduced by subsequent investigators. This term is more descriptive and appropriate and will therefore be used in this discussion. The process of pervaporation essentially involves selective sorption of a liquid mixture into the membrane, selective diffusion or flow through the membrane, and then desorption into a vapor phase.

As early as 1906, Kahlenberg [2] reported a qualitative study on the separation of a mixture of a hydrocarbon and an alcohol through a rubber membrane. The first quantitative investigation was conducted by Hagerbaumer [3] in 1955 with a microporous Vycor glass membrane and a high-pressure drop across it, to separate organic liquid-liquid mixtures, specifically azeotropic compositions. Definite separation was observed, suggesting a potential of its application to an industrial operation. Later, Heisler et al. [4] in 1956 and Binning et al. [1, 5] in 1958 utilized this hybrid operation of separation of a liquid-liquid mixture into a vapor mixture through nonporous polymeric films. Binning's studies were rather extensive, and he was able to obtain a high degree of separation and also high permeation rates. It has been recognized since then that, like gaseous permeation, the excellent single-state selectivity and the low operating cost due to high thermal efficiency and simplicity of the operation are major advantages over other conventional equilibrium separation processes such as distillation and extraction. A comparison of the separation of a water-isopropanol mixture by distillation and pervaporation is shown in Fig. 7.1. Furthermore, the fast permeation rate in liquid permeation indicates a technological advantage over gaseous permeation.

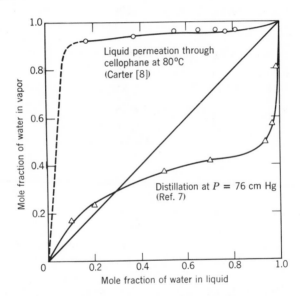

Fig. 7.1 Comparative separation of water-isopropanol mixtures by distillation and pervaporation.

The reasons why the operation has not yet become entrenched, in spite of the above-mentioned advantages, are the high equipment costs, difficulties in fabricating the unit, and stability of the membrane. In view of the present fast-developing technology of synthetic polymer membranes, the operation of liquid permeation should, however, compete with other conventional operations, particularly for separation of hydrocarbon mixtures, azeotropic mixtures,

mixtures with close boiling points and similar solubilities in a solvent, and mixtures of water pollutants.

1 PERMEATION OF PURE LIQUIDS

The permeation of pure components is of fundamental importance and deserves somewhat detailed treatment.

Solubility, Diffusivity, and Concentration Profile

Originally Binning et al. [5] assumed the existence of the so-called "two phases" inside of the membrane, that is, the "solution phase" on the incoming side of the film and the "vapor phase" on the other side, so that the permeating molecules experience their main resistance of permeation only in the vapor phase. More generally and phenomenologically, Li and Long [6] described the process in terms of ordinary diffusion with a concentration-dependent diffusivity $D(C)$, that is, the molecular flux J^* in the one dimensional space is

$$J^* = -D(C) \frac{dC}{dx} \ , \tag{7.1}$$

where C is the concentration of solvent molecules in the polymer film, and distance x is measured from the incoming side of the film. Then, Fick's second law at steady state says

$$\frac{d}{dx} \ D(C) \frac{dC}{dx} = 0 \ . \tag{7.2}$$

The most popular functional form of $D(C)$ is

$$D(C) = D_0 \ \exp \ (aC) \ . \tag{7.3}$$

Here, the parameters D_0 and a are functions of temperature and the chemical and physical nature of the polymer and solvent molecules, such as crystallinity, degree of plasticization, or Flory-Huggins interaction parameter.

Substituting Eq. 7.3 into Eq. 7.2 and integrating with boundary conditions $C=C_1$ at $x=0$ and $C=C_2$ at $x=l$, the flux is obtained as (see Fig. 7.2):

$$J^* = \frac{D_0}{al} \ [\ \exp \ aC_1 \ - \ \exp \ aC_2 \ ] \ , \tag{7.4}$$

and the concentration profile becomes

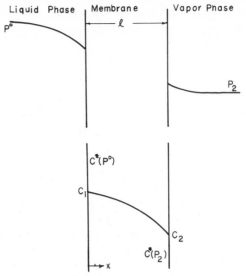

Fig. 7.2 Driving force diagram of pervaporation process.

$$C = \frac{1}{a} \ln \left[\exp aC_1 - \frac{x}{l} (\exp aC_1 - \exp aC_2) \right] \quad . \tag{7.5}$$

Suppose that the thermodynamic equilibrium exists at both interfaces:

$$C_1 = C^* (P^0) ,$$

$$C_2 = C^* (P_2) .$$

Here, $C^*(P)$ is a function of pressure with some parameters depending on temperature and the nature of the polymer and permeate. Also P^0 is the saturated vapor pressure of the liquid at the prevailing temperature, and P_2 is the pressure in vapor phase.

Then, one may express Eqs. 7.4 and 7.5 in terms of P^0 and P_2 by substituting the above equilibrium relationships for C_1 and C_2, respectively.

Finally, the permeability Q becomes

$$Q = \frac{J^* l}{\Delta P} = \frac{D_0}{a \, \Delta P} \left[\exp aC_1 - \exp aC_2 \right] ,$$

where $\Delta P = P^0 - P_2$.

It is well known that for the static equilibrium relationship $C^*(P)$, most polymer-solvent systems obey either

$$C^*(P) = P \exp \sigma C$$

or Henry's law: $\hspace{12cm}$ (7.6)

$$C^*(P) = S^*P \quad ,$$

where σ and S^* are constant parameters that depend only on the temperature and the nature of the system, as they were orginally expressed by the Flory-Huggins interaction parameter.

In particular, if Henry's law is assumed, one obtains the flux:

$$J^* = \frac{D_0}{al} \left[\exp aS^*P^0 - \exp aS^*P_2 \right] \tag{7.7}$$

and the concentration profile:

$$C = \frac{1}{a} \ln \left[\exp aS^*P^0 - \frac{x}{l} (\exp aS^*P^0 - \exp aS^*P_2) \right] . \tag{7.8}$$

Equation 7.8 was used by Li and Long [6] to obtain computed concentration profiles. Finally, the permeability becomes

$$Q = \frac{D_0}{a \, \Delta P} \left[\exp aS^*P^0 - \exp aS^*P_2 \right] . \tag{7.9}$$

Indeed, the validity of the concentration dependence of diffusivity in the form of Eq. 7.2 was first confirmed experimentally by Richman and Long [9] using the microradiographic technique in unsteady state, and later more comprehensively by Kim [10] employing a simpler but more accurate method under steady-state conditions. In contrast to Li and Long's work, Kim measured the actual concentration profile in a stacked film packet without assumption of any equilibrium relationship, and he was able successfully to fit the measured profiles with Eq. 7.3 as shown in Fig. 7.3. To give an idea of the magnitude of the parameters in Kim's equations, some sample values are shown in table 7.1.

Kim also was able to establish that the concentration at the ingoing interface is not in static equilibrium, but that a dynamic steady-state equilibrium exists with the liquid phase. The discrepancy of the concentration from the static equilibrium value was pronounced, particularly in the system of Nylon 6 and dioxane. It is interesting that the numerical value of a for the system was unusually high, indicating a high degree of plasticization.

It is also important to note that for a high numerical value of a, the

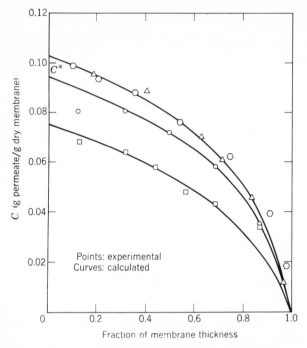

Fig. 7.3 Concentration profile for pervaporation of water through Nylon-6; effect of thickness. □−0.0195 cm (7.7 mils); ○−0.0444 cm (17.5 mils); △−0.0868 cm (34.2 mils); −0.1285 cm (50.6 mils). Reprinted from Ref. 10, by courtesy of Marcel Dekker, Inc.

Table 7.1 Parameters in Eqs. 7.4 and 7.5[a]

Membrane	Permeate	Temp ($^{\circ}$C)	$\dfrac{a}{\text{(g dry memb.}}$ g permeate)	$D_0 \times 10^8$ (cm^2/sec)	$\dfrac{C^*/P^0}{\text{(g permeate}}$ g dry memb.)
Nylon 6	Water	35	27.7	0.20	0.101
Nylon 6	Dioxane	35	86.3	14.5	0.0113
Cellulose Acetate	Water	35	17.4	9.2	0.006
Polyethylene	Dioxame	35	39.5	9.0	0.039
Polyethylene	Hexane	35	47.4	26.8	0.043
Polyethylene	Benzene	35	41.9	102.0	0.044

[a]Reference 10.

concentration profile is almost flat in the upstream section of the membrane, whereas it is very steep in the downstream section. This suggests that Binning's "two-phase" model is the most appropriate interpretation of the pervaporation process.

Effect of Temperature on Permeation Rate

According to Kim [10], D_0 increases very rapidly as the temperature increases, but its dependency on a and static sorption $C^*(P^0)$ is not uniform, perhaps due to structural changes in the polymer. Indeed, Li and Long [6] reported that the permeation rate of paraffins and naphthenes through polypropylene film increases rapidly with temperature, but the rate changes abruptly around the glass-transition temperature as shown in Fig. 7.4, where the product of molecular flux and thickness is plotted against the reciprocal absolute temperature.

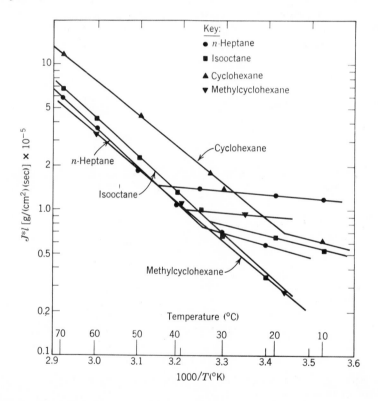

Fig. 7.4 Permeation rates for paraffins and naphthenes in pervaporation through 1-mil annealed polypropylene film [6]. Courtesy of the American Institute of Chemical Engineers.

Effect of Pressure on Permeation Rate

Pressure influences the permeation rate in two respects: implicitly, in properties of the polymer and liquid itself such as solubility and the parameters a and D_0, and explicitly, in the integrated form of the permeation rate, Eq. 7.7. Since solubility of liquid permeate in a polymer is not so sensitive to hydrostatic pressure beyond the saturation pressure, the effect of upstream pressure would not be appreciable. In addition, the pressure effect on the parameters a and D_0 would be negligible, considering that the saturated vapor pressure of ordinary liquids is usually below or around atmospheric pressure, and such pressures are too low to cause a change in the microstructure of the polymer. Explicit effect of downstream pressure, P_2, would not be large either, due to large numerical values of a for most liquid solvents, unless the pressure approaches the saturated vapor pressure as illustrated for the system of Nylon 6 and water in Fig. 7.5. For instance, if the downstream pressure P_2 were to rise from zero to a value such that P_2/P^0 becomes 0.30, the permeation rate at P_2 would still be 90% of that into a vacuum. Normal operating conditions are likely to be well below the 0.30 ratio. This argument was confirmed by experiments of Binning [5].

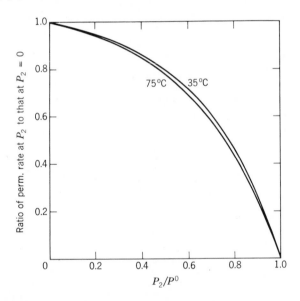

Fig. 7.5 Effect of down stream pressure in pervaporation for the system nylon-water [10]; Henry's law is assumed to be valid.

Effect of Nature of Polymer and Liquid

Liquid permeation is a combination of phenomena of sorption, diffusion, and desorption in series just like gaseous diffusion. Hence, a high permeation can be achieved for a system of high solubility and high diffusivity.

It is well known that a system that has similar solubility parameters will exhibit a high mutual solubility and therefore a high permeability. Thus, Li and Long [6] reported that hydrocarbons permeate faster through a hydrophobic film like polypropylene, than do polar liquids such as water and methanol, although the permeation rates are reversed through Mylar film. They also found that the aromatic ring has a strong effect on Saran film. Cook [11] was able to obtain a higher permeation rate of water through Nylon 6 by treating the surface of the film with nitric acid, which was interpreted as a polization of surface polymer molecules.

Crystallinity of the polymeric film is also an important parameter for permeation. Li and Long [12] found that the exponential factor a of diffusivity increases rapidly as crystallinity increases, but that the amount of permeate dissolved in the film decreases.

As discussed before, plasticization influences the premeation greatly, which is expressed by the magnitude of the parameter a. The liquid solvent loosens the microstructure of the polymer and thus enchances the diffusivity and therefore, the permeation rate, especially if it swells the polymer. It should, however, be noted that the effect is not always favorable for the separation of a liquid mixture, becuase it may reduce the separation factor. The effect of annealing with liquid solvents has also been studied [6, 13]. In general, solvent-annealed polyethylene film showed a greater sorption and a higher permeation rate.

A good deal of the technology of pervaporation, as applied to organic compounds and of preparation of membranes, is contained in the patents issued to Binning and his co-workers. They are of sufficient interest to warrant the attention of the reader [18]. A good summary of membrane separations reported in the literature was presented by Friedlander and Rickles [14].

2 PERMEATION OF LIQUID MIXTURES

The simplest case of permeation of a liquid mixture is where sorption and permeation of one species of the mixture do not influence those of other species. This is called "ideal permeation." In this case the total permeation rate and degree of separation of a mixture can be calculated directly from permeation data of its pure constituents, just as in the case of gaseous permeation. The actual permeation of real liquid mixtures is, however, far from the ideal permeation due to possible modification of the microstructure of the polymer film by liquid solvents, such as plasticization and redistribution of crystallite size in the film, as well as due to mutual interaction among liquid permeants within the film. Experimental studies [1, 5, 6, 8, 11, 12, 15] show that observed sorption and total permeation rates are often much larger, but degrees of separation smaller, than those for ideal mixtures. On the other hand, there are some reports in which a reverse situation is observed [6, 8].

An example of enhanced total permeation rates, but lowered separations, presented by Fels and Huang [15] is illustrated in Fig. 7.6 and 7.7 for the system benzene-n-hexane-polyethylene. As shown, the total permeation rate is enhanced tremendously and passes through a maximum as concentration of

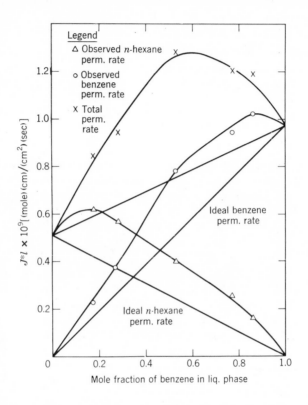

Fig. 7.6 Enhancement of permeation rates for system benzene-n-hexane-polyethylene at 308°K and $P_2 = 0$ (data from Ref. 15).

benzene increases (Fig. 7.6). But the separation is reduced very much compared to that in the ideal permeation, particularly at high concentrations of benzene (Fig. 7.7). Evidently there exists a maximum enrichable concentration beyond which the separation can not be achieved, just as in distillation with an azeotropic mixture. In the system benzene-isopropyl alcohol-polyethylene, the permeation rate of benzene is, however, retarded by the presence of isopropyl alcohol, and the degree of separation is lower than that of the ideal mixture. This is as illustrated in Fig. 7.8 and 7.9.

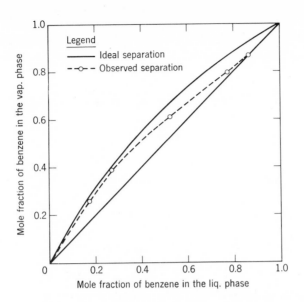

Fig. 7.7 Separation of mixtures of benzene and n-hexane with polyethylene (see Fig. 7.6); Example of decreased separation due to solvent and polymer interaction [15].

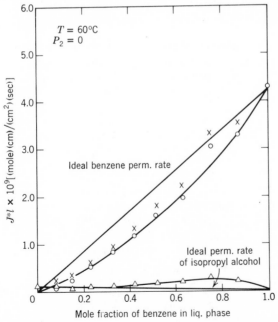

Fig. 7.8 Separation of benzene-isopropyl alcohol with polyethylene: system interaction leads to retarded permeation rates, data from [8]. $\circ$: observed benzene perm rate; $\triangle$: observed perm rate of isopropyl alcohol; $\times$: observed total perm rate.

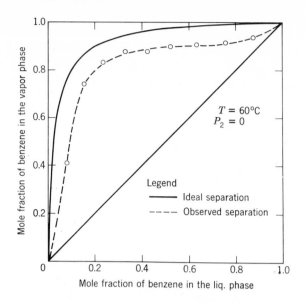

Fig. 7.9 Deviation of degree of separation from ideal separation for system shown in Fig. 7.8, data from [8].

3 SORPTION, DIFFUSION, AND PERMEATION: LIQUID AND VAPOR

In this section the discussion shall be confined to a binary mixture. However, its extension to multicomponent mixtures can easily be made.

If ordinary diffusion is assumed for the liquid permeation, the molecular fluxes, J_A^* and J_B^*, for a binary mixture can be expressed as

$$J_A^* = - D_A(C_A,C_B) \frac{dC_A}{dx} \quad ,$$

$$J_B^* = - D_B(C_A,C_B) \frac{dC_B}{dx} \quad , \tag{7.10}$$

where C_A and C_B are concentrations of species A and B, respectively, in the polymer film, and $D_A(C_A,C_B)$ and $D_B(C_A,C_B)$ diffusivities of A and B, respectively. These are, in general, functions of concentrations, as well as temperature and the nature of the polymer and permeants. Then, Fick's second law at steady state becomes

$$\frac{d}{dx} D_A (C_A,C_B) \frac{dC_A}{dx} = 0 \quad ,$$

$$\frac{d}{dx} D_B (C_A, C_B) \frac{dC_B}{dx} = 0 \ . \tag{7.11}$$

If functional forms of $D_A(C_A, C_B)$ and $D_B(C_A, C_B)$ are known, Eq. 7.11 can be integrated with boundary conditions, $C_A = C_{A1}$, $C_B = C_{B1}$ at $x = 0$, and $C_A = C_{A2}$, $C_B = C_{B2}$ at $x = l$, to find the fluxes:

$$J_A^* = J_A'(C_{A1}, C_{B1}, C_{A2}, C_{B2};l) \ ,$$

$$J_B^* = J_B'(C_{A1}, C_{B1}, C_{A2}, C_{B2};l) \ , \tag{7.12}$$

and the concentration profiles in the membranes:

$$C_A = C_A'(C_{A1}, C_{B1}, C_{A2}, C_{B2};l, x) \ ,$$

$$C_B = C'(C_{A1}, C_{B1}, C_{A2}, C_{B2};l, x) \ . \tag{7.13}$$

Suppose that sorption relationships at interfaces between concentrations in the membrane and partial pressures in the bulk liquid and vapor phase are known in the form:

$$C_{i1} = C_i(\gamma_A P_A^0 x_A, \gamma_B P_B^0 x_B) \ ,$$

$$C_{i2} = C_i(f_A P_2 y_A, f_B P_2 y_B). \quad i = A, B \ .$$

Here, P_i^0 is the vapor pressure of pure component i at the temperature of the system T, and P_2 is the total pressure in the vapor phase. Also x_i and y_i are mole fractions of component i in the liquid and vapor phase, respectively, and γ_i is the activity coefficient of i in the liquid phase, and f_i the fugacity coefficient of i in the vapor phase at T and P_2. If these relationships are substituted into Eq. 7.12 and rearranged, permeabilities can be conveniently defined in analogy with gas-phase permeabilities:

$$Q_i = \frac{J_i^* l}{P_i^0 (x_i - \text{Pr}_i \ y_i)}, \quad i = A, B \ , \tag{7.14}$$

where the generalized pressure ratio of i component, Pr_i, is defined as

$$\text{Pr}_i = \frac{P_2 f_i}{P_i^0 \gamma_i} \ . \tag{7.15}$$

Consequently, permeabilities Q_i's are, in general, functions of concentrations in both bulk phases, x_i and y_i, as well as P_i^0, P_2, γ_i, f_i and the respective properties of the polymer and permeate molecules. But Binning [5] found that Q_i's are essentially independent of the thickness of the membrane.

The above-mentioned procedure to estimate permeabilities from a study of diffusivities and sorption relationship is quite formal. However, the present knowledge on diffusivities and sorption properties is very limited, both experimentally and theoretically. Recently, Fels and Huang [15] attempted to correlate their experimental data of diffusivities using the free-volume theory, and Tombalakian and Markarian [16] introduced a set of constant linear phenomenological coefficients analogous to that in reverse osmosis. It appears that their results have not yet been well established. Hence, most present experimental work has dealt with the permeation rates themselves, rather than through study of diffusivities and sorption. Usually the permeation rates are measured with varying concentration of the liquid phase and vacuum in the vapor phase.

In the actual operation of liquid permeation, the boundary layer effect may not be so negligible as in gaseous permeation. In this case, one can conveniently introduce the effective permeability Q_i^e such that

$$\frac{1}{Q_i^e} = \frac{(r_i^l + r_i^v)}{l} + \frac{1}{Q_i} \quad , \tag{7.16}$$

where r_i^l and r_i^v are resistances to the transport of component i through boundary layers in the liquid phase and the vapor phase, respectively. They are, in general, functions of flow rates of both streams, permeabilities, and properties of mixtures such as diffusivity, viscosity, and density.

Effect of Nature of Polymer and Liquids on Separation

In permeation of one species of a real liquid mixture, strong interaction with that of the other sepcies always takes place. This is due to the structural change in the polymer caused by liquid permeants and also due to mutual interaction between permeant molecules themselves, as discussed previously. In many cases, sorption of a less-soluble species is enhanced by a higher-soluble species, and the total sorption of the mixture often exceeds that of the higher-soluble pure species as illustrated in Fig. 7.10. So does the total permeation rate as shown in Fig. 7.6 and 7.8. Consequently, the degree of separation is reduced below that of the ideal mixture and, at best, equal to the ideal separation.

Effect of plasticization of the polymer by permeants is believed to be the main reason for the strong interaction. The plasticization increases the total permeation rate, but it reduces the degree of separation. Hence, there exists an optimum degree of plasticization for a practical purpose.

Li and Long [12] summarized various means of preparing the polymer film to obtain an optimum degree of plasticization in the polymer film and also to

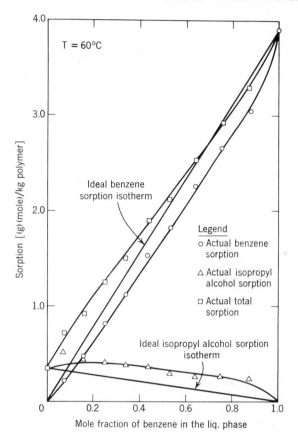

Fig. 7.10 Sorption of liquid mixture benzene-isopropyl alcohol in polyethylene, data from [8].

ensure the stability of the film, such as control of crystallinity, annealing, heat treatment, irradiation, adding a third compound, and preparing a graft copolymer film.

Effect of Temperature on Sorption, Permeation, and Separation

In many cases, the total sorption and total permeation rate increase as temperature increases as shown in Fig. 7.11 and 7.12, but the separation is reduced as shown in Fig. 7.13. However, the degree of separation for the system of benzene-*n*-hexane-polyethylene increases again beyond a certain concentration, as the temperature increases [15]. On the other hand, the reverse situation in total sorption and degree of separation was reported for the system of water-isopropyl alcohol-cellophane as shown in Fig. 7.14 and 7.15.

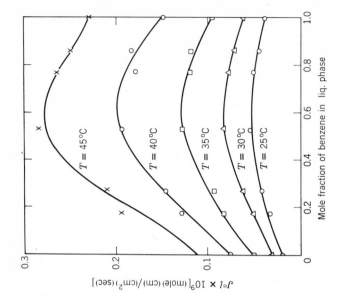

Fig. 7.12 Temperature effect on permeation rate of mixtures of *n*-hexane-benzene through polyethylene, data from [15].

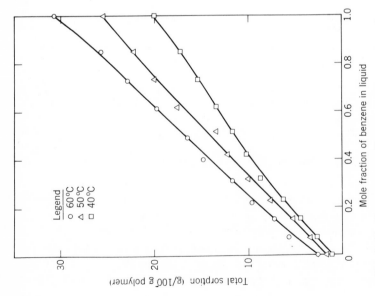

Fig. 7.11 Temperature effect on total sorption of mixtures of benzene-isopropyl alcohol in polyethylene, data from [8].

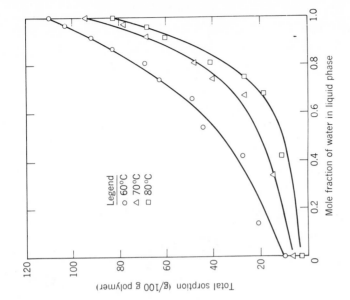

Fig. 7.14 Temperature effect on total sorption of water-isopropanol mixtures in cellophane, data from [8].

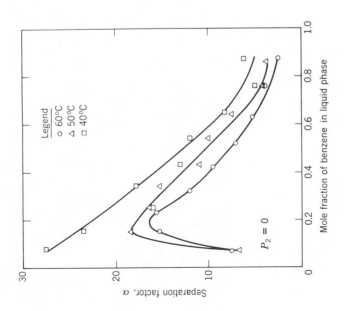

Fig. 7.13 Temperature effect on separation of benzene-isopropyl alcohol mixtures with polyethylene, data from [8].

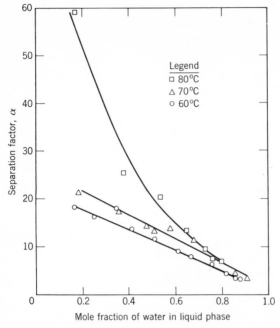

Fig. 7.15   Temperature effect on separation of system shown in Fig. 7.14, data from [8].

4 DESIGN OF A SEPARATION UNIT FOR LIQUID MIXTURES

A unit for pervaporation can be designed in exactly the same manner as in gaseous permeation, discussed in Chapter IV, such as the Weller-Steiner Cases I and II, or the cocurrent and countercurrent operations. However, the difference from gaseous permeation in the computational viewpoint lies in that permeabilities Q_i and separation factor a^* are strongly dependent on concentrations in the liquid and vapor phases. Also the boundary-layer effect, particularly in the liquid phase near the interface, may not be negligible.

Since most operations in pervaporation can be effectively conducted in vacuum in the vapor phase, and also because the permeation rates are not so sensitive to the downstream pressure in the vapor phase, the discussion in this section will be confined to two typical continuous operations. One scheme considers complete mixing; the other is based on plug flow in the liquid phase with the vapor phase under vacuum conditions. This is somewhat analogous to the Weller-Steiner Cases I and II, respectively, in gaseous permeation.

The same convention will be used for notation in this section as in gaseous permeation, that is, the species A refers to a higher-permeable species, and B represents the less-permeable species, such that the separation factor a is defined as

$$a = \left(\frac{y_A}{1-y_A}\right)\left(\frac{1-x_A}{x_A}\right) \; ,$$

and a is always greater than unity. It is convenient to introduce the ideal separation factor a^* and the reduced permeability Q'_A such that

$$a^* = \frac{Q_A P^o_A}{Q_B P^o_B} \qquad (7.17)$$

and

$$Q'_A = \frac{\text{permeability of component A in a mixture}}{\text{permeability of pure component A}}$$

$$= \frac{Q_A(x_A)}{Q_A(1)} \; . \qquad (7.18)$$

As mentioned earlier, a^* and Q'_A depend strongly on the concentration of x_A.

Operation with Complete Mixing in the Liquid Phase

When the liquid phase is completely mixed and the vapor phase is under a vacuum, the separation factor a becomes identical with the ideal separation factor a^*.

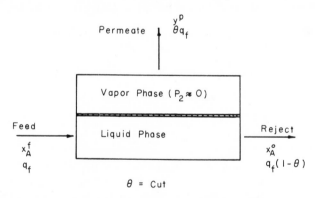

Fig. 7.16 Diagram for operation with complete mixing in liquid phase.

The material balance over the separation unit (see Fig. 7.16) leads to

$$\theta q_f y^p_A = \frac{Q_A \, P^o_A \, S \, x^o_A}{l} \; ,$$

$$\theta q_f (1 - y_A^p) = \frac{Q_B \ P_B^0 \ S(1 - x_A^0)}{l} \ .$$

After rearranging we get

$$\frac{y_A^p}{1 - y_A^p} = a* \ \frac{x_A^0}{1 - x_A^0} \qquad (7.19)$$

and

$$x_A^f = (1 - \theta) \ x_A^0 + \theta y_A^p \ ; \qquad (7.20)$$

the total membrane area required S can be expressed as a product of two coherent groups of parameters as follows:

$$S = \left(\frac{q_f \ l}{Q \ (1) P_A^0} \right) \left(\frac{\theta y_A^p}{Q_A' x_A^0} \right) \ , \qquad (7.21)$$

where $a*$ and Q_A' are values at $x_A = x_A^0$ and Q_A (1) is defined by Eq. 7.18. It is important to note that the first term on the right-hand side of Eq. 7.21 is just the membrane area required for q_f moles of pure component A to permeate through. Therefore, one may conveniently define a dimensionless area S' such that

$$S' = S \left(\frac{Q_A (1) P_A^0}{l \ q_f} \right) \ . \qquad (7.22)$$

Then, Eq. 7.21 reduces to

$$S' = \frac{\theta y_A^p}{Q_A' x_A^0} \ . \qquad (7.23)$$

It becomes apparent that S' depends only on the desired degree of separation and fraction of recovery, but not on q_f. Also, the relative quantity S' is very similar to the so-called "number of transfer units," which is widely used in the design of a continuous fractionation column.

It should be noted that in the operation with complete mixing in the liquid phase there exists a minimum stripping concentration x_{AM}^0 for a given feed composition x_A^f. It is given by

$$\frac{x_A^f}{1 - x_A^f} = a* \ \frac{x_{AM}^0}{1 - x_{AM}^0} \ ,$$

where a^* is evaluated at $x_A = x_{AM}^0$. Beyond this concentration no further stripping can be achieved by this operation.

Operation with Plug Flow in the Liquid Phase

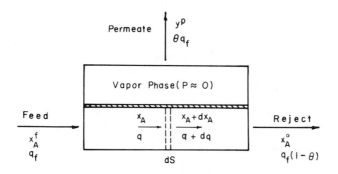

Fig. 7.17   Diagram for plug-flow operation in liquid phase.

Mass conservation over a differential area dS (see Fig. 7.17) leads to

$$d(qx_A) = - \frac{Q_A P_A^0}{l} \, dS \, x_A \, ,$$

$$d[q(1-x_A)] = - \frac{Q_B P_B^0}{l} \, dS \, (1-x_A) \, ,$$

where Q_A and Q_B are functions of x_A. Rearranging these equations, one can obtain a more convenient system of ordinary differential equations as follows:

for the flow rate:

$$\frac{dq'}{dx_A} = \frac{q'[x_A + (1-x_A)/a^*]}{x_A(1-x_A)(1-1/a^*)} \, , \qquad (7.24)$$

and for the dimensionless membrane surface area S':

$$\frac{dS'}{dx_A} = \frac{q'}{Q_A' x_A(1-x_A)(1-1/a^*)} \, , \qquad (7.25)$$

where

$$q' = \frac{q}{q_f} \, .$$

Hence, the problem is reduced to solving the system of ordinary differential equations with initial conditions:

$$q' = 1, \quad S' = 0 \quad \text{at } x = x_A^f \; ,$$

and to find

$$q' = 1 - \theta \quad \text{and} \quad S' \quad \text{at } x_A = x_A^0 \; .$$

Then, the mole fraction in the permeate stream y_A^p is

$$y_A^p = \frac{1}{\theta} \; [x_A^f - (1 - \theta)x_A^0] \; .$$

In general, the analytical solution for the system of ordinary differential equations, Eqs. 7.24 and 7.25, is not possible, because the functional forms of Q_A' and a^* with respect to x_A (Eqs. 7.17 and 7.18) may not be simple, or they are more often given in tabulated fashion. Hence, the solution can be obtained more effectively by an appropriate scheme of numerical integration like the Runge-Kutta-Gill method, with a suitable numerical scheme of interpolation such as the Lagrange-Aitken method [17] if Q_A' and a^* are available in tabular forms.

Computer programs for the two cases are presented in Appendix A.

Comparison of the Two Operations — Sample Calculations

As an illustration, numerical calculations were made for the two operations, using the system, benzene-n-hexane-polyethylene at $T = 308°K$ from Fels and Huang [15]. The results are presented in Fig. 7.18 and 7.19. Here, the boundary-layer effect is assumed to be negligible. From Fig. 7.18, it is seen that the definite advantage of plug-flow operation becomes pronounced as the cut increases. It is particularly interesting that the total areas required for a given cut are almost identical for the two operations (see Fig. 7.19).

Nomenclature

Symbol	Meaning
C	Concentration in the membrane phase
C^*	Concentration in the membrane phase in equilibrium with a bulk phase (see Fig. 7.2)
D	Diffusivity of a permeant in the membrane phase
D_0	Constant defined in Eq. 7.3

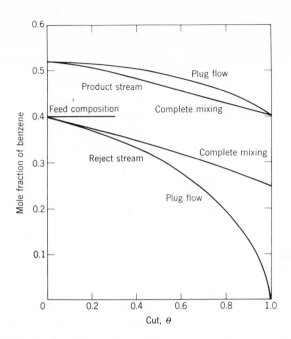

Fig. 7.18 Numerical results for pervaporation of system *n*-hexane-benzene through polyethylene at 308°K [15]; computer data.

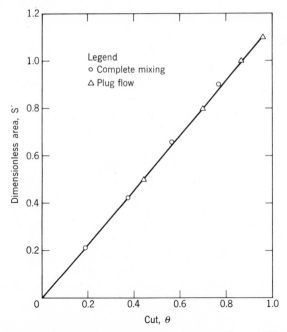

Fig. 7.19 Total area requirements for pervaporation of system shown in Fig. 7.18.

J^*	Molecular flux of a permeant in the membrane phase
P^0	Vapor pressure of a liquid
P	Pressure
P_1	Upstream pressure on liquid phase
P_2	Downstream pressure in the vapor phase
ΔP	Pressure drop across the membrane
Pr	Generalized pressure ratio defined in Eq. 7.15
Q	Permeability of a permeant
Q^e	Effective permeability
Q'	Reduced permeability of species A
S	Total area of the membrane required
S^*	Henry's constant
S'	Dimensionless area, Eq. 7.22
T	Temperature, absolute
a	Constant defined in Eq. 7.3
f	Fugacity coefficient
l	Thickness of membrane
q	Flow rate in liquid phase
r^l	Resistance to transport through the boundary layer in the liquid phase
r^v	Resistance to transport through the boundary layer in the vapor phase
x	Distance measured from upstream side of membrane
x_i	Mole fraction in liquid phase (i = A, B,. . .)
x^0	Mole fraction in reject stream
x^f	Mole fraction in feed stream
y	Mole fraction in vapor phase
y^p	Mole fraction in product stream
a	Separation factor defined as $a = \dfrac{y_A}{(1-y_A)} \dfrac{(1-x_A)}{x_A}$
a^*	Separation factor defined by Eq. 7.17
σ	Constant in Eq. 7.6
θ	Cut
γ	Activity coefficient

Subscripts

A	Referes to higher permeable sepcies A
B	Refers to less permeable species B
AM	Refers to minimum concentration of species A
f	Refers to feed
i	Refers to a species

p	Refers to permeate
1	Refers to the upstream side of the membrane
2	Refers to the downstream side of the membrane

Superscripts

f	Refers to feed
p	Refers to permeate

References

1. R. C. Binning and F. E. James *Petr. Refiner,* **39**, 214 (1958).
2. L. Kahlenberg, *J. Phys. Chem.,* **10**, 141 (1906).
3. K. Kammermeyer and D. H. Hagerbaumer, *A.I.Ch.E. J.,* **1**, 215 (1955).
4. E. G. Heisler, A. S. Hunter, J. Siciliano, and R. M. Treadway, *Science,* **124**, 77 (1956).
5. R. C. Binning, R. J. Lee, J. F. Kennings, and E. C. Martin, *Ind. Eng. Chem.,* **53**, 45 (1961).
6. N. N. Li and R. B. Long, *A.I.Ch.E. J.,* **15**, 73 (1969).
7. *International Critical Tables,* McGraw-Hill, New York, 1933.
8. J. W. Carter and B. Jagannadhaswamy, Proceedings "Symposium on the Less Common Means of Separation," *Inst. Chem. Engs,* **1964**, 35.
9. D. Richman and F. A. Long, *J. Amer. Chem. Soc.,* **82**, 509 (1960).
10. S. N. Kim and K. Kammermeyer, *Sepn. Sci,* **5**, 679 (1970).
11. R. L. Cook, M.S. thesis, University of Iowa, 1971.
12. N. N. Li and R. B. Long, in *Progress in Separation and Purification,* Vol. 3, E. S. Perry and C. J. Van Oss, Eds., Wiley, New York, 1970, p. 153.
13. R. F. Baddour, A. S. Michaels, H. J. Bixler, R. P. de Fillipi, and J. A. Barrie, Div. of Ind. and Chem., 143rd Meeting, A.C.S., Los Angeles, March 1963.
14. H. Z. Friedlander and R. N. Rickles, *Anal. Chem.,* **37**, 27A (1965).
15. M. Fels and R. Y. M. Huang, *J. Macromol. Sci. – Phys. B,* **5**, 89 (1971).
16. A. S. Tombalakian and G. K. Markarian, *Can. J. Chem. Eng.,* **50**, 131 (1972).
17. M. Abramowitz, and I. A. Stegun, *Handbook of Mathematical Functions,* Dover, New York, 1965.
18. R. C. Binning and J. T. Kelley, U.S. Patent 2,913,507 (1959). R. J. Lee and R. C. Binning, U.S. Patent 2,923,749 (1960). R. C. Binning, J. E. Jennings, and E. G. Martin, French Patent 1,198,940 (1959). R. C. Binning and R. J. Lee, U.S. Patent 2,923,751 (1960); U.S. Patent 2,953,502 (1960). J. F. Jennings and R. C. Binning, U.S. Patent 2,956,070 (1960). R. C. Binning and J. M. Stuckey, U.S. Patent 2,958,657 (1960). R. C. Binning and W. F. Johnston, Jr., U.S. Patent 2,970,106 (1961). R. C. Binning, U.S. Patent 2,981,680 (1961). R. C. Binning, J. F. Jennings, and E. G. Martin, U.S. Patent 2,985,588 (1961); U.S. Patent 3,035,060 (1962).

Chapter VIII

SEPARATIONS IN LIQUID PHASE

There are many processes that fall into the category of liquid-phase operation. However, they can be distinguished on the basis of a major difference, that is, whether or not an electric potential is used as a driving force. On this basis it will be convenient to cover strictly liquid-phase processes as such, and present a separate treatment for the so-called electromembrane processes.

Thus, the present chapter will deal with reverse osmosis, ultrafiltration, dialysis, and liquid membranes, as the most important representatives of liquid-phase separations. Before the individual processes are discussed it is necessary to consider the phenomena involved in the generalized flow through membranes in the liquid phase.

1 TRANSPORT THROUGH MEMBRANES

Even when there is no external electric potential applied, the simple diffusion equation may not explain the observed membrane-transfer processes. This is due to the permselective property of membranes. For example, if a membrane permits transfer of only small ions and rejects large ions, Donnan equilibrium occurs (see Chapter III). In reverse osmosis, the membrane rejects solute and passes only the solvent. In facilitated diffusion and active transport (see also Chapter III), the simple diffusion equation totally fails to explain the experimental observations. Another example is the case of electrically charged membranes. Some membranes allow only cations to pass, and the others allow only anions. All of these peculiar phenomena are caused by the membrane-diffusant interactions.

In this chapter, for the sake of convenience, we will divide the discussion of membrane transport into two groups: one for electrolyte solutions and the other for nonelectrolyte solutions. However, even this differentiation msut be treated with caution. In a system containing only one solute that is an electrolyte, the constraint of electric neutrality makes this system behave the same as a nonelectrolyte system. It appears that the reverse-osmosis process as reported in the literature has been treated as a nonelectrolyte system.

The basic difference in the two systems, electrolyte solution and nonelectrolyte solution, lies in the fact that the transport equation for the electrolyte system involves an extra term representing the electric potential, which is generated by the presence of ions.

Transport of Nonelectrolytes

As discussed in Chapter II, transport of a chemical species in a nonelectrolyte solution through a membrane can generally be described by the abstract theory of nonequilibrium thermodynamics. When the solution is a liquid phase, the concentration gradients of solutes and the solution pressure are important forces, and they primarily induce the fluxes of the solutes and the solvent. Each flux can then be associated with the various forces using a finite number of the linear phenomenological coefficients. Some of the coefficients defined in this way can be eliminated by employing the Onsager reciprocal theorem. However, this phenomenological theory does not offer any explanation of the mechanism of transport. Thus, two models shall be considered to elucidate the mechanism. Both have received attention by various investigators in one form or another.

In the first model, the membrane is viewed as a homogeneous medium with a finite thickness. When the membrane is associated with a solution, the membrane phase itself is considered as a solution in which the "component" of the membrane segment is very sluggish and almost stationary. Thus, the well-established solution theory can be applied directly to the system. In the present discussion it will be designated as the *diffusion model.* This model is very convenient, particularly in the case where the molecular size of a chemical species is of the same order as that of another species from which it is to be separated, as, for instance, in reverse osmosis. If the two molecular sizes are quite different as in ultrafiltration, a porous membrane with a suitable average pore size is commonly chosen to effect the separation. For this situation it will be convenient to use what would be the second model, which could be called a *pore model.* Then, the membrane will be considered to be porous with a finite thickness. Also, it is assumed that the fluxes of chemical species whose molecular sizes are much smaller than the average pore size result primarily from the convective pore flow, whereas other species of much larger sizes are filtered out. However, the flux of the species of an intermediate size is hindered by interaction with the membrane wall and therefore is usually less than what would be expected from convective pore flow. The possibility of interaction between intermediate-size molecules and the membrane wall can lead to complications, so that the pore model may not be as simple as one might infer from a casual inspection. Unless the two molecular sizes are very different, the diffusion model is often more convenient, even in ultrafiltration, if an appropriate set of parameters such as diffusivity and distribution coefficient is chosen.

Diffusion Model

Here the membrane shall be considered as a homogeneous medium with a finite thickness l. In a rigorous sence, all actual membranes are heterogeneous to some extent. However, in most cases, we can treat the membrane as a homogeneous medium, if we wish, by choosing a sufficiently large value in a λ'-scale evaluation scheme as is often done in statistical mechanics. This conceptual parameter λ' is strictly arbitrary and may be any magnitude, depending on how detailed the observer wishes to interpret a given medium and its associated phenomena. For example, if one wants to characterize a gas phase as a homogeneous medium, then the value of λ' should be greater than the mean free path of the molecules. Similarly, the λ'-scale in a liquid phase can not be smaller than, say, 10Å. In the case of membrane phenomena, the value of λ' is much larger than the pore size or any other characteristic dimension of the heterogeneity.

If the membrane is a composite or anisotropic (see Section 2) in scale, we shall assume that the scale is in the order of l and therefore much larger than λ', as in most practical membranes. When a fluid is associated with such a membrane, the homogeneity implies that intensive thermodynamic variables and dynamic behaviors of the fluid in the actual membrane are effectively averaged in the λ' scale and then assigned to every local point in the membrane. Thus, the variables defined in this way may be expected to be smooth and analytic in the mathematical sense.

With these postulates, the interpretation can be undertaken by means of the diffusion model. Then, the effective diffusion of a molecular species in such a homogeneous membrane is due to the local gradient of the averaged chemical potential as in a homogeneous medium, so that the chemical potential $\overline{\mu}_i$ of the species i at temperature T and pressure $\overline{P}$ can be written as

$$\overline{\mu}_i = RT \ln \overline{a}_i + \overline{v}_i \ (\overline{P}\text{-}P^*) + \overline{\mu}_i^* \ (P^*) \ , \qquad (8.1)$$

where R is the gas constant, and $\overline{a}_i$ and $\overline{v}_i$ are the activity and the partial molar volume, respectively, of the species i in the membrane. Also, $\mu_i^*(P^*)$ is the chemical potential at a standard state at temperature T and pressure P^*.

The interaction taking place at the solution-membrane interface can be considered as follows. Suppose that a membrane is immersed in a homogeneous solution at a uniform temperature T and pressure P. The solution consists of solute i of activity a_i and partial molar volume v_i, and solvent B of activity a_B and molar volume v_B. If the membrane is in equilibrium with the solution, the chemical potentials of both the solute and solvent must prevail uniformly in the two phases:

for the solute species i:

$$RT \, ln \, \frac{\overline{a_i}}{a_i} = -v_i(\overline{P}-P) = -v_i\pi_s \quad , \tag{8.2}$$

and for solvent B:

$$RT \, ln \, \frac{\overline{a_B}}{a_B} = -v_B(\overline{P}-P) = -v_B\pi_s \quad . \tag{8.3}$$

Here the quantities without the bar refer to the solution phase, and π_s is the so-called swelling pressure. The partial molar volumes of both solute and solvent are assumed to be the same in both phases. Eliminating the swelling pressure in Eqs. 8.2 and 8.3, one may obtain the distribution coefficient K_i of molar concentrations of i in the two phases:

$$K_i \equiv \frac{\overline{C_i}}{C_i} = \frac{\gamma_i}{\overline{\gamma_i}} \left(\frac{\overline{\gamma_B}\overline{C_B}}{\gamma_B C_B} \right)^{\frac{v_i}{v_B}} \tag{8.4}$$

The activity coefficients γ_j and $\overline{\gamma_j}$ are defined as

$$a_j = C_j\gamma_j; \quad \overline{a_j} = \overline{C_j}\overline{\gamma_j}, \quad j = i, \, B \, . \tag{8.5}$$

The above treatment, resulting in Eq. 8.4, represents a generalized expression for the thermodynamic equilibrium between the solution and the membrane.

When this concept is applied specifically to *osmosis*, the following analysis can be used. Consider two solutions separated by a membrane. Both solutions contain the same solvent B, and the membrane permits the passage of the solvent only. If the system is in equilibrium at temperature T, the uniformity of the chemical potential of the solvent requires

$$P^{\mathrm{I}} - P^{\mathrm{II}} = -\frac{RT}{v_B} \, ln \, \frac{a_B^{\mathrm{I}}}{a_B^{\mathrm{II}}}$$

$$= \pi^{\mathrm{I}} - \pi^{\mathrm{II}} \quad . \tag{8.6}$$

The superscripts I and II refer to the two solution phases, and the so-called osmotic pressure π is conveniently introduced as

$$\pi = -\frac{RT}{v_B} \, ln \, a_B \quad , \tag{8.7}$$

as specified by Eq. 3.24 in Chapter III. As indicated by Eq. 8.6, if the two solution pressures are the same, the solvent transports from the dilute solution (of higher a_B but lower π) to the concentrated solution (of lower a_B but higher π), until $a_B^I = a_B^{II}$ or $\pi^I = \pi^{II}$. This phenomenon is considered to be *regular osmosis*. However, if the two external pressures are set at the initial state such that

$$(P^I - P^{II})_0 > (\pi^I - \pi^{II})_0 \quad ,$$

the solvent diffuses from the concentrated solution to the dilute solution until the equilibrium relationship, Eq. 8.6, is satisfied. This phenomenon is called *reverse osmosis*.

When the solution is very dilute, Eq. 8.7 is simplified by the Taylor expansion to a variation of Eq. 3.24:

$$\pi = RT\sum_i C_i \quad ,$$

where the summation is over all solutes present in the solution. This summation equation is the well-known van't Hoff equation.

The driving force of transport for a molecular species through a quasihomogeneous membrane is the local gradient of the chemical potential. Thus, the flux of component j at a uniform temperature T in one-dimensional space with respect to the stationary membrane is given by the basic diffusion equation:

$$N_j = - \frac{\overline{D_j}\overline{C_j}}{RT} \frac{d\overline{\mu}_j}{dz} \tag{8.8}$$

with $\overline{D}_j$ being the effective diffusion coefficient at a distance z in the membrane from the ingoing interface.

For the solvent (j = B), it is convenient and instructive to always consider an *imaginary* solution of molar concentration C_B and chemical potential μ_B. At each point in the membrane these imaginary values are in equilibrium with actual membrane concentration $\overline{C}_B$ and chemical potential $\overline{\mu}_B$. The solution is in thermodynamical and in mechanical equilibrium with the point. It corresponds to the assumption of equilibrium at both the membrane-solution interfaces. The gradient of the chemical potential in the membrane can then be transformed into

$$\frac{d\overline{\mu}_B}{dz} = \frac{d\mu_B}{dz}$$

$$= RT \frac{d}{dz} \ln a_B + v_B \frac{dP}{dz}$$

$$= v_B \left(\frac{dP}{dz} - \frac{d\pi}{dz} \right) . \tag{8.9}$$

Substituting the gradient into Eq. 8.8, the flux of the solvent becomes

$$N_B = - \frac{\overline{D}_B \overline{C}_B v_B}{RT} \left(\frac{dP}{dz} - \frac{d\pi}{dz} \right)$$

$$= - \frac{\overline{D}_B \overline{C}_B v_B}{RT} \frac{d\mathcal{P}}{dz} , \tag{8.10}$$

where $\mathcal{P} = P - \pi$. Integration of the equation over the membrane thickness l at steady state yields (see Eq. 3.26):

$$N_B = \frac{Q_B}{l} \Delta \mathcal{P} = \frac{Q_B}{l} (\Delta P - \Delta \pi) . \tag{8.11}$$

Note that Q_B is then the flow coefficient for solvent B, based on pressure differences, where

$$\Delta \mathcal{P} = \mathcal{P}^I - \mathcal{P}^{II}$$

$$\Delta P = P^I - P^{II}$$

$$\Delta \pi = \pi^I - \pi^{II}$$

$$Q_B = \frac{1}{\Delta \mathcal{P}} \int_{\mathcal{P}^{II}}^{\mathcal{P}^I} \frac{\overline{D}_B \overline{C}_B v_B}{RT} d\mathcal{P} . \tag{8.12}$$

Superscripts I and II refer to the ingoing and outgoing interfaces, respectively, as shown in Fig. 8.1. It should be stressed that in deriving the equation the existence of thermodynamical and mechanical equilibria was assumed at both the ingoing and outgoing interfaces.

For the solute ($i = A$), it can be shown, for a dilute solution, that in the practical range of operating pressure [5]:

$$\left| \left(\frac{\partial \overline{\mu}_A}{\partial \overline{C}_A} \right)_{P,T} \frac{d\overline{C}_A}{dz} \right| \gg \left| \left(\frac{\partial \overline{\mu}_A}{\partial P} \right)_{\overline{C}_A, T} \frac{dP}{dz} \right| .$$

If the solution is ideal ($\overline{\gamma}_A = 1$), Eq. 8.8 can, therefore, be approximated to the simple Fick diffusion equation:

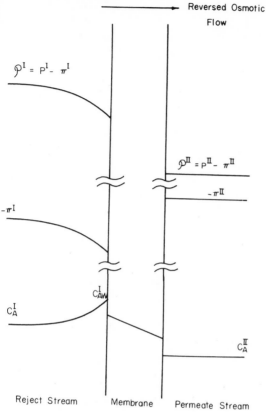

Fig. 8.1 Schematic of transport of nonelectrolyte through membrane

$$N_A = - \overline{D}_A \; \frac{d\overline{C}_A}{dz} \tag{8.13}$$

or, in the integrated form at steady state:

$$N_A = \frac{Q'_A}{l} \, (C_A^I - C_A^{II}) \; , \tag{8.14}$$

where

$$Q'_A = \frac{K_A}{\overline{C}_A^I - \overline{C}_A^{II}} \int_{\overline{C}_A^{II}}^{\overline{C}_A^I} \overline{D}_A \; d\overline{C}_A \; .$$

Note that Q'_A represents the flow coefficient of solute A based on a concentration gradient. Again, equilibrium was assumed at both interfaces between the membrane and the solution, with the distribution coefficient K_A defined by Eq. 8.4. Then the solute rejection SR, can be defined as

$$SR = \frac{C_A^I - C_A^{II}}{C_A^I} = 1 - \frac{C_A^{II}}{C_A^I} \quad . \tag{8.15}$$

In order to evaluate the performance of a membrane in *reverse osmosis*, it is customary to use the solute rejection SR along with the production rate Q_B/l instead of the parameters Q_B/l and Q'_A/l. If the transport by molecular diffusion in the product stream is negligible compared to the convective diffusion, the ratio of the two fluxes becomes

$$\frac{N_A}{N_B} \simeq \frac{C_A^I}{C_A^{II}}$$

$$= \frac{Q'_A(C_A^I - C_A^{II})}{Q_B \, \Delta P} \quad ,$$

where C^{II} is the total molar concentration in the product stream. Hence, the solute rejection can be given in terms of Q'_A and Q_B as

$$SR \cong \frac{Q_B \, \Delta P}{Q_B \, \Delta P + Q'_A C^{II}} \quad . \tag{8.16}$$

Pore Model

If the pore radius r of a porous membrane is much larger than both the molecular sizes of the solvent and the solute in a solution, the total volume flux through the membrane can be well estimated by the Hagen-Poiseuille equation (see Eq. 3.17) in the following form:

$$N'_v = \frac{Q_v}{l} \, \Delta P \quad , \tag{8.17}$$

where

$$Q_v = \frac{GK}{\mu} = \frac{Gr^2 \epsilon}{8\mu} \quad , \tag{8.18}$$

and μ is the viscosity of the solution. The geometrical factor G of the porous membrane depends on porosity, tortuosity, pore-size distribution, anisotropy, and so on.

On the other hand, the separation can be easily achieved by sieve action, if the radius r is much larger than the molecular size of the solvent but much smaller than that of solute. The total volume flux N'_ν becomes just that of the solvent in this case. However, the flow mechanism is greatly complicated by the molecular interaction between the solute molecule and molecules of the membrane if the pore radius r is larger than, but comparable to, the molecular size of the solute. Indeed, experimental data [1] show that in this case significant separation has been observed, and the separation generally depends on the gradient contrary to that predicted by Eq. 8.17. Furthermore, the pore sizes estimated by the equation for various kinds of solutes are not constant but vary with the size of the solute and its concentration. This is primarily because of the physical-molecular interaction between the solute molecule and the membrane, such as van der Waals interaction or electrostatic interaction (if any), as well as because of the viscous friction. In order to explain the situation, Ferry [2] considered the probability for a solute molecule to strike a pore at the ingoing interface, and Faxén [3] estimated the frictional inter-action with the membrane. Spiegler [4], and later Merten [5], on the other hand, developed the so-called finely porous model more fundamentally, by using a hydrodynamic model with frictional coefficients. However, it appears that these theoretical models need further extensive study, theoretically and experimentally, in order to be useful for practical purposes.

Katchalsky [6] started with the linear thermodynamic theory of irreversible processes and defined three phenomenological coefficients to describe the total volume and solute fluxes N_ν and N_B for a dilute binary solution as

$$N'_\nu = L_\nu \ (\Delta P - \sigma \ \Delta \pi) \ , \tag{8.19}$$

$$N_A = C_{Am} \ (1 - \sigma)N'_\nu + L_A \ \Delta C_A \ , \tag{8.20}$$

where the coefficients L_ν, L_A, and σ are called the filtration coeffieient, the solute permeability, and the Staverman reflection coefficient, respectively. Also, C_{Am} is the log-mean molar concentration of the solute between that in the feed stream C_A^I and that in the product stream C_A^{II}, and $\Delta C_A = C_A^I - C_A^{II}$. The reflection coefficient σ is known as a measure of coupling between the total volume flux and the solute transmission. Thus, $\sigma = 0$ if the solute is identical with the solvent (e. a., isotope), and $\sigma = 1$ if there is no coupling. If the pore size is much larger than both molecules sizes of the solvent and the solute, one may express the phenomenological coefficients in Eqs. 8.19 and 8.20 in terms of the diffusion model, Eqs. 8.11 and 8.14, and the pore model, Eq. 8.17. The solvent and solute permeation fluxes in this case can be stated as

$$N_B = \frac{Q_B}{l} \ (\Delta P - \Delta \pi) + C_{Bm} \ \frac{Q_\nu}{l} \ \Delta P \ , \tag{8.21}$$

$$N_A = \frac{Q'_A}{l} \ \Delta C_A + C_{Am} \ \frac{Q_\nu}{l} \ \Delta P \ , \tag{8.22}$$

where C_{Am} and C_{Bm} are the log-mean concentrations of the two solutions. If the solution is very dilute, one can approximate the above equations to

$$N'_v \cong \frac{N_B}{C_{Bm}} = \frac{Q_B + C_{Bm}Q_v}{l} \left(\Delta P - \frac{Q_B}{Q_B + Q_v C_{Bm}} \Delta \pi \right) ,$$

$$N_A = \frac{1}{l} \left(Q'_A + \frac{Q_B Q_v C_{Am} RT}{Q_B + Q_v C_{Bm}} \right) \Delta C_A + C_{Am} \left(1 - \frac{Q_B}{Q_B + Q_v C_{Bm}} \right) N'_v .$$

Thus, one can identify the phenomenological coefficients as follows:

$$L_v = \frac{Q_B + C_{Bm}Q_v}{l} , \tag{8.23}$$

$$\sigma = \frac{Q_B}{Q_B + C_{Bm}Q_v} , \tag{8.24}$$

$$L_A = \frac{1}{l} \left(Q'_A + \frac{Q_B C_v C_{Am} RT}{Q_v + Q_v C_{Bm}} \right) . \tag{8.25}$$

Recently, the phenomenological coefficients in more general cases were extensively discussed by Manning [7] in terms of an energy-barrier model and also a friction-coefficient model.

Transport of Electrolytes

It is commonly accepted that the driving force in diffusion of a nonelectrolyte is just the gradient of its chemical potential, which generally consists of the activity and the pressure gradient in an isothermal system. The diffusion of an ionic species is, however, subject to an additional force, the electric-potential gradient. Indeed, it is well known that, in a dilute solution of hydrochloric acid even in the absence of an external electric potential, the diffusion of the faster-moving hydrogen ion, or hydronium ion, is depressed by the electrical potential generated by the slow-moving chloride ion. On the other hand, the effective mobility of the chloride ion is enhanced by the potential created by the hydrogen ion. Thus, the overall diffusion rate of the hydrochloric acid as a whole assumes an averaged value between the two individual diffusion rates of the hydrogen ion and the chloride ion.

If there exists in a solution only one electrolyte that consists of one cation and one anion, the electric potentials created by both ions balance each other, and the electrolyte as a whole behaves as a nonelectrolyte in the absence of the external electric potential. One may consider the electrolyte as one single species in this case, rather than as two species. A single ordinary chemical potential may then be assigned to the electrolyte without considering the electric potential, and effective behaviors as a whole, such as diffusion and sorption, may then be studied. This procedure cannot, however, be applied directly to the solution that contains more than one electrolyte. It has commonly been observed that the diffusion and sorption of an electrolyte in the membrane, even in dilute solutions, are altered greatly by the presence of other electrolytes, from the behavior that would be observed in a solution of a single electrolyte. Also, the diffusion of an electrolyte occurs often in the direction from a dilute solution to a concentrated solution, even in the absence of a strong osmotic effect. These "anomalous" phenomena can hardly be understood in terms of the ordinary chemical potential and the averaged properties without considering the electrical potential. The phenomena become particularly pronounced when the solution is associated with a permselective membrane, charged or neutral, with respect to one of the ionic species.

The transport process of an electrolyte through a membrane can be explained either by the diffusion (homogeneous) or the pore (heterogeneous) model, as in the previous section. In most cases, the homogeneous model has proven to be simple and adequate, except in the case of osmotic flow. We shall, therefore, discuss the transport of electrolytes in light of the homogeneous model exclusively. Also, the most general theory of nonequilibrium thermodynamics as it applies to flow through membranes has already been discussed in Chapter II.

Within the diffusion model one can distinguish between dilute and concentrated solution behavior. As applied here, only the electrokenitic phenomena in a *dilute* solution will be discussed, as this is the situation that is relevant to the membrane process. For a detailed discussion of both dilute and concentrated solution in membranes, the reader should consult the elegantly written books by Helfferich [8] and Tuwiner [9]. The historical development of the theory of electrically charged membranes has already been reviewed in the Prologue.

It is worthwhile to consider the matter of *thermodynamic equilibria* for electrolyte solutions and their significance in several cases involving Donnan equilibria. When an ion in a solution is subject to an electric as well as a chemical potential, the ion cannot distinguish the origin of the potential and simply responds to the net resultant force. Hence, it is customary to define a total chemical potential μ_i^e of an ionic species i such that

$$\mu_i^e = \mu_i + z_i \mathcal{F} \phi , \qquad (8.26)$$

where z_i is the electrochemical valence of the i species, ϕ is an electric potential, and $\mathcal{F}$ is the Faraday constant (96,500 C/equiv.). The ordinary chemical potential μ_i is given by Eq. 8.1. When two homogeneous phases are in equilibrium, the total chemical potential μ_i^e must be the same in both phases. In addition, the following electroneutrality must be maintained:

$$\sum_i z_i C_i = 0 \quad , \tag{8.27}$$

where C_i is the volumetric molar concentration, and the summation is to be carried over all the ionic species present in each phase.

Donnan Exclusion

Consider a cation-exchange membrane of capacity $\overline{C}_R$ eq/cc, immersed in a homogeneous electrolyte solution. The solution contains one electrolyte that comprises a cation A of valence z_A and an anion Y of valence z_Y. One mole of the electrolyte dissociates in the solution to give ν_A moles of ion A and ν_Y moles of ion Y. Then the following relation holds:

$$z_A \nu_A + z_Y \nu_Y = 0 \ , \tag{8.28}$$

and the partial molar volume of the electrolyte v_{AY} in the solution is by definition

$$v_{AY} = v_A \nu_A + v_Y \nu_Y \quad . \tag{8.29}$$

If the two phases, the solution and the membrane phase, are in thermodynamic equilibrium at temperature T and the solution pressure P, the total chemical potentials must be

$$\mu_i^e = \overline{\mu}_i^e \ , \quad i = A, Y \ . \tag{8.30}$$

Here, the bar refers to the membrane phase and the quantity without the bar to the solution phase. From Eqs. 8.26 and 8.1, the Donnan potential E_{Don} across the membrane-solution boundary becomes

$$E_{\text{Don}} \equiv \overline{\phi} - \phi$$

$$= \frac{1}{z_i \mathcal{F}} \left(RT \ln \frac{a_i}{\overline{a}_i} - \pi_s v_i \right) \ , \quad i = A, Y \tag{8.31}$$

where $\pi_s = \overline{P} - P$ is the so-called swelling pressure. Here, the molar volumes of both ionic species are assumed to be the same in both phases. The system must also satisfy the electroneutrality in each phase:

$$z_A C_A + z_Y C_Y = 0 \ ,$$

$$z_A \overline{C}_A + z_Y \overline{C}_Y - \overline{C}_R = 0 \ . \tag{8.32}$$

The activity coefficients γ_i and $\overline{\gamma}_i$ are defined as usual:

$$a_i = \gamma_i C_i; \quad \overline{a}_i = \overline{\gamma}_i \overline{C}_i \ . \tag{8.33}$$

Substituting Eqs. 8.32 and 8.33 into Eq. 8.31 yields

$$\left(\frac{\overline{C}_Y}{C_Y} \right)^{\nu_Y} = \left(\frac{|z_Y|\, C_Y}{\overline{C}_R + |z_Y|\,\overline{C}_Y} \right)^{\nu_A} \left(\frac{\gamma_\pm}{\overline{\gamma}_\pm} \right)^{\nu} \exp\left(- \frac{\pi_s \nu_{AY}}{RT} \right), \qquad (8.34)$$

where $\nu = \nu_A + \nu_Y$, and the mean activity coefficients, $\gamma_\pm$ and $\overline{\gamma}_\pm$, are defined by

$$\left(\gamma_\pm \right)^{\nu} = \left(\gamma_A \right)^{\nu_A} \left(\gamma_Y \right)^{\nu_Y},$$

$$\left(\overline{\gamma}_\pm \right)^{\nu} = \left(\overline{\gamma}_A \right)^{\nu_A} \left(\overline{\gamma}_Y \right)^{\nu_Y}. \qquad (8.35)$$

This corresponds to the definitions of the mean activities:

$$a_\pm^{\nu} = a_A^{\nu_A}\, a_Y^{\nu_Y},$$

$$\overline{a}_\pm^{\nu} = \overline{a}_A^{\nu_A}\, \overline{a}_Y^{\nu_Y}. \qquad (8.36)$$

If the concentrations in both phases are very dilute ($\gamma_\pm \cong \overline{\gamma}_\pm \cong 1$) and the contribution by swelling is negligible, Eq. 8.35 can be simplified to give

$$\left(\frac{\overline{C}_Y}{C_Y} \right)^{\nu_Y} = \left(\frac{|z_Y|\, C_Y}{\overline{C}_R + |z_Y|\,\overline{C}_Y} \right)^{\nu_A}. \qquad (8.37)$$

Inspection of Eqs. 8.37 and 8.32 shows that the ratio of the concentration of anion Y to that of cation A in the cation-exchange membrane is smaller than that in the solution. For dilute solutions and a high capacity ($\overline{C}_R \gg C_Y$) in particular, the concentration $\overline{C}_Y$ becomes negligible. Thus, the Donnan potential E_{Don} has the effect of excluding the ion Y from the membrane. Similarly, one may show that a cation is repelled from the anion-exchange membrane. An ion whose charge has the opposite sign to that of an ion-exchange membrane is called a counterion, and an ion of the same sign is termed a co-ion.

In general, the Donnan exclusion of a co-ion from an ion-exchange membrane becomes more effective with the increase of the capacity and the degree of cross-linking of the membrane polymer, and of the valence of the co-ion. It is, however, less pronounced at a higher concentration of the electrolyte solution and for a larger valence of the counterion. The exclusion is also less effective when there occurs association of ionic pair formation between the counterion and the fixed ionic group, and when mobile ions form a complex [8].

If there are more than two ionic species in the solution, Eq. 8.32 still applies to each ionic species. But the electroneutrality conditions require that

$$\sum_i z_i C_i = 0$$

$$\sum_i z_i \overline{C}_i + \omega \overline{C}_R = 0. \qquad (8.38)$$

The summation is to be carried over all of the ionic species. The sign indicator ω is -1 for the cation-exchange membrane and $+1$ for the anion-exchange membrane. If the effect of swelling is not significant, Eq. 8.31 can be rearranged into

$$\left(\frac{a_1}{\overline{a}_1} \right)^{\frac{1}{z_1}} = \left(\frac{a_2}{\overline{a}_2} \right)^{\frac{1}{z_2}} = \cdots = \left(\frac{a_i}{\overline{a}_i} \right)^{\frac{1}{z_i}} . \qquad (8.39)$$

Donnan Equilibrium with Charged Membranes

Now consider *two electrolyte solutions* that are separated by a high-capacity cation-exchange membrane. The two solutions contain two counterions (cations), 1 of valence z_1 and 2 of valence z_2, and a common co-ion (anion) Y of valence z_Y. The two solutions are so dilute $(\overline{C}_R \gg C_Y)$ that the passage of the co-ion through the membrane is almost negligible due to the Donnan exclusion. The system is kept in thermodynamic equilibrium under a uniform temperature. Similarly to the previous example, the uniformity of the total chemical potentials of species 1 and 2 in both solutions in terms of activities a_i requires

$$\phi^{II} - \phi^{I} = \frac{1}{z_i \mathcal{F}} \left[RT \, \ln \frac{a_i^{I}}{a_i^{II}} - (P^{II} - P^{I}) v_i \right], \quad i = 1,2 \qquad (8.40)$$

where the superscripts I and II refer to the two solutions. If the pressure effect on the transport of solute is negligible (which is true in most dilute solutions), Eq. 8.40 yields

$$\left(\frac{a_1^{I}}{a_1^{II}} \right)^{\frac{1}{z_1}} = \left(\frac{a_2^{I}}{a_2^{II}} \right)^{\frac{1}{z_2}} = \text{const} \qquad (8.41)$$

subject to the electroneutrality conditions:

$$z_1 C_1^{I} + z_2 C_2^{I} = |z_Y| C_Y^{I} \ ,$$

$$z_1 C_1^{II} + z_2 C_2^{II} = |z_Y| C_Y^{II} \ , \qquad (8.42)$$

In order to examine the implications of Eqs. 8.41 and 8.42, consider a simple case in which both solutions are ideal $(\gamma_i^{I} \cong \gamma_i^{II} \cong 1, \ i = 1,2)$ and $z_1 = z_2$. Then the system of equations is simplified to

$$\frac{C_1^{I}}{C_1^{II}} = \frac{C_2^{I}}{C_2^{II}} = \frac{C_Y^{I}}{C_Y^{II}} = \text{const.} \qquad (8.43)$$

Hence, the counterions 1 and 2 are distributed in equilibrium, regardless of their initial distribution, such that the ratios of their concentrations in the two solutions satisfy the above relationship. The equilibrium distribution depends, however, on the initial distribution of the co-ion, since the ideal permselective membrane does not permit the passage of the co-ion. Suppose that at the *initial* state the concentration of counterion 1 in the solution II is larger than in the solution I, that is, $(C_1^I/C_1^{II})_0 < 1$. But this ratio is larger than that of the concentrations of the co-ion $(C_Y^I/C_Y^{II})_0$. As the system approached the equilibrium state, the ratio $(C_Y^I/C_Y^{II})_0$ remains constant. However, the counterion 1 diffuses from solution I to solution II in order to achieve the equilibrium ratio, Eq. 8.43, even if the concentration of 1 in the former solution is smaller than in the latter solution. Thus, the cation is "pumped" through the cation-exchange membrane from the dilute solution into the concentrated solution.

The driving force for diffusion of 1 is due to the electric potential generated by the distribution of concentration of anion Y. Therefore, a large driving force can be achieved by controlling the anionic concentrations in both solutions, which often exceeds the force due to the difference of its own concentrations between the two solutions. One may also consider a similar anion-exchange process with an anion-exchange membrane. This phenomenon has recently been proposed to be used in the separation of metallic ions from an aqueous solution using a cation-exchange membrane [10]. The continuous ion-exchange process is often called the Donnan dialysis. In the actual case, the osmotic transport of water and the co-ion leakage through the ion-exchange member, however, considerably reduce the driving force. Also, it is commonly observed that the mass transfer is very slow, and Donnan equilibrium can hardly be achieved in a short time.

Donnan Equilibrium with Neutral Membranes

Finally, consider another example of Donnan equilibrium between two solutions separated by a membrane. In analogy to the previous example, the two solutions contain two cations, 1 of valence z_1 and 2 of valence z_2, and a common ion Y of valence z_Y. But the membrane is not electrically charged. The ionic size of 2 is so large that the permeation of 2 is almost negligible compared to that of 1 and Y. The two solutions are also so dilute that they behave like an ideal solution, and the osmotic pressure does not directly affect their transport through the membrane. If the system is in thermodynamic equilibrium under a uniform external pressure and temperature, the electrochemical potentials of ionic species 1 and Y must be balanced. Similarly to the previous example, one may obtain the equilibrium distribution of concentrations of 1 and Y:

$$\left(\frac{C_1^I}{C_1^{II}}\right)^{\frac{1}{z_1}} = \left(\frac{C_Y^I}{C_Y^{II}}\right)^{\frac{1}{z_Y}} = \text{const} \qquad (8.44)$$

subject to the electroneutrality conditions:

$$z_1 C_1^I + z_2 C_2^I + z_Y C_Y^I = 0 ,$$

$$z_1 C_1^{II} + z_2 C_2^{II} + z_Y C_Y^{II} = 0 . \tag{8.45}$$

If the concentration of 2 in the solution II is zero, that is, $C_2^{II} = 0$, and $z_1 = -z_Y = 1$, the equation reduces to

$$\left(\frac{C_1^{II}}{C_1^I} \right)^2 = 1 + \frac{z_2 C_2^I}{C_1^I} , \tag{8.46}$$

which is essentially the same as Eq. 3.34 in Chapter III. The latter equation was derived from the uniformity of the mean activity. As seen from the equation, the equilibrium ratio, C_1^{II}/C_1^I, is always greater than 1. It is, thus, another example of a "pump" that can transport the smaller cation 1 from the dilute solution to the concentrated solution. The energy of the pump is supplied by the electric potential generated by the concentration distribution of the larger cation 2. The efficiency increases with the valence of 2, the concentration of 2 in the solution I, and the permselectivity of the membrane. Indeed, Vromen [11] attempted to dialyze nitric acid from the aqueous solution of ferric nitrate and free nitric acid, and found that the acid concentration in the diffusate is often higher than in the feed liquor, which could not be understood in terms of the ordinary chemical potential (or concentration gradient).

Diffusion of Electrolytes

First, it is appropriate to present a generalized discussion of diffusion in membrane and solution systems. To treat diffusion in membranes we use the diffusion model for nonelectrolytes, where the membrane is considered to be a homogeneous medium on the λ scale. If the membrane is electrically charged, the fixed ionic group is thus assumed to be immobile and uniformly distributed throughout the membrane. Also, the membrane is formed by colloidal ion-exchange particles embedded in an inert binder, and the average particle size is considered to be smaller than λ. When a fluid is associated with such a membrane, we shall, somewhat arbitrarily, distinguish the dynamic behavior of a molecular species or an ionic species due to its individual motion, from that due to the motion in a group, such as the cluster transport or pore flow. The group behavior has been considered as a single effective behavior in the previous section. The distinction between individual and group motion is particularly helpful in clarifying the phenomenon of electro-osmosis which is discussed later.

Here, we shall only deal with the case in which an external electric potential is absent, and therefore no net electric current density is involved. The transport process under the external electric potential is discussed in conjunction with electrodialysis in Chapter IX. The review of the membrane potential created during the diffusion process of ionic species through a membrane will also be postponed to Chapter IX.

The electrolyte solution under discussion will be restricted to a dilute solution, therefore all electrolyte solutes are completely dissociated into ionic forms. Furthermore, only cross-coupling of individual fluxes arising from electric potential and convective flow will be considered. The more general case is treated in Chapter II. In a concentrated solution, the ionic migration is retarded by the effects of the relaxation field and of electrophoresis [12]. The former effect essentially results from the relative motion of one type of ion (e.g., cation) with respect to that of the other type (e.g., anion). The effect is absent in the diffusion process caused by the ordinary chemical potential. However, the validity of the Nernst-Einstein relationship will be assumed. It signifies the equivalence of ionic mobility due to ordinary chemical potential and electric potential. This relationship may be expected to be valid, particularly in the ionic transport through most practical ion-exchange membranes, where the average pore size is so small that the formation of the so-called ionic cloud is severely restricted. The molecular theory of the general electrokinetic phenomenon in a concentrated solution was treated comprehensively by Conway [12], and the macroscopic theory was reviewed by Newman [13]. The formalism of Helfferich [8], treating the electrokinetic phenomenon associated with such a membrane, will serve as the basis of discussion.

Then, the molar flux N_j of a neutral species j with respect to the stationary membrane at a uniform temperature T can be expressed as follows:

$$N_j = - \frac{\overline{D_j}\overline{C_j}}{RT} \nabla \overline{\mu_j} + \overline{C_j}N_v'$$

$$= - \overline{D_j} [\nabla \overline{C_j} + \overline{C_j}\nabla \ln \overline{\gamma_j} + \frac{\overline{v_j}\overline{C_j}}{RT} \nabla \overline{P}] + \overline{C_j}N_v' \ , \qquad (8.47)$$

where ∇ is the gradient operator and N_v' is a bulk volumetric flux. Compared to Eq. 8.10, this equation more generally includes the effect of the gradient of the activity coefficient $\overline{\gamma_j}$, as well as that of the bulk transport. Similarly, the flux N_i of an ionic species i of valence z_i under an electric potential $\overline{\phi}$ is given by

$$N_i = - \frac{\overline{D_i}\overline{C_i}}{RT} \nabla \overline{\mu_i^e} + \overline{C_i}N_v'$$

$$= - \overline{D_i}(\nabla \overline{C_i} + \overline{C_i}\nabla \ln \overline{\gamma_i} + \frac{\overline{v_i}\overline{C_i}}{RT} \nabla \overline{P} + \frac{z_i\overline{C_i}\mathcal{F}}{RT} \nabla \overline{\phi}) + \overline{C_i}N_v' \ . \qquad (8.48)$$

This equation corresponds in form to the general equation, Eq. 2.16.

Thus, the ionic flux constitutes the ordinary diffusional flux due to the activity, the pressure-diffusional flux, and the electric transference due to the electric potential, as well as the convective flux. If the concentration of the ionic species is extremely small and the gradient of the activity coefficient is negligible, and also if the membrane is so dense that the convective flux is insignificant, Eq. 8.48 reduces to the Nernst equation:

$$N_i = - \overline{D}_i \left(\nabla \overline{C}_i + \frac{z_i \overline{C}_i \mathcal{F}}{RT} \nabla \overline{\phi} \right) . \qquad (8.49)$$

In addition, the system must satisfy the condition of electroneutrality:

$$\sum_i z_i \overline{C}_i + \omega \overline{C}_R = 0 , \qquad (8.50)$$

where

$$
\begin{aligned}
\omega &= 1 &&\text{for anion-exchange membrane} \\
&= 0 &&\text{for neutral membrane} \\
&= -1 &&\text{for cation-exchange membrane,}
\end{aligned}
$$

and also the conservation of the electric charge:

$$I = \mathcal{F} \sum_i z_i N_i . \qquad (8.51)$$

The summation is to be carried out over all ionic species present in the membrane, and I is the net electric current density $[C/(cm^2)(sec)]$.

Since, in most practical cases, the volume of the membrane is much smaller than that of the associated solution, the steady or quasi-steady state is usually maintained in the membrane:

$$\nabla \cdot N_i = 0 \quad \text{or} \quad N_i = \text{const.} \qquad (8.52)$$

At the membrane-solution interface, the local thermodynamic equilibrium is also usually maintained (see Chapter II for a more general case). When this is the case, the boundary conditions at the interface for the system of ordinary differential equations of ionic fluxes, Eqs. 8.48 through 8.51, become the Donnan equilibrium relationship, Eq. 8.39. The general integration of the system of differential equations for a solution of an arbitrary number of ionic solute species is very complicated, even in the case where $\overline{D}_i$ and $\overline{C}_i$ are constant, and requires some sophisticated mathematical manipulation [14]. We shall confine ourselves to some simple cases that are relevant to practical membrane processes.

The *diffusion in charged membranes*, as a more specific case, leads to the following analysis. Consider two electrolyte solutions separated by a cation-exchange membrane of capacity $\overline{C}_R$ without an external electric potential. Both solutions contain one electrolyte that comprises a cation A and an anion Y. The concentration of the electrolyte in the membrane is so small that the gradients of the activity coefficients are almost zero, and also the membrane is so dense that the bulk flow is negligible. The pressure diffusion of the solute is assumed to be negligible. Therefore, the Nernst-Planck equation, Eq. 8.49, is valid for the transport of each ionic species:

$$N_A = -\overline{D}_A \left(\frac{d\overline{C}_A}{dx} + z_A \overline{C}_A \frac{\mathcal{F}}{RT} \frac{d\overline{\phi}}{dx} \right), \qquad (8.53)$$

$$N_Y = -\overline{D}_Y \left(\frac{d\overline{C}_Y}{dx} + z_Y \overline{C}_Y \frac{\mathcal{F}}{RT} \frac{d\overline{\phi}}{dx} \right) \qquad (8.54)$$

subject to the electroneutrality condition:

$$z_A \overline{C}_A + z_Y \overline{C}_Y = \overline{C}_R \quad , \qquad (8.55)$$

and the conservation of the electric charge:

$$\frac{I}{\mathcal{F}} = z_A N_A + z_Y N_Y = 0 \ . \qquad (8.56)$$

Eliminating the electric potential from Eqs. 8.53 and 8.54, using the conservation of the electric charge, Eq. 8.56, one obtains

$$N_A = -\frac{\overline{D}_A \overline{D}_Y (z_A^2 \overline{C}_A + z_Y^2 \overline{C}_Y)}{(\overline{D}_A z_A^2 \overline{C}_A + \overline{D}_Y z_Y^2 C_Y)} \frac{d\overline{C}_A}{dx}$$

$$= -\overline{D}_e \frac{d\overline{C}_A}{dx} \ . \qquad (8.57)$$

Here $\overline{D}_e$ is an effective diffusivity defined by the equation. Thus, the ionic flux of A is influenced by the electric potential in the presence of the anion Y and assumes an average diffusivity $\overline{D}_e$ between $\overline{D}_A$ and $\overline{D}_Y$. This interionic effect, which slows down the faster-moving ion and speeds up the slower-moving ion, is the direct consequence of the conservation of the electric charge. For the diffusion of an electrolyte of equi-valence ions through a neutral membrane, the effective diffusivity is a harmonic mean of diffusivities of the two ions. If the cation-exchange membrane possesses a high capacity and concentrations in both solutions are relatively dilute, that is, $\overline{C}_R \gg C_Y$, the concentration of the co-ion Y in the membrane phase is negligible. In this case, the ionic flux N_A is simplified to

$$N_A = - \overline{D}_Y \frac{d\overline{C}_A}{dx} . \tag{8.58}$$

The equation shows that the flux of counterion A through a cation-exchange membrane of a high capacity is primarily controlled by the diffusivity of co-ion Y. It also implies that the ionic flux (leakage) of the co-ion N_Y obeys the simple Fick law:

$$N_Y = - \overline{D}_Y \frac{d\overline{C}_Y}{dx} . \tag{8.59}$$

This relationship is obtained by substituting Eqs. 8.55 and 8.56 into Eq. 8.58. The above-mentioned behavior of an electrolyte in a cation-exchange membrane can similarly be expected in an anion-exchange membrane.

Donnan Dialysis with Charged Membranes

Suppose that two electrolyte solutions are separated by a *cation-exchange membrane* of a high capacity $\overline{C}_R$, similar to the previous example. The two solutions, however, contain two cations, 1 and 2, and a common co-ion (anion) Y. The concentration of the fixed ionic (anionic) group in the membrane phase is so high that the passage of the co-ion Y is almost not permitted by the Donnan exclusion. The fluxes of the counterions, N_1 and N_2, obey the Nernst-Planck equation, Eq. 8.49, subject to

$$z_1 \overline{C}_1 + z_2 \overline{C}_2 = \overline{C}_R \quad \text{electroneutrality,} \tag{8.60}$$

$$z_1 N_1 + z_2 N_2 = 0 \quad \text{no electric current.} \tag{8.61}$$

In analogy to the previous example, one may obtain the flux N_1 in the differential form:

$$N_1 = - \frac{\overline{D}_1 \overline{D}_2 (z_1^2 \overline{C}_1 + z_2^2 \overline{C}_2)}{(\overline{D}_1 z_1^2 \overline{C}_1 + \overline{D}_2 z_2^2 \overline{C}_2)} \frac{d\overline{C}_1}{dx} . \tag{8.62}$$

If diffusivities $\overline{D}_1$ and $\overline{D}_2$ are independent of concentrations of both ions, integration of the differential equation at steady state gives the following situations:

when $z_1 \neq z_2$ and $z_1\overline{D}_1 \neq z_2\overline{D}_2$,

$$N_1 = \frac{\overline{D}_1\, a(1 - \beta)}{l\,(1 - a\beta)} \left[(\overline{C}_1^I - \overline{C}_1^{II}) + \frac{\overline{C}_R\beta(1 - a)}{z_1(1 - \beta)(1 - a\beta)} \right.$$

$$\left. \ln \left(\frac{\overline{C}_1^I + \dfrac{a\beta\overline{C}_R}{z_1(1 - a\beta)}}{\overline{C}_1^{II} + \dfrac{a\beta\overline{C}_R}{z_1(1 - a\beta)}} \right) \right]; \qquad (8.63)$$

when $z_1 = z_2$ but $\overline{D}_1 \neq \overline{D}_2$,

$$N_1 = \left(\frac{\overline{D}_1}{l} \right) \left(\frac{a\overline{C}_R}{z_1(1 - a)} \right) \ln \left(\frac{\overline{C}_1^I + \dfrac{a\overline{C}_R}{z_1(1 - a)}}{\overline{C}_1^{II} + \dfrac{a\overline{C}_R}{z_1(1 - a)}} \right) ; \qquad (8.64)$$

and when $z_1 = z_2$ and $\overline{D}_1 = \overline{D}_2$,

$$N_1 = \frac{\overline{D}_1}{l} (\overline{C}_1^I - \overline{C}_1^{II}) ; \qquad (8.65)$$

where

$$a = \frac{\overline{D}_1}{\overline{D}_2} \qquad \text{the diffusivity ratio,}$$

$$\beta = \frac{z_2}{z_1} \qquad \text{the valence ratio.}$$

It is interesting that the flux N_1 for $z_1 = z_2$ and $\overline{D}_1 = \overline{D}_2$ is the same as the asymptotic solution of N_1 when the concentration of the counterion in the membrane phase is very small. The relationship between the concentration of the solution and that of the membrane at both interfaces is governed by the Donnan equilibrium as follows:

$$\frac{\overline{C}_1^i}{C_1^i} = \left(\frac{\overline{C}_R - z_1\overline{C}_1^i}{z_2 C_2} \right)^{\frac{1}{\beta}} \left(\frac{\gamma_1}{\overline{\gamma}_1} \right) \left(\frac{\overline{\gamma}_2}{\gamma_2} \right)^{\frac{1}{\beta}}, \qquad i = \text{I,II} . \qquad (8.66)$$

Here the effect of swelling in the membrane, if any, is included in the activity coefficients, $\overline{\gamma}_1$ and $\overline{\gamma}_2$. Thus, one may express the flux N_1 in terms of the concentrations of the counterion 1 in both solutions by substituting the boundary conditions, Eq. 8.66, into the integrated forms of the flux N_1, Eqs. 8.63 through 8.65. The flux N_2 can then be obtained from the conservation of the electric charge, Eq. 8.61.

As explicitly seen in the boundary conditions, the flux N_1 depends not only on the concentrations of cation 1 in both solutions, but also on the concentrations of co-ion Y in the solutions. The boundary conditions also indicate a greater (or smaller) driving force than would be expected in dialysis of nonelectrolyte solutes. The diffusion of the counterion 1 is even possible in the direction from a solution of its small concentration to a higher concentrated solution, as mentioned previously. The above discussions represent idealized situations. In actual dialysis, several other phenomena, such as osmotic flow, leakage of co-ion, and especially concentration polarization, complicate the transport mechanism: This situation will be discussed in a later section (Section 4).

Donnan Dialysis with Neutral Membranes

Finally, consider two electrolyte solutions that are separated by a *neutral membrane*. Both solutions contain two cations, 1 and 2, and a common anion Y, again as in the previous example. However, the ionic size of cation 2 is so large that the permeation of the cation is negligible compared to that of cation 1 and anion Y. The diffusions of cation 1 and anion Y are assumed to obey the Nernst-Planck equation, Eq. 8.49, for the case where $i = 1$ and Y. Electroneutrality and conservation of the electric charge require

$$z_1 \overline{C_1} + z_Y \overline{C_Y} = 0 \quad \text{electroneutrality,} \tag{8.67}$$

$$z_1 N_1 + z_Y N_Y = 0 \quad \text{no electric current.} \tag{8.68}$$

Eliminating the term involving the electric potential in Eq. 8.49 using Eqs. 8.67 and 8.68, the flux N_1 can be written as

$$N_1 = -\frac{\overline{D_1}\overline{D_Y}(z_1 - z_Y)}{\overline{D_1}z_1 - \overline{D_Y}z_Y} \frac{d\overline{C_1}}{dx} . \tag{8.69}$$

If $\overline{D_1}$ and $\overline{D_Y}$ are constant throughout the membrane, one may integrate the equation at steady state to obtain

$$N_1 = \frac{\overline{D_1}}{l} \frac{\overline{D_Y}(z_1 - z_Y)}{(\overline{D_1}z_1 - \overline{D_Y}z_Y)} (\overline{C_1}^{I} - \overline{C_1}^{II}) . \tag{8.70}$$

At the membrane solution interfaces, the following Donnan equilibrium is usually maintained:

$$\frac{\overline{C}_1^i}{C_1^i} = \left(\frac{z_1 C_1^i + z_2 C_2^i}{z_1 \overline{C}_1^i} \right)^{\frac{z_1}{|z_Y|}} \left(\frac{\gamma_1}{\overline{\gamma}_1} \right) \left(\frac{\gamma_Y}{\overline{\gamma}_Y} \right)^{\frac{z_1}{|z_Y|}}, \quad i = \text{I,II.} \quad (8.71)$$

The flux N_1 can then be expressed in terms of concentrations, C_1^I and C_1^{II}, by combining the boundary condition, Eq. 8.71, and the integrated form of the flux equation, Eq. 8.70.

Osmotic Flow

In the previous examples involving electrolytes, it was assumed that the osmotic flow (solvent flow) is negligible. In most practical processes, a porous membrane is commonly used to obtain a high flux of a certain ionic species. In that case, the effect of the osmotic flow cannot, however, be neglected, as it sometimes improves the performance of a membrane process, but more often it has an adverse effect.

As discussed in Section 1, p. 126, the osmotic flow can be viewed in two ways. In the diffusion model, the flux is primarily due to the molecular diffusion through a quasihomogeneous membrane according to the local gradient of the chemical potential. On the other hand, the osmotic flux in the pore model results from convective pore flux. The two points of view are in good agreement, particularly in the case where the associated solutions are very dilute. Thus the osmotic flow will be considered as a result of the convective pore flow. This model is particularly convenient to explicitly explain the phenomena of electro-osmosis and anomalous osmosis.

Suppose that two electrolytic solutions of the same concentration are separated by a membrane under an *external electric potential*. The membrane is porous and electrically neutral. The average pore size is much smaller than the thickness of the so-called electric double layer on the inner membrane surface. In the pore, the concentration of a cation is stoichiometrically balanced with that of an anion. As a cation is transported in the pore by the electric potential, the cation experiences a frictional force by the surrounding solvent and exchanges its momentum. The momentum absorbed by the solvent exactly balances with that transferred from an anion, which is moving in the opposite direction. Therefore, the external electric potential has no effect on the solvent flow. However, if the membrane is electrically charged, the counterion is much more concentrated in the pore than is the co-ion. Thus the momenta transferred from both ions to the solvent do not balance each other, and the solution receives a net force in the same direction as the counterion is moving. Then, the solution in the pore as a whole behaves like a charged species and is driven in the direction by the electric potential in the absence of a pressure gradient. The net transport of the counterion is, therefore, enhanced with respect to the stationary membrane, although the migration of the co-ion is retarded.

Schmid [15] assumed that the concentration of the solution in the pore of a small size (say, 50 Å) is uniform in the scale of the order of the average pore size, and the electric surplus charged per unit pore volume is $\omega \overline{C}_R \, \mathcal{F}/\epsilon$, where ϵ is the porosity of the porous membrane. The force, $(\omega \overline{C}_R \, \mathcal{F}/\epsilon)'(d\overline{\phi}/dx)$ resulting from the surplus charge under the electric potential, therefore, uniformly acts on the unit volume of the solution in the pore. In analogy to the electric transport of an ionic species, the volumetric convective flux in the presence of the pressure gradient can be written as

$$N_v' = D_v \left(- \frac{d\overline{P}}{dx} + \frac{\omega C R_R \, \mathcal{F}}{\epsilon} \cdot \frac{d\overline{\phi}}{dx} \right). \tag{8.72}$$

where D_v is the bulk solution permeability, which in the Poiseuille model is related to K in Eq. 3.19 as

$$D_v = \frac{K}{\mu} = \frac{\epsilon r^2}{8\mu} \; .$$

The equation (8.72) has proved to be valid in most practical ion-exchange membranes in which the capacity and the degree of cross-linking are not high [16]. Also the average pore size is smaller than the thickness of the electric double layer (say, 100 Å) but larger than the ionic size.

If, on the other hand, the average pore size is very large (say, 1000 Å or more) the electric double layer is formed on the inner membrane surface. The outer layer consists of the charged membrane surface, and the charged inner layer is mobile with the solution in the pore. Thus, the external electric-potential force primarily acts on the inner layer rather than on the volume of the solution and counteracts the hydrodynamic drag force by the fixed outer layer. In that case, the following equation is valid [17]:

$$N_v' = D_v \left(- \frac{d\overline{P}}{dx} + \frac{2D\zeta}{\pi r^2} \frac{d\overline{\phi}}{dx} \right), \tag{8.73}$$

where D is the dielectric constant of the solvent and ζ is the difference of the electric potential across the electric double layer. The double layer can also be formed on the neutral membrane surface. In this case, the charged outer layer consists of a sluggish liquid film adhering to the membrane surface. For pure water, the film is negatively charged if the surface is weakly acidic, and positively charged for the weakly basic surface. Therefore, the water is transported under the electric potential to the cathode for the negatively charged film, but to the anode for the positively charged film. In an aqueous electrolyte solution, the ζ potential is very sensitive to the concentration of the electrolyte and the sign of ζ may be reversed [17]. Owing to the interaction of a mobile ion and the membrane surface or the fixed outer layer, adsorption or complex formation can occur. This effect is pronounced particularly for a multivalent ionic species. Indeed, when a solution containing TiO^{2+} is associated with a cation-exchange membrane under an external electric potential, the osmotic flow is in the same direction as the co-ion moves [8].

When both the concentration and the pressure gradients exist as well as the electric potential, the diffusional flux of ionic species must also be considered in order to understand the osmotic flux. The situation is, therefore, very complicated. However, the following qualitative explanation can be rendered.

When two nonelectrolyte solutions are separated by a membrane, the solvent flux is in the direction from the dilute solution to the concentrated solution under a uniform external pressure. The flux is proportional to the difference of the total concentration of solutes in the two solutions. This osmosis is said to be "normal." However, when two electrolyte solutions are divided by a charged membrane, the osmotic flow considerably deviates from what would be observed in normal osmosis, and often its direction is reversed. If the counterion diffuses much faster than the co-ion, for example, the counterion builds up its space charge on the outgoing membrane surface and creates an electric potential. This potential in part drives back the solution in the pore from the outgoing membrane surface to the ingoing surface. Thus, it enhances the normal osmotic flux, if the counterion transports from the concentrated solution to the dilute solution. This enhancement is called *positive anomalous osmosis.* However, if the co-ion transports faster than the counterion, the osmotic flow is generally retarded, and the direction is even reversed. This is called *negative anomalous osmosis.* Thus, a transient rise or depression in the capillary on the side of the concentrated solution is commonly observed in the measurement of the osmotic pressure with electrolyte solutions. A rather comprehensive analysis of the anomalous osmotic phenomenon on the basis of Eq. 8.48 was made by Schlögle [18].

If there exists physical or chemical interaction between a mobile ion and the membrane, such as adsorption of solutes and complex formation between the fixed ionic group and electrolyte solutes, the above-mentioned generalization is not valid. In the diffusion of acid through a neutral membrane, the hydrogen ions are apparently adsorbed on the membrane wall, and the membrane behaves like a charged medium. Thus, anomalous osmotic flow can be observed even with a neutral membrane [11]. If the mobile ion tends to be hydrated, the apparent anomalous osmotic flux may, of course, be observed.

With highly concentrated solutions, the effect of the electric potential is smeared out, and the electrolyte behaves like a nonelectrolyte. In this case, the osmosis becomes approximately normal.

2 REVERSE OSMOSIS

Reverse osmosis is the separation process of a solution by the preferential permeation of the solvent through a membrane with application of an external pressure. The pressure is usually much larger than the difference of the osmotic pressures across the membrane. The solvent is thus "squeezed out" from the concentrated solution into the dilute solution. (The basic mechanism was discussed in Chapter III). The molecular size of the solute to be separated by this process is assumed to be in the same order of magnitude as that of the solvent. Its sister process, ultrafiltration, involving solutes of larger molecular sizes, will be discussed in the next section.

At the present time, the reverse-osmosis process is perhaps the most active field among membrane processes, especially in the desalination of saline and brackish waters. The current widespread attention to the process can be credited to the discovery of cellulose acetate as a highly salt-rejecting membrane by Reid and his co-workers [19], and later to the successful preparation of the so-called "anisotropic" or "asymmetric" cellulose acetate membrane by Loeb and Sourirajan [20]. The asymmetric membrane consists of a very thin but dense layer and a supporting sublayer of larger pore size. The separation occurs mainly during the passage of the solution through the thin layer, and therefore a high flux can be obtained without sacrificing the degree of salt rejection. The research projects of both groups were primarily supported by the Office of Saline Water, U.S. Department of the Interior. Further innovative preparations of asymmetric polyamide and cellulose acetate membranes, in hollow-fiber form as well as composite structure, are likely to increase the application possibilities of the process.

In 1955, Hagerbaumer [21] extended reverse-osmosis application to the separation of hydrocarbon liquid mixtures using microporous Vycor glass and accomplished a reasonable degree of separation. These findings may indicate a potential for an industrial operation in the future. The process of reverse osmosis has found successful use in the food and pharmaceutical industries, pulp and paper processing, and waste-water treatment, as well as desalination of saline and brackish waters.

Only a brief summary of the aspects relevant to the practical application of the process will be presented. The presentation is conducted mainly in terms of the desalination of saline water, where the process of reverse osmosis is most effectively applied and where a great deal of research has thus been done. More detailed discussions can be found, for example, in the books by Lonsdale [22] and by Sourirajan [23].

Reverse-Osmosis Membranes

Obviously, an effective membrane is of great importance in any membrane-separation process. The specific requirements for a reverse-osmosis membrane are attainment of a high flux in the liquid phase, high selectivity, that is, effective salt rejection, and mechanical ruggedness to withstand substantial pressure differentials across the membrane. It is surprising that these aims have indeed been accomplished, primarily through the development of the so-called asymmetric polymer membranes.

The most important membranes today are cellulose acetate and polyamide, both with asymmetric structures. Cellulose acetate membranes are usually employed in flat-sheet form, whereas the polyamide membrane is available mainly as a hollow-fiber configuration; however, progress is under way to prepare cellulose acetate in a hollow-fiber configuration. The cellulose acetate membrane is the one that so far has been most widely used. The performance of the two membranes is best illustrated in Fig. 8.2, which was prepared by Lonsdale [22]. The wide range of values for a given membrane in the figure

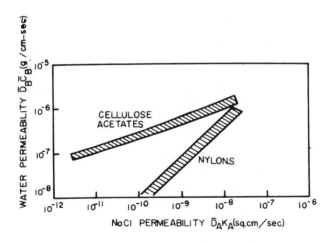

Fig. 8.2   Water permeability versus NaCl permeability in celluslose acetate membrane [22].

represents changes in chemical composition, and different conditions of its preparation and measurement. Nevertheless, the performance of the cellulose acetate membrane is superior to that of the polyamide membrane in its present state of development. However, the latter membrane in a hollow-fiber form has the high mechanical strength to withstand a high pressure gradient. In addition, it is chemically stable, especially with respect to high pH, compared to the anisotropic cellulose acetate membrane. The configuration of hollow fibers also provides compactness and some operational advantages over other configurations like the plate-and-frame and the spiral-wound module as described in Chapter XIV. The polyamide hollow fiber has, therefore, already found many uses in moderate-size installations.

Membranes of other cellulose derivatives in anisotropic forms have been examined for desalination [22, 23]. Among them are ethyl cellulose, cellulose acetate butyrate, cellulose acetate methacrylate, other highly substituted cellulose acetates, and the so-called thin-film composite membrane. The thin-film composite membrane consists of a thin film of cellulose acetate and a porous support membrane like the Loeb-Sourirajan type of anisotropic membrane, but is prepared in a different way. As these membranes were tested in desalination, a high salt rejection was often obtained but the water flux was generally lower than the anisotropic cellulose acetate. A number of other potentially useful membranes have reportedly been prepared and tested, such as the ion-exchange membrane [24], the dynamically formed membrane [25], the graphic oxide membrane [26], and porous Vycor glass [27]. However, their performances in desalination have so far been rather poor compared to the anisotropic cellulose acetate and the polyamide hollow fiber. A valuable summary of published performance data has been prepared by Walch [28]: these are presented in Chapter XIV. The research and development efforts that have gone into membrane preparation and polymer tailoring are also treated in Chapter XIV.

Performance of Cellulose Acetate Membranes

As already mentioned, both the anisotropic cellulose acetate and the polyamide membranes have found extensive use in desalination. Most of the published studies on the membrane performance have, however, been limited to the cellulose acetate configuration. Therefore, the following discussion is confined to the cellulose acetate membrane. The performance of the polyamide membrane is reported to be very similar to the cellulose acetate in many respects, but the salt rejection is more sensitive to pH.

The membrane performance greatly depends on the method of its preparation. The acetyl content, chemical composition of the casting solution, evaporation condition, and annealing temperature are important factors that influence the performance of the cellulose acetate membrane. The effect of these parameters is discussed in Chapter XIV. Only the effects of operating parameters and the general rejection behaviors will be treated here.

Effect of Operating Temperature

As the operating temperature of a reverse-osmosis unit increases, both permeabilities for the saline water, Q'_A of the solute and Q_B of the solvent, which are defined in Eqs. 8.12 and 8.14, will increase. However, the salt rejection decreases with the temperature, since the apparent activation energy for salt is greater than that of water. The activation energy of the water permeation is in the range of 5 to 6 kcal/mole and the energy for the salt transport in the order of 7 to 8 kcal/mole. A high temperature generally enhances the membrane compaction and deterioration, and therefore shortens the life of the membrane. The membrane compaction shall be discussed in greater detail in conjunction with the effect of the operating pressure.

Effect of Operating Pressure

The solute permeability Q_B decreases, but the salt rejection is enhanced, as the operating pressure increases. The dependence of Q_B on the pressure is reduced at a high annealing temperature as discussed in Chapter XIV. At constant pressure, the thin, dense layer grows in thickness slowly, Q_B being gradually reduced as time elapses. But the salt rejection remains almost constant. This membrane compaction, thus, deteriorates the performance and shortens the life of the membrane. The typical compaction rate is shown in Fig. 8.3. The rate is slower at a lower pressure and operating temperature. The compaction can often be estimated by an empirical equation [22]:

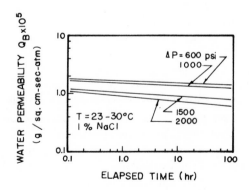

Fig. 8.3 Compaction curves for anisotropic cellulose acetate membrane annealed at 80°C [22].

$$N_B(t) = N_B(\tau) \left(\frac{t}{\tau} \right)^m, \quad t \geqslant \tau, \quad (8.74)$$

where t is time, and m and τ are constants that depend on the operating pressure and temperature, the annealing temperature and the physical and chemical nature of the membrane and the solution. The constant parameter m is often called the compaction slope. The magnitude of the slope is usually in the order of -0.03 to -0.1 for operating pressures of 500 to 1500 psi with the typical system of anisotropic cellulose acetate and saline water. The compaction rate can be retarded to some extent by operating the osmotic unit at a low temperature and pressure, and also by modifying the membrane, such as using a filler in the membrane.

Effect of Nature of Feed Stream

The most important factors influencing the membrane life are hydrolysis, solvation, deterioration by microorganisms, and fouling, as well as the membrane compaction.

Cellulose acetate is slowly hydrolyzed in water, which reduces the acetyl content in the membrane and therefore its performance. Vos et al. [30] conducted a kinetic study on the hydrolysis of the cellulose acetate of 39.8% acetyl content. The study showed that the hydrolysis rate exhibits a minimum at pH 4.5 to 5.0 in the temperature range of 23 to 95°C. The minimum rate decreases as the temperature is lowered. Polar organic compounds with low molecular weights, such as alcohols, ketones, and amides, always plasticize or swell cellulose acetate as was shown in the chapter on pervaporation (Chapter VII). Perchlorate salt and zinc dichloride also solvate cellulose acetate. Some microorganisms have been reported to digest the membrane enzymatically. Chlorination or addition of dilute cupric sulfate in the proper range of pH has proved to be an effective method to prevent the enzymatic deterioration of the membrane [22].

Fouling is always a troublesome problem in desalination. It is essentially caused by precipitation of iron oxide compounds, MnO, $CaCO_3$ and $CaSO_4$, as well as organic molecules in the saline water. The precipitation can, however, be minimized by using certain reducing agents or precipitation inhibitors, or by controlling pH in the solution. The organic material can, on the other hand, be removed with active carbon; or the precipitates can be defouled by depressurizing, by backflushing, or by using detergents or acid washing. A high flow rate of the feed stream also helps to minimize fouling.

Rejection of Solutes

The separation of a solution through a membrane is essentially due to the different molecular interaction between permeant molecules and the membrane through adsorption, diffusion, and desorption. The different selectivity of solutes in a solution for a given membrane can, therefore, be characterized by the molecular size and shape and the intermolecular interaction potential. Thus, Loeb [31] summarized the different rejections of electrolytes and organic solutes in an aqueous solution through the cellulose acetate membrane in terms of those fundamental molecular variables.

The different rejections of aqueous electrolyte solutions can generally be related to the electromotive potential, the unhydrated ionic radius, the ionic charge, and the dielectric constant. Indeed, Sourirajan [23] found that the order of the rejections is as follows: for *anion:* citrate > tartrate = sulfate > acetate > chloride > bromide > nitrate > iodide > thiocyanate; and for *cation:* Mg, Ba, Sr, Ca > Li, Na, K. The order is very similar to the lyotropic series of electrolytes, that is, the series of their decreasing ability to remove a lyophilic substance from a colloidal solution. According to the Glueckauf model [32], the rejection increases with the valence and the unhydrated ionic radius. However, Blunk [33] found that, when a monovalent anion is associated with hydrogen or ammonium, the tendency is reversed as follows: HI > HBr > HCl > HF; $NaClO_4$ > NH_4ClO_4 > $HClO_4$, and $NaNO_3$ > NH_4NO_3 > HNO_3. On the other hand, most divalent or trivalent electrolytes are relatively well rejected.

Considering the molecular size and shape of the solute, the following tendency of rejection of organic solutes can be easily expected: The rejection increases with the molecular weight in the homolog series, and for isomers, tertiary > iso > secondary > primary. For polar substances of low molecular weights, aldehyde > alcohol > amine > acid. It is worthwhile to note that negative rejection of phenyl compounds has been observed, that is, the permeation of the solute is faster than that of water [34].

Design and Operation of Reverse-Osmosis Unit

When two solutions are separated by an osmotic membrane, the fluxes of the solvent B and the solute A are given by Eqs. 8.11 and 8.14. These equations describe the resistance to transport of A and B only through the membrane. In an actual reverse-osmosis unit the overall mass transfer of a chemical species from one solution to another solution is greatly influenced by the concentration polarization and the longitudinal backmixing, as well as the permeabilities through the membrane. The concentration polarization is the phenomenon that the solute A is accumulated near the ingoing side of the membrane due to the slow permeation rate of A through the osmotic membrane compared to that of the solvent B. The polarization severely reduces the solvent flux and enhances the solute leakage. Also, the accumulated solute often deteriorates the membrane.

As demonstrated in Chapter XIII, the concentration polarization depends in a rather complicated manner on the ratio of the two fluxes, the longitudinal velocity, the channel width or radius of the unit, and the diffusivity of the solute in the solution. In general, the degree of polarization decreases, as the longitudinal velocity and the diffusivity increase. But a large channel width or radius and a high ratio of solvent flux to solute flux enhance the polarization. The concentration buildup can thus be reduced by using a thin channel and high feed-flow rate. Also, various devices to promote local mixing have been proposed to minimize the buildup. Typical examples are the so-called turbulence promotor, an ultrasonic device, and the zig-zag path in the plate-and-frame type of modules.

Four types of commercial reverse-osmosis modules are available in the present market. They are the plate-and-frame, the spiral-wound, the tube, and the hollow-fiber types. However, those commercial units can more conveniently be classified according to the degree of longitudinal backmixing. The extreme cases are, of course, the complete mixing and the nonmixing, or plug flow. The backmixing is caused by molecular diffusion and eddy motion. The mixing depends on the geometry of the osmosis module, the stream velocity and fluid properties such as diffusivity and viscosity. In general, the degree of mixing is enhanced as the ratio of the longitudinal dimension to the channel width and the longitudinal stream velocity decrease. Therefore, a low degree of backmixing, or even plug flow, can likely be attained in a hollow-fiber configuration with a high stream velocity. On the other hand, complete mixing is apt to take place in a short tube of large radius.

Thus, a module of closely packed hollow fibers possesses definite operational advantages. In such a module, the adverse effects of concentration polarization and backmixing are very small compared to other types of modules and the efficiency is therefore high. However, the pressure loss due to hydrodynamic friction in the small channels of hollow fibers can be considerable, and it reduces the available driving force. Hence, one should seek an optimum condition in design and operation of the module by balancing the two opposing effects. The principle of design and operation of the reverse-osmosis module is discussed in greater detail in Chapter XIII.

3 ULTRAFILTRATION

Ultrafiltration is a membrane process of separating a large molecular weight (say, larger than 500) of solute from a solution by an external pressure. The molecular sizes of the solvent and other solutes, if any, in the solution is much smaller than the large solute. The process under discussion here also includes the so-called membrane filtration, which isolates very fine particles from a colloidally dispersed mixture.

The characteristics of the membrane, the transport mechanism, and other operational features are essentially the same as in its sister process—reverse osmosis. The process differs from reverse osmosis only in the molecular size of the solute to be separated, and thus some authors consider it as a special case of the latter process. The concentration of the solute in the feed solution in ultrafiltration is usually quite small, and the solution exhibits a very low osmotic pressure. Therefore, the process is operated at a low external pressure (5 to 100 psi), whereas the operating pressure for reverse osmosis is usually in the range of 500 to 2000 psi. Figure 8.4 schematically illustrates this behavior in comparison with reverse osmosis [35]. In an efficient unit of ultrafiltration,

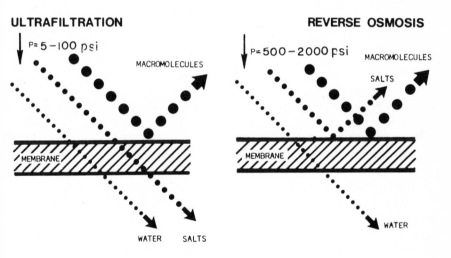

ULTRAFILTRATION　　　　　　　　　　　**REVERSE OSMOSIS**

P = 5−100 psi

MACROMOLECULES

MEMBRANE

WATER　　SALTS

P = 500−2000 psi

MACROMOLECULES

SALTS

MEMBRANE

WATER

Fig. 8.4 Comparison of flow schemes in reverse osmosis and ultrafiltration [35].

a porous membrane is commonly employed. The pore size is large enough that the solvent molecules experience little resistance during permeation through the membrane, but the solute is almost completely rejected or retained. However, the adverse effect of the concentration polarization on the ingoing side of the membrane is very severe compared to that in reverse osmosis. It is not unusual that the solute is so highly polarized that it even forms a gel layer on the membrane surface, and the overall solvent transport is considerably retarded. This phenomenon is mainly due to the fact that the diffusivity and the solubility of large molecules are generally very small. The concentration polarization associated with the large molecular size of solute is discussed in greater detail in Chapter XIII.

Simplicity of construction, short time of operation, and low operating cost are the main advantages of ultrafiltration over other conventional separation processes, such as evaporation, solvent extraction, selective precipitation, or chromatographic separation. The process is particularly profitable in the separation of a solution containing traces of thermally and chemically labile substances like food products, drugs, and other biologicals.

The development of the ultrafiltration process was elegantly reviewed by Michaels [38]. Michaels and Porter summarized their recent valuable experiences associated with this process in a series of articles; also, a bibliography of 414 items is available from Amicon Corporation [36].

Ultrafiltration Membranes and Transport Mechanism

As mentioned earlier, a porous membrane of suitable pore size is effectively used in ultrafiltration. The solute of the large molecular size is almost completely retained and the solvent and other solutes of smaller size, if any, permeate through the membrane by convection with little resistance. Plugging of pores by the larger solute molecules has, however, been troublesome with some porous membranes. It greatly reduces the solvent transport and often shortens the membrane life irreversibly. The problem of plugging has recently been resolved by employing the anisotropic (asymmetric) membrane. This kind of membrane is highly permeable to the solvent as discussed in Section 2. Also, it withstands relatively high pressures, shear stress, and abrasion with a solution containing large solute molecules for a considerable period of time.

In reverse osmosis, the cellulose acetate membrane is extensively used. However, this membrane is not always the best for ultrafiltration. When the process is associated with an organic solvent, cellulose acetate is readily plasticized or swollen. At low and high pH values, it is easily hydrolyzed in an aqueous solution. Its glass-transition temperature is relatively low, and the operating temperature is limited to a low value (50-60°C). Furthermore, cellulose acetate is usually subject to attack by enzymes and microorganisms. Hence a variety of new improved synthetic membranes, rather than cellulose acetate, have been introduced in ultrafiltration. Those membranes that are available on the market were described in considerable detail by Michaels [115], Lonsdale [22], and van Oss [116].

In particular, it is worthwhile to note that the UM-series of membranes marketed by Amicon Corporation is essentially amphoteric membranes (also see Chapter XIV). In such membranes, both acidic and basic ionic groups are present. When a nonionic macromolecule is separated from an aqueous solution containing a small electrolyte solute, the amphoteric membrane is expected to exhibit a high selectivity with respect to the electrolyte solute due to its highly hydrophilic character. However, the mobility of the small electrolyte solute is generally retarded. This phenomenon is called *ion retardation* [8]. The retarded mobility is, to some extent, improved by adjusting the pore size and also by preparing the membrane in an asymmetric form. On the other hand, an ion-exchange membrane containing only an acidic ionic group or a basic group is more effective in separation of an ionic macrosolute from a solution containing a nonelectrolyte solute. The ionic macromolecule is repelled from the membrane by the Donnan exclusion, and, therefore, a high retention can be achieved.

As has been observed in the reverse-osmosis membrane, the solute rejection increases with the external pressure gradient and then approaches an asymptotic value at a sufficiently high pressure. The asymptotic value is generally not unity because of increased coupling between the solute flux and solvent permeation by convection. When the molecular size of the solute is of the same order of magnitude as the average pore size of the membrane, the solute rejection often passes through a maximum and then decreases with the pressure gradient. This behavior is partially due to both the elastic distortion of pores of the membrane and the possible elongative distortion of the large but soft solute molecules induced by pressure and shear stress [38]. The solute rejection is also slightly enhanced by the solute concentration, which indicates the existence of a frictional interaction between solute molecules and the wall of fine pores in the membrane.

4 DIALYSIS

In a broad sense, *dialysis* refers to a process separating one chemical species from another species in a liquid solution through a semipermeable membrane. The driving force is the difference of the chemical potentials across the membrane. In this section, the process will, however, be confined to the case where the external pressure and electric potential are absent. The process driven by an external pressure has already been discussed in Sections 2 and 3 on reverse osmosis and ultrafiltration, respectively, and the process driven by an external electric potential will be treated in the next chapter. Thus, only concentration-driven processes are covered here.

The permselectivity of a chemical species relative to another species in a membrane largely depends on the ratio of their molecular sizes in the absence of a strong specific interaction. Therefore, the process of dialysis has been effectively applied to separate a small crystalloid from a large colloid in a solution. The recovery of caustic soda from an aqueous mixture of caustic and hemicellulose in the manufacture of viscose rayon is a typical example. However, the permeation flux of the smaller species with the driving force of the concentration gradient alone is often found to be too small to be of economic significance. Processes driven by an external pressure such as reverse osmosis and ultrafiltration are more attractive in such cases. Still, dialysis has been an important process in the separation of a chemical species that is extremely labile to an excessive pressure or electric potential. Also, dialysis of one ionic solute from another ionic solute has recently been recognized again to be effective and competitive with other processes. It was demonstrated in Section 1 that a large driving force can be achieved in a system where Donnan equilibrium prevails. In this case, a considerable permeation flux can be obtained without an external pressure or an electric potential. The recovery of acid from the spent liquor in the metal-refining industry is a classical example. Presently, explorative research on the process has been extended to the recovery of a precious metal in a waste stream [10]. In desalination plants, the process is a potential candidate for removing a polyvalent electrolyte prior to actual desalination. Such polyvalent electrolytes tend to form troublesome scale on reverse-osmosis membranes or on the inside surface of evaporators. Dialysis in which an electrolyte solute is separated from other electrolytes is often called Donnan dialysis, or Donnan softening in the desalination plant. The process is also considered a continuous ion-exchange process, if the membrane is electrically charged.

For the sake of convenience the dialysis process will be classified into two categories: ordinary dialysis and Donnan dialysis. In ordinary dialysis, a nonelectrolyte is separated from another nonelectrolyte or from a mixture containing an electrolyte solute. If the mixture contains more than one electrolyte, the interionic interaction among ions of the same sign is insignificant in ordinary dialysis. The driving force for transport of a solute through the membrane is predominantly the concentration gradient. In Donnan dialysis, on the other hand, the interionic interaction plays an important role, and the Donnan equilibrium is maintained at the membrane-solution interface. The term *Donnan dialysis* is usually limited only to dialysis of electrolyte solutes with an ion-exchange membrane. However, the term will be extended to all cases where the Donnan equilibrium has a significant effect. In Section 1 it was demonstrated that Donnan equilibrium can prevail even with a neutral membrane, if the membrane is permselective with respect to one of the ions present in the solution.

Ordinary Dialysis

Suppose that two dilute solutions, designated by I and II, are separated by a membrane at a uniform temperature and external pressure. Solution I consists of solvent B and two solutes. The molecular size of the one solute, designated by A, is much smaller than that of the other solute, but almost of the same size as solvent B. Therefore, the passage of the larger solute through the membrane is negligible compared to those of solute A and solvent B. On the other hand, solution II consists of solute A and solvent B only. The concentration of solute A in solution I is larger than in solution II. Thus, the solute A is transported from the former solution to the latter, while the solvent transfer is usually in the opposite direction by osmotic pressure. Then, the molar flux of solute A is given by Eq. 8.14 and the solvent flux B by Eq. 8.11. The abnormal osmosis was discussed in Section 1.

Dialysis Membrane and Osmotic Flux

The earlier membranes used in dialysis were exclusively cellulosic. However, the newly developed synthetic membranes for reverse osmosis and ultrafiltration can also be used directly in dialysis. The amphoteric membrane described in the section on ultrafiltration membranes (Section 3) can, for example, be effectively employed to separate a nonionic macrosolute from an aqueous solution containing a small electrolyte solute. On the other hand, a cation- or anion-exchange membrane may be very efficient for isolation of an ionic macromolecule from a solution of a small nonelectrolyte solute.

A high *positive* osmotic flux, which occurs from the diffusate stream to the dialysate stream generally has an adverse effect. It dilutes the concentration of the dialysate stream and therefore reduces the available driving force. When the dialysate is a waste stream, its dilution is an additional burden. It is therefore extremely important to select an appropriate membrane that shows low *positive* or *negative* osmotic transport. When the solute to be diffused is ionic, the osmotic transport can be controlled to some extent by choosing a weakly electrically charged membrane. In dialysis of an aqueous caustic solution with a polyvinyl alcohol membrane, for example, the sodium ion is adsorbed by an alcoholation reaction, and the membrane is electrically charged positive. Since the mobility of the hydroxyl ion is high compared to the unbound sodium ion, the osmosis is positive [9], and therefore the membrane is not suitable. In such dialysis, a weak anion-exchange membrane with an amine group, however, shows a favorable negative water flow [11]. Similarly, a weak cation-exchange membrane is effective in controlling the osmotic flow in acid recovery. However, the mobility of the electrolyte solute to be diffused is generally reduced in such ion-exchange membranes. An electrically charged membrane with suitable ionic strength is therefore required.

Overall Dialysis Coefficient

As in other membrane processes in the liquid phase, a significant effect of the concentration polarization near the membrane-solution interface is commonly observed. The polarization in dialysis, however, differs slightly from that in reverse osmosis or ultrafiltration in which the osmotic flux (the solvent flux) is important. In these osmotic processes, the solution accumulates near the ingoing side of the membrane, and the polarized boundary layer retards the osmotic flux but enhances the solute flux. The concentration polarization is, thus, important only near the ingoing side of the membrane.

On the other hand, in dialysis in the absence of the nonpermeable macrosolute, the solute A to be diffused depletes near the ingoing side of the membrane, but accumulates at the outgoing side of the membrane. The solute concentration is thus polarized on both sides of the membrane. The polarized boundary layers retard the solute flux N_A and also the solvent flux N_B, which occurs usually in the direction opposite to the solute flux. If the nonpermeable macrosolute is, however, present in the feed stream, the macrosolute accumulates greatly on the ingoing side of the membrane. This polarized layer further reduces the solute flux N_A, but the solvent flux is generally enhanced. Such concentration polarization of the macrosolute is so severe that even a gel layer is often formed on the membrane surface, as was mentioned in the discussion of ultrafiltration. The resistance across the layer frequently exceeds that across the membrane itself and thus plays a major role in the overall mass transfer. This case is treated in greater detail in Chapter XIII.

The flux of solute A to be diffused in the presence of such concentration polarization is customarily expressed in terms of bulk concentrations in the two streams as

$$N_A = U_A(C_A^I - C_A^{II}) , \qquad (8.75)$$

where U_A is called the overall dialysis coefficient. The osmotic flux in dialysis is usually not so significant compared to the solute flux N_A, and the solvent is almost stagnant. In this case, the overall dialysis coefficient U_A can be approximated by

$$\frac{1}{U_A} = \frac{l}{Q_A'} + \frac{1}{k_A^I} + \frac{1}{k_A^{II}} . \qquad (8.76)$$

Here, k_A^I and k_A^{II} are mass-transfer coefficients through the ingoing and the outgoing boundary layers, respectively, which are defined in Chapter XIII. The mass-transfer coefficients, k_A^I and k_A^{II}, increase with the circulation velocities in the corresponding streams, and also with diffusivity of solute A in the solution. However, the coefficients decrease as the solute flux N_A and the channel width increase. For a high solute flux, k_A's are usually functions of the two bulk concentrations.

In a continuous dialysis with a thin channel (say, 2 cm width) in the absence of the macrosolute, the thickness of the boundary layer is often not so sensitive to the circulation velocity in the usual range of 0.4 to 2 cm/min. The thickness assumes a constant value in the order of 0.6 to 0.8 cm [11, 40]. In this case, the mass-transfer coefficient k_A is in the order of

$$k_A = \frac{D_{AB}}{\delta} \cong (120{\sim}170) \; D_{AB} \; \frac{cm}{sec} \; ,$$

which is usually in the same order of magnitude as Q'_A/l.

Figure 8.5 shows the influence of concentration on the overall dialysis coefficient U_A in dialysis of aqueous acids and of sodium hydroxide, two systems that were extensively studied by Vromen [11]. The membrane was

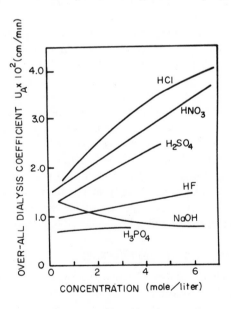

Fig. 8.5 Overall dialysis coefficient of acids and base through Nalfilm D-30. Reprinted by permission from Ref. 11. Copyright 1962 by the American Chemical Society.

Nalfilm D-30 described by Nalco Company as a vinyl type of polymer film. Its thickness was 0.0132 cm for acids and 0.0112 cm for sodium hydroxide. As the liquor concentration increases, the coefficient U_A is markedly enhanced, particularly for the strong acids. Vromen postulated the enhancement as results of the increase of the diffusivity in the boundary film D_{AB} and permeability Q'_A with concentration. But the increase of Q'_A is mainly responsible for such steep enhancement for a high concentration. In dialysis of strong acids, the hydrogen ion is apparently absorbed on the membrane surface, and the membrane behaves like an electrically charged membrane. Thus, Vromen observed a considerable abnormality in the osmotic flow, as mentioned earlier in Section 1.

Principle of Design and Operation of Dialyzer

Three types of commercial dialyzers have been available in the market: the tank, the plate-and-frame, and the tube types. In the tank-type dialyzers, flat bags of membranes are simply immersed in a tank. The feed liquor is circulated through the tank outside the bags. The Cerini dialyzer [41] is a typical example. This type of dialyzer has some disadvantages. It requires a large space for the equipment and also a heavy thick membrane in order to ensure its strength. Furthermore, such a configuration is not effective in minimizing the effect of the concentration polarization on both sides of a membrane bag. On the other hand, the resistance across the polarized boundary layer can be considerably reduced in the plate-and-frame and the tube types of dialyzer by employing a high circulation rate. In addition, a high efficiency can be achieved by designing and operating the dialyzer in such a condition that a countercurrent plug flow can be maintained. In the plate-and-frame-type dialyzer in particular, very thin membranes can be safely used so that a high flux of the solute may be obtained. One may also design a zig-zag path in the plate in order to promote local turbulence and thus further reduce the boundary-layer resistance. However, the complicated design of such a plate-and-frame type is a definite disadvantage. The usage of the tube type of dialyzer has been very limited because of the difficulty in preparing thin but stable membranes in the tube shape, particularly in the shape of hollow fibers. This type of dialyzer can, however, be expected to be very promising in the near future, considering the current rapid development of hollow-fiber technology.

Donnan Dialysis with Neutral Membranes

Suppose that two dilute solutions, designated by I and II, are separated by a neutral membrane. Both solutions consist of two cations, 1 and 2; a common anion Y and a common solvent B. The ionic size of cation 2 is so large that its permeation is extremely small compared to those of cation 1 and anion Y. The concentration of cation 2 in solution I is much greater than in solution II. Thus, cation 2 is to be dialyzed, and cation 1 to be diffused, from solution

I to solution II through the membrane. As shown in Section 1, the permeation of cation 1 is greatly enhanced by the electric potential created by the slower-permeating cation 2. On the other hand, the diffusion of cation 2 is further retarded by the faster-moving cation 1. Even pumping of cation 1 by the electric potential can occur from its less concentrated solution to its more concentrated solution as demonstrated by Eq. 8.44. In dialysis of such a solution, the ionic flux N_1 of cation 1 is given by Eqs. 8.70 and 8.71, and the flux N_2 of cation 2 can be approximated by the simple Fick equation, Eq. 8.14. Also, the solvent flux is given by Eq. 8.11. As indicated in these equations, the efficiency of such dialysis increases with the electrochemical valence of the macro-ion 2 as well as with the permselectivity of the membrane. But the *positive* osmotic flux and the leakage of the macro-ion reduce the efficiency.

Vromen [11] attempted to recover free nitric acid from a waste stream containing ferric salt. The experimental results are presented in Table 8.1. As shown in this table, the apparent overall dialysis coefficients $U_{1 (app)}$ of the free

Table 8.1   Vromen's Dialysis Experiment[a]

Stream	Flow Rate (liter/min)	Iron Concn., C_2 [g(mole)/liter]	Free Acid Concn., C_1 [g(mole)/liter]	Apparent Overall Dialysis Coeff.[b] (cm/min)
		Experiment 1		
Water	0.070			
Feed	0.058	0.516	0.384	$U_{1(app)} = 6.6 \times 10^{-2}$
Dialysate	0.066	0.346	0.0698	$U_{2(app)} = 0.55 \times 10^{-2}$
Diffusate	0.062	0.112	0.298	
		Experiment 2		
Water	0.042			
Feed	0.034	0.516	0.384	$U_{1(app)} = 35 \times 10^{-2}$
Dialysate	0.042	0.285	0.00868	$U_{2(app)} = 0.56 \times 10^{-2}$
Diffusate	0.034	0.173	0.371	

[a] Countercurrent experiment with 3350 cm^2 of the Nalfilm D-30.
[b] Calculated from the log-mean concentration, $(\Delta C_i)_{lm}$, that is, $U_{i(app)} = N_i/(\Delta C_i)_{lm}$, $i = 1,2$. Data from [11].

acid are indeed markedly enhanced complared with its value 1.6×10^{-2} cm-/min in the absence of the ferric ion (see Fig. 8.6). The coefficient is higher by four times in Experiment 1 and even by 20 times in Experiment 2. For such dialysis in the absence of significant concentration polarization, the flux can be obtained from Eqs. 8.70 and 8.71 as

$$N_1 = U_1 \left[\left((C_1^I + 3C_2^I)C_1^I \right)^{1/2} - \left((C_1^{II} + 3C_2^{II})C_1^{II} \right)^{1/2} \right], \quad (8.77)$$

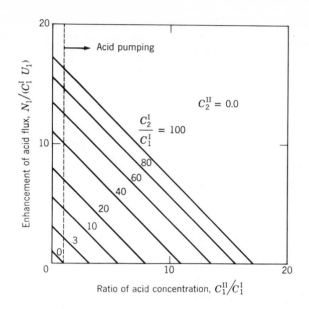

Fig. 8.6 Enhancement of acid flux by the presence of ferric ion in dialysis with a neutral membrane (calculated from Eq. 8.77).

where U_1 is the dialysis coefficient of the pure acid. Here, it is assumed that the coefficient is affected only to a limited extent by the presence of the ferric ion, since the concentration of the macro-ion in the membrane phase is expected to be very small. Substituting the arithmetically average values of the experimental concentrations in the two streams into Eq. 8.77, the enhancement can be estimated as 4.5 times in the first experiment and 17 times in the second experiment. The agreement with experimental data is very good, particularly in the first experiment. Figure 8.6 illustrates the enhancement of the acid flux in the presence fo the ferric ion. The curves are computed from Eq. 8.77.

Donnan Dialysis with Electrically Charged Membranes

This particular type of dialysis is essentially a continuous ion-exchange process. One ionic species in one solution is continuously exchanged, by dialysis, with another ionic species of the same sign in another solution, separated from the former solution by an ion-exchange membrane. It has already been pointed out in Section 1 that in this dialysis an electric potential is created by the difference of the concentrations of an ionic species across the membrane. In a suitable condition, the potential, in turn, greatly enhances the permeation rate of another ionic species. Even "pumping" of the other ionic species is possible from its less concentrated solution into its more concentrated solution. This phenomenon has been a popular subject of research among physical chemists and biologists since Donnan [42] first formulated his well-known thermodynamic principle fifty years ago. Particularly, Sollner has published over 100 articles on the subject. However, it was Wallace [43] who recognized the Donnan dialysis as a potentially effective large-scale, continuous separation process of an ionic species. Its advantages over other conventional separation processes are evident. Donnan dialysis is a continuous process, simple in operation but reliable and also requiring only low energy consumption. The scaleup of a dialysis module to meet a desired capacity is easy to accomplish. Explorative research by Wallace and his co-workers [10] has been started for the following promising applications:

• removal of ^{137}Cs and mercury from a dilute waste

• removal of ^{90}Sr from a dilute waste with a complexing agent

• separation of silver and copper with a complexing agent

• concentration of uranium from dilute processing solutions

• water softening: removal of calcium and magnesium

• pH adjustment

• removal of zinc from textile waste

• recovery of copper from industrial wastes and low-grade ore

• recovery of nickel, cadmium, and chromium from electroplating wastes

• stripping common pollutants such as mercury, cadmium, copper, and lead from industrial and mining waste

However, studies on the relatively new Donnan dialysis process have been limited at most to the pilot-plant scale as of now.

Most applications of dialysis concern the removal or recovery of metal ions by exchange with an acid. The metal ions permeate through a cation-exchange membrane from the feed stream to the diffusate stream, and hydrogen ions of the acid transport in the opposite direction. Thus, the feed stream is to be stripped of the metal ions by the hydrogen ions. The ion-exchange rate may be accelerated by adding a complexing agent to the stripping stream that can form a complex with the metal ion. However, discussion will be confined only to the case where the metal ions are removed from a solution by an acid without a complexing agent.

Minimum Charge of Acid

When the two dilute streams, separated by an ideal cation-exchange membrane, are under thermodynamic equilibrium, the equilibrium distribution of the two ions across the membrane is given by Eq. 8.41. Suppose that the feed stream contains only the metal ions and the stripping stream contains only the acid at the initial state. Then, the minimum charge of the acid required for a given fraction recovery of the metal ion

$$f = \frac{W_{1E}}{W_{1F}} = \frac{q_E C_{1E}}{q_F C_{1F}} \qquad (8.78)$$

can be derived by combining the overall material balance and Eq. 8.41 with the electroneutrality condition as

$$\frac{W_{2S}}{W_{1F}} = \frac{f}{\beta} \left[\left(\frac{f}{1-f} \right)^{\beta} \left(\frac{q_S}{q_F} \right)^{1-\beta} + 1 \right] . \qquad (8.79)$$

Here, the subscripts S, F, and E refer to the stripping, feed, and diffusate (extract) streams, respectively, and the subscripts 1 and 2 refer to the metal ion and the hydrogen ion, respectively. Also, W and q are the molar solute flow rate and the total volumetric flow rate. It is assumed in the derivation that the *ideal* membrane does not permit the leakage of the co-ion (anion) nor the occurrence of osmotic flow. Figure 8.7 illustrates the effect of the ratio of

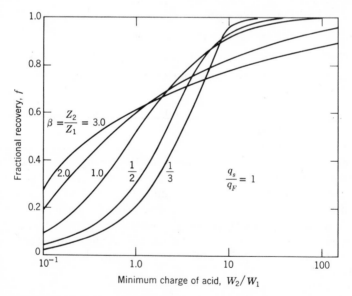

Fig. 8.7 Effect of ratio of chemical valences on the minimum charge of acid (calculated from Eq. 8.79).

electrochemical valences on the minimum charge of acid. As indicated in this equation, the minimum charge of the acid is independent of the ratio q_S/q_F of the flow rate of the feed stream to that of the stripping stream, provided that the electrochemical valence ratio $\beta = z_2/z_1$ is unity. However, the minimum charge for a polyvalent metal ion ($\beta < 1$) increases with the ratio q_S/q_F, as indicated in Fig. 8.8. In other words, the maximum attainable fraction recovery of the polyvalent metal ion for a given charge of the acid decreases with the

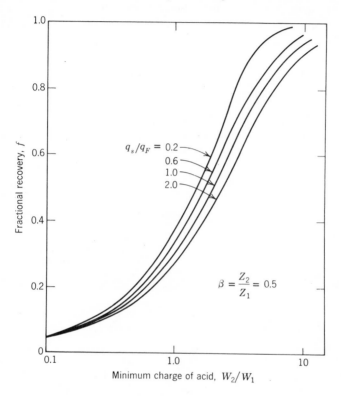

Fig. 8.8 Effect of q_S/q_F on the minimum charge of acid (calculated from Eq. 8.79).

ratio q_S/q_F. Hence, a very low value of q_S/q_F is commonly used to recover a polyvalent metal ion. When the feed stream flows too rapidly and the stripping stream too slowly, the diffusate stream is saturated with the metal ion, and thus its recovery is reduced.

In practice, thermodynamic equilibrium is, however, hardly attained for a finite membrane area. Also, the actual amount of the acid required is more than the minimum charge due to leakage of the co-ion, occurrence of osmotic flow, concentration polarization, and backdiffusion. Indeed, Wallace and co-workers [44] were able to achieve an impressive recovery of uranyl ion with an excessive amount of nitric acid that was larger by 2 to 10 times that of the minimum charge.

Ionic Flux and Osmotic Flow

In such ion-exchange dialysis, the ionic flux N_1 of the metal ion 1 is given by Eqs. 8.63 to 8.65 in terms of ionic concentrations in the membrane phase. Substitution of the boundary condition, Eq. 8.66, into these equations leads to an expression in terms of concentrations of the solution phases. The ionic flux N_2 of the hydrogen ion is then obtained from the charge conservation, Eq. 8.61. Also, the leakage of the co-ion (anion) N_Y can be approximated by the simple Fick equation (see Section 1):

$$N_Y = \frac{\overline{D}_1 K_Y}{l} (C_Y^{I} - C_Y^{II}) .$$

(8.80)

The osmotic flux has already been discussed in Section 1.

When an excessive amount of acid is used in such dialysis, the following condition is usually satisfied:

$$\frac{D_2}{D_1} > 1 \gg \frac{z_1^2 \overline{C}_1}{z_2^2 \overline{C}_2}$$

(8.81)

or, in terms of the solution concentrations:

$$\frac{C_1}{C_1 + C_2} \ll 1 \quad \text{if } z_1 = z_2 = 1,$$

(8.82)

$$\frac{8 \overline{C}_R C_1}{C_2^2} \ll 1 \quad \text{if } z_1 = 2 \text{ and } z_1 = 1.$$

(8.83)

Then, both Eqs. 8.63 and 8.64 can be approximated by Eq. 8.65. Substituting Eq. 8.66 into Eq. 8.65, one may obtain, when $z_1 = z_2 = 1$,

$$N_1 = \frac{\overline{D}_1 K_{12} \overline{C}_R}{l} \left(\frac{C_1^{I}}{C_T^{I}} \right) \left[1 - \left(\frac{C_1^{II}}{C_1^{I}} \right) \left(\frac{C_T^{I}}{C_T^{II}} \right) \right] ,$$

(8.84)

$$K_{12} = \left(\frac{\gamma_1}{\overline{\gamma}_1} \right) \left(\frac{\overline{\gamma}_2}{\gamma_2} \right) ,$$

(8.85)

$$C_T = C_1 + K_{12} C_2$$

(8.86)

and, when $z_1 = 2$ and $z_2 = 1$,

$$N_1 = \left(\frac{\overline{D}_1 K_{12} \overline{C}_R^2}{l} \right) \left(\frac{C_1^{I}}{(C_2^{I})^2} \right) \left[1 - \left(\frac{C_1^{II}}{C_1^{I}} \right) \left(\frac{C_2^{I}}{C_2^{II}} \right)^2 \right] ,$$

(8.87)

$$K_{12} = \left(\frac{\gamma_1}{\bar{\gamma}_1}\right)\left(\frac{\bar{\gamma}_2}{\gamma_2}\right)^2 . \tag{8.88}$$

In order to derive Eq. 8.87, further approximation was made by the Taylor expansion with respect to the small term $8\bar{C}_R C_1/C_2^2$. Thus, in the presence of an excessive amount of the acid, the electric potential created by the concentration distribution of the hydrogen ion enhances the driving force for transport of the metal ion, but does not significantly influence the mobility. The ion-exchange rate is primarily determined by the diffusivity of the metal ion in the membrane phase but almost independent of the mobility of the hydrogen ion. In dilute solutions where ionic species effectively exhibit their ideal electrochemical behaviors, the distribution coefficient K_{12} is usually close to unity. Then, the value C_T approximately equals the total cation concentration in the solution when $z_1 = z_2$. Figures 8.9 and 8.10 demonstrate the enhancement of the driving force for the ion-exchange rate.

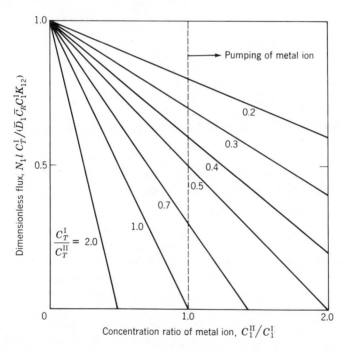

Fig. 8.9 Ion-exchange dialysis of a monovalent metal ion with an excessive amount of acid (calculated from Eq. 8.84).

With an excess of acid in the stripping stream, the solvent flux is primarily due to the osmotic pressure gradient. In this case, the solvent permeates through

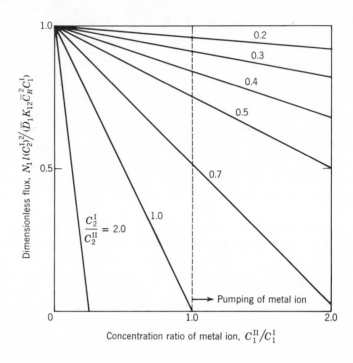

Fig. 8.10 Ion-exchange dialysis of a divalent ion with an excessive amount of acid (calculated from Eq. 8.87).

the membranes from the feed (dialysate) stream into the stripping (diffusate) stream and thus dilutes the concentration of the latter stream.

Design and Operation of a Dialysis Module

Essential features in design and operation of ion-exchange dialysis are the same as for the ordinary dialysis. Modules such as the plate-and-frame, spiral, tube, and hollow-fiber configurations can be employed directly in this dialysis. Also, the concentration polarization and backdiffusion are important factors that must be carefully considered in design and operation. The metal ion is depleted, but the hydrogen ion is accumulated, on the ingoing side of the membrane. On the other hand, the concentrations of the two ions are polarized in the opposite way on the outgoing side of the membrane. Thus, the concentration polarization reduces the driving force for ion exchange.

As mentioned earlier, a high flow rate of the feed stream but a low flow rate of the stripping stream are usually employed to obtain a high recovery of the metal ion. Therefore, the concentration polarization near the interface between the membrane and the stripping (diffusate) stream is very important. The backdiffusion is also expected to be considerable. Another serious adverse effect, due to the extremely low flow rate of the stripping stream, is dilution of the concentration of the stripping (diffusate) stream by the osmotic flux. The net driving force for the ion-exchange rate is thus greatly reduced. Therefore, selection of suitable operating conditions is essential.

5 LIQUID MEMBRANES

Most of the natural and artificial membranes are solid. Fluid membranes might have been conceived by earlier investigators, but they have not come into existence until recently. It is easy to form a fluid membrane; however, it is difficult to maintain and control the membrane in a separation process. In order to avoid easy breakups, some type of reinforcement is necessary to support such a weak membrane structure.

Liquid Membranes with Support

The first successful liquid membrane was discovered by Martin [45] in his study of desalination by reverse osmosis. A few parts per million of polyvinyl methyl ether were added to the saline feed, which was capable of forming a membrane at the interface between a cellulose acetate membrane and a saline solution. A tremendous increase was noted in permselectivity accompanied by a small decrease in flux. This *in situ* formation of the liquid membrane is equivalent to having two membranes in series resulting in a change of flux and selectivity.

The explanation of the increased permselectivity was given by Kesting [46, 47] based on the hypothesis of the existence of a liquid membrane. Kesting and co-workers [48] also extended the liquid-membrane study in reverse osmosis using various surfactant membranes. Below the critical micelle concentration of the surfactant, both fluxes of salt and water decreased by increasing the concentration of surfactant due to the additional resistance of the surfactant layer. However, beyond the critical micelle concentration, neither flux changed significantly, implying the complete formation of the surfactant membrane under the cellulose acetate membrane. There was a considerable variation in the total flux and permselectivity of different liquid membranes. It was believed that the ratio of the hydrophillic group to the hydrophobic group and the structure of the micelles were primarily responsible for these changes.

Liquid Membranes without Support

Without substrate membrane support, a liquid surfactant membrane alone can also be operational when the membrane forms tiny droplets suspended in a liquid phase. Li [49-51] discovered a separation technique using liquid surfactant membranes in emulsified solutions. He placed aqueous solution of surfactant on the bottom of a separation column. On top of the surfactant solution, an organic solvent occupied a space, through which tiny droplets rose up. The feed solution was introduced at the bottom of the column as shown in Fig. 8.11. As the surfactant droplets containing feed pass through the solvent phase, selective permeation takes place through the thin liquid membrane. As the droplets reach the top of the solvent phase, they are enriched with less-permeating components and coalesce to form a new phase of product solution.

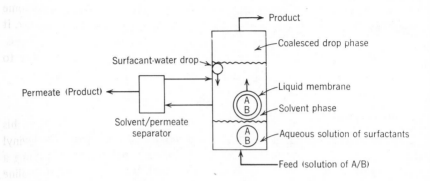

Fig. 8.11 Schematic of liquid-membrane operation [50]. Courtesy of the American Institute Chemical Engineers.

The liquid membrane is believed to consist of two surfactant layers with a very thin water layer sandwiched in-between. This water layer presents the major part of permeation resistance to the hydrocarbons. Unlike solid membranes, liquid membranes go through changes in their thickness. Because of film drainage, the water layer becomes thinner and thinner with time, thus causing the variation in permeability and selectivity. Assuming that the permeability is a product of diffusivity and solubility, we can immediately see that the solubility is the controlling variable. Although diffusivities of different hydrocarbons are about the same, the solubilities of hydrocarbons in water vary greatly [52].

A study [51] of the toluene permeation rate through surfactant liquid membrane showed that the rate of permeation increased exponentially with temperature due to higher drainage rates at higher temperatures. However, the selectivity changed little with temperature implying that the thickness of liquid membrane really became reduced. The permeation rates of toluene, heptane, and the mixture of both are given in Table 8.2. The variation of membrane structure is evident as a function of time. It should be noted that the thickness of membrane is not included in the permeation rate.

Table 8.2 Permeation Rate through Liquid Membrane[a]

	Surfactant Solution = 0.1% Saponin	
Feed	Permeation rate $\times$ 10^6 [g/(cm^2)(sec)]	Sample Timing (min)
Toluene	29.9	6
	58.8	10
Heptane	0.168	10
	0.396	18
1 Toluene / 1 Heptane	18.6	6
	32.9	10

[a]Reference 50.

Separation of benzene and hexane by a liquid membrane was also reported recently [53]. As benzene is far more soluble in water than hexane, benzene can permeate through the liquid surfactant film much more quickly. The separation factor varied as a function of feed composition as well as of the surfactant concentration.

Besides the application in blood oxygenators (see Chapter X), the liquid membrane also finds its usefulness in waste-water treatment [54]. Contaminated water is brought into contact with tiny droplets of selected reagents coated by a surfactant liquid membrane. Contaminants may be adsorbed, extracted, permeated, and then converted by reaction on the inside of the droplet. The emulsion is separated from the treated water and may then be demulsified and recycled for regeneration of the liquid-membrane emulsion, or may be disposed of by incineration.

Also, some laboratories (for instance General Electric's research center in Schenectady, New York) are working on the removal of hydrogen sulfide from gasified coal by a liquid-membrane technique.

Liquid Membranes in Gas Separation

Gas separation does not have to be limited necessarily to solid membrane processes. It is possible to apply the liquid-membrane technique to gas separations. Liquid-membrane processes may offer better separations at higher flow rates resulting in lower overall costs.

Some fundamental studies were carried out by Cook [55] recently on the determination of separation characteristics of liquid surfactant membranes for various gases. A side-arm burette was employed as the main piece of apparatus. A liquid surfactant film was generated at the bottom of the burette and pushed upward by a different gas than the gas above it. The liquid film was stopped in the center of the burette with two pure gases on either side of the membrane. Both ends of the burette were then sealed. The liquid membrane allowed permeation of both gases in opposite directions. Owing to different rates of permeation, the liquid membrane moved either up or down until it reached an equilibrium position, when the concentrations of the gases were equal on either side. By reading the membrane position as a function of time, the permeabilities of gases were claculated as shown in Table 8.3.

Table 8.3   Permeabilities of Gas Pairs through Liquid Membranes[a]

System A-B	2% Ivory Film $Q_A \times 10^9$	$Q_B \times 10^9$	2% Duponol Film $Q_A \times 10^9$	$Q_B \times 10^9$
$CO_2 - N_2$	4.47	0.48	12.30	0.26
$CO_2 - O_2$	7.42	0.88	12.97	0.75
$CO_2 - He$	11.20	3.39	19.99	1.03
$CO_2 -$ ethane	8.6	0.38	9.59	0.41
$CO_2 -$ propane	5.98	0.20	12.26	0.33
$He - N_2$	5.79	1.80	6.43	2.80
$He -$ ethane	4.36	3.56	3.27	2.83
$He -$ propane	4.61	2.14	3.95	2.32
$O_2 - N_2$	2.84	1.27	2.51	1.13
$O_2 -$ ethane	6.91	6.72	2.62	2.49
$O_2 -$ propane	1.40	0.94	1.25	0.74
Ethane $- N_2$	2.71	1.24	2.20	1.04
Propane $- N_2$	1.67	1.13	3.27	2.56
Ethane $-$ propane	1.84	1.30	2.02	1.35

[a] Reference 55.

The differences in permeabilities are mainly caused by the differences of gas solubilities in water. Some pairs of gases showed large separation factors, promising that liquid membranes may become an alternative separation technique. Since the liquid membrane can be made very thin, the flux can be very high. Also, the liquid-membrane technique requires no sweep gas, which is an added advantage.

6   COUPLING WITH CHEMICAL REACTIONS

When chemical reactions are coupled with membrane permeation, many interesting phenomena can occur. On one hand, the chemical reactions merely augment the mass-transport rate through the membrane, resulting in so-called facilitated transport or active transport (see Chapter III). On the other hand, however, the chemical reactions take place either inside or outside of the membrane and products are permeated out of the membrane, thus eliminating the conventional separation steps before and/or after the reaction. When the chemical reactions are carried out outside of the membrane, the membrane functions as a boundary of the reaction zone. However, if the reactions are carried out inside the membrane, the membrane acts as more than a partition. For this reason, such a membrane will be called a *reactive membrane*. Conceptually, this kind of reactive membrane represents a revolution of unit operations and reaction engineering. Normal chemical processes involve several separation or purification operations before and after the main chemical reactions in the reactor. In a reactive membrane, all of these series treatments are compacted in a small space. Therefore, the reactive membrane may be viewed as a combination of several different unit operations equipment plus a chemical reactor.

The preparation of a reactive membrane requires "loading" a selected membrane with appropriate chemical reagents, or catalysts or enzymes. The liquid membrane technique easily allows the encapsulation of the desired chemicals in the membrane. However, the most promising technique is the immobilization of enzymes on membranes. There has been much recent development in this area from many different fields of discipline. Especially, the study of immobilized enzymes has become an attractive subject in biochemistry. It is beyond the scope of this book to go into details and compile all the publications in this area. Instead, short summaries and references are presented. First, the facilitated transport will be discussed as an example of reactive membrane operation. Then, a summary of immobilized enzyme studies will be covered, followed by the discussion of the concept of a liquid-membrane reactor and of product removal by membranes. All of these examples show how the membrane processes may be coupled with chemical reactions to yield various facilitated phenomena.

Facilitated Transport

Facilitated transport describes a phenomenon where the rate of flow of a given species through a membrane is greater than would be expected from basic diffusional behavior (see Chapter III). The term was created by a physiologist to denote observed transport across cell membranes far in excess of that which was calculated to occur from permeability considerations.

A rather comprehensive treatise on what is termed "facilitated diffusion" is contained in a book by Stein [56] that is concerned with molecular movement across cell membranes. It is evident that observations of a facilitated or augmented rate of transport date back to at least 1933, when Höber and Ørskov [57] noted unusually high transport of urea in their study of the rate of hemolysis of red cells. The first reports of facilitated gas transport, specifically oxygen, appear to be by Wittenberg [58] and by Scholander [59] in studies of oxygen movement through films of hemoglobin solution. These observations may have suggested the concept of a "loaded" membrane for accelerated gas transport with nonliving membranes.

The first such use may be that described by Robb et al. [60] in a U.S. Patent, where a thin film of liquid is confined within a suitable membrane structure—porous or swollen membrane—so that the liquid film becomes the primary separation barrier.

A study by Ward and Robb [61] on facilitated transport of carbon dioxide through an immobilized film of aqueous cesium and potassium bicarbonate-carbonate solutions proved that the permeability of CO_2 could indeed be enhanced appreciably. The salt solutions were contained in a porous cellulose acetate film that was able to sorb 60% water, and presumably a like amount of solution that was incorporated by soaking the film in the solution. The permeability measurements were made by using a backing of 0.5-mil silicone rubber to prevent the displacement of the solution by the applied gas pressure.

In order to catalyze the bicarbonate-forming reaction, the experiments were also carried out with additions of carbonic anhydrase and sodium arsenite. Although the anhydrase proved ineffective, except for an initial transient increase, the arsenite additive was effective. The reported results are presented in Table 8.4.

Table 8.4 Facilitated Transport

10^9 *Permeabilities* (std cc)(cm)/(sec)(cm^2)(cm Hg)		
	CO_2	O_2	*Separation Factor*
Water only	210	9.5	22
Membrane + water	40	2	20
Memb. + satd. Ce bicarb.	75	0.05	1500
Memb. + satd. Ce bicarb. + 0.5N Na arsenite	214	0.052	4100

A later study by Ward [62] covered the transport of nitric oxide through ferrous chloride dissolved in formamide. The liquid film was contained between silicone-rubber films, and a porous polyethylene backing was used on the downstream side. The cell design and the apparatus diagram are shown in Fig. 8.12 and 8.13.

The permeabilities were thus measured with a variable-pressure method (see Chapter XII). Appreciably greater transport of nitric oxide was obtained with the solution as compared to a pure formamide film. However the main objective of the study was the establishment of the transport mechanism. The analysis is rather involved and requires a lengthy mathematical treatment. Thus, the reader should consult the original treatise [61].

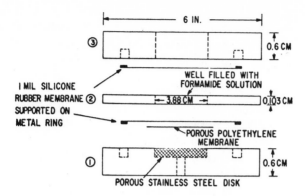

Fig. 8.12 Cell design for facilitated transport [62]. Courtesy of the American Institute of Chemical Engineers.

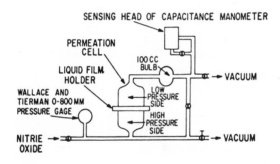

Fig. 8.13 Flow diagram for facilitated transport αperiments [62]. Courtesy of the American Institute of Chemical Engineers

There is one further study by Ward and Neulander [63] that deserves to be mentioned. It deals with the possibility of using an immobilized liquid membrane for separating sulfur dioxide from power-plant-stack gases, specifically from carbon dioxide. The "liquid" membrane was prepared by casting films of a mixture of a 5% water solution of hydroxymethyl cellulose and polyethylene glycols (the solvent for SO_2) with Carbosil addition. A microporous layer of Teflon was used as backing. Difficulties were encountered with breaking-down of the film under pressure. Although a rather good permeability of 3.2×10^{-3} (standard units) was obtained for SO_2, the separation factor of SO_2/CO_2 was only about 15, and thus considered not sufficiently high for stack gas cleanup.

An indication of the potential interest in the subject of facilitated transport is represented by an abstract of a paper given at a symposium on "Reactions in Multiphase Systems," March 15-18, 1971 in Tutzing, Germany. There,

Sauer [64] reported on "coupling of mass transport and chemical reactions" in steady-state membrane systems. The study was concerned primarily with devising a model interpretation of the process. A valuable journal review by Schultz et al. [117] represents the most up-to-date developments.

Immobilized Enzyme Membranes

If enzymes can be fixed to a membrane surface or incorporated into a membrane matrix, the study of enzyme kinetics will make tremendous progress in both theory and application. Enzymes in natural cell membranes are mostly bound, thus giving rise to heterogeneous reactions. On the other hand, most enzyme reactions in industry are carried out in a homogeneous phase. This requires constant loss of enzymes, and the efficiency and degree of control are very low. The immobilized enzyme membrane allows easy removal from the solution after the completion of reaction. It can be used again and again if so desired. Of course, it should be noted here that the immobilized enzyme does not have to be in the form of a membrane in general. It may be in a particulate form.

There are basically four different methods of attaching enzymes to membranes. These procedures are well outlined in a recent review article by Carbonell and Kostin [65]. The first method is to bind enzymes covalently to an inert support. Lilly and his co-workers [66-70] used cellulose as supporting material. Covalent binding was achieved between enzymes and polysaccharides by using cyanogen halides as activators [71-77]. Other types of covalent attachment of enzymes to various polymeric materials [78-84] have also been reported. It is notable that inorganic materials, mostly glass, were successfully used to bind enzymes covalently [85-89]. However, the life of enzyme activity was shorter than with other media.

The second method of binding enzymes to a membrane is cross-linking of the protein molecules by using bifunctional reagents. Collodion membranes were used to cross-link with papain by Goldman et al. [90, 91]. Also, silica particles [92] and cellophane sheets [93] were tried to bind enzymes.

The third approach of immobilizing enzymes is to have physical adsorption of enzymes on solid carriers. Various enzymes were adsorbed on porous glass [94-96] and other solids [97-100] retaining good enzymatic activity.

The fourth method is to embed enzymes mechanically into either a solid matrix [101-103] or a liquid film [104]. Even hollow fibers [105] can be used to entrap enzyme solutions.

The use of immobilized enzyme membranes makes it easy to separate enzymes from the reacting solution for the termination of reaction and recovery of enzymes. Enzymes can be reused, thus yielding a better design of continuous enzyme processes. When the immobilized enzymes are employed in place of homogeneous enzymes, often the enzyme kinetics change due to the structural changes upon binding. However, a great deal of insight of reaction kinetics can be studied by the enzyme membranes [90, 91, 106-110] that could not be revealed by the homogeneous reactions.

Liquid-Membrane Reactor

With the development of the surfactant liquid-membrane system, it became evident that a catalyst could be contained within such a membrane. The advantages of having a catalytic liquid-membrane reactor are numerous. It avoids the excess handling of the catalyst, requires no solid recovery phases, and partially separates the product from the reactants. The concept of the liquid-membrane reactor and some preliminary experiments are well demonstrated by Ollis et al. [111] in their study of acetaldehyde synthesis. The schematic diagram of a liquid catalytic membrane is shown in Fig. 8.14, where reactants permeate, react, and then the product is preferentially extracted out

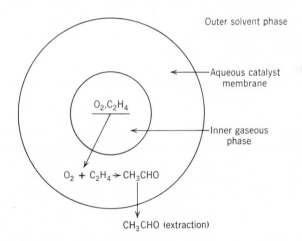

Fig. 8.14   Schematic diagram of liquid catalytic membrane [111]. Courtesy of the American Institute of Chemical Engineers.

into the solvent phase. Figure 8.15 presents a flow diagram for the continuous process. After the reaction, the product is separated from the solvent by distillation. A number of different solvents were tried to ease the distillation load. However, several problems remain to be solved such as membrane stability, optimum catalyst concentration, and solvent loss.

Product Removal by Membranes

When a selective membrane is in contact with a reacting solution and is preferentially permeable to one of the products, the thermodynamic equilibrium of the reaction involved will be shifted so as to compensate for the loss of this product from the reaction zone. The following three cases will serve as

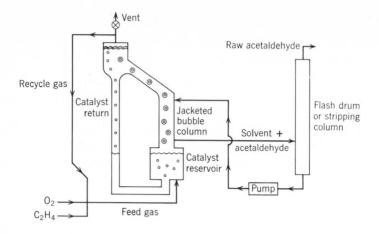

Fig. 8.15 Flow diagram for multiphase reactor and product recovery [111]. Courtesy of the American Institute of Chemical Engineers

examples of this type. Each case concerns a different type of reaction. Also they involve different mechanisms of permeation, namely, pervaporation and reverse osmosis.

The first example relates to a method of completing those chemical reactions that evolve water as one of the products [112]. The water-permeable membrane eliminates the moisture from the reaction mixture by pervaporation. Thus, it drives the reaction to completion. After the reaction, the nonpermeated reaction products are removed for further separation and purification of the final product.

Jennings and Binning [112] were successful in demonstrating the above idea through an illustrative chemical reaction. Esterification of *n*-butanol and acetic acid was carried out in a permeation device. The permeation membrane was hydrolyzed polyvinyl acetate, which removed continuously the water that had been formed by the reaction. At 155°C, it took 60 hr to complete the esterification. The permeate consisted of 95% water.

The second example also shows how pervaporation can be used in removing reaction products from the dissolved catalyst in hydrocarbon conversion [113]. The alkylation reaction between isobutane and ethylene was conducted at 35°C and 600 psig. The catalyst was ferric pyrophosphate hydrate - BF_3 complex, which was retained within the reaction zone while the reaction products were being pervaporated through a film of polyethylene. The downstream side was kept at 50 mm Hg. Thus, the permeated chemicals were rapidly removed. The permeated hydrocarbons contained only 0.1% of BF_3, which is much less than the BF_3 concentration of 3% in the reaction solution.

The third example involves reverse osmosis to separate the reaction product. The following reaction was studied by Gosser et al. [114]:

$$RuH_2(PPh_3)_4 \rightleftharpoons RuH_2(PPh_3)_3 + PPh_3,$$

where the triphenylphosphine (PPh_3) ligand is an unwanted by-product. By removing the PPh_3 from the $RuH_2(PPh_3)_3$ in reverse-osmosis equipment, enhancement of the catalytic activity was achieved, since the equilibrium of the above chemical reaction could be shifted to the right.

It is evident that the application of membrane processes in conjunction with chemical reactions presents a wide-open area of research. It can serve to drive a reaction to completion as well as purify the product.

Nomenclature

Symbol	Meaning
A	Solute
B	Solvent
C_i	Molar concentration of solute A
C_{Am}	Log-mean molar concentration of A across the membrane
C_{Bm}	Log-mean molar concentration of B across the membrane
$\overline{C}_R$	Capacity of an ion-exchange membrane (*mequiv*/liter)
C_T	Total concentration defined by Eq. 8.86
$\overline{D}_e$	Effective diffusivity defined by Eq. 8.57
D_i	Diffusivity of molecular species i
E_{Don}	Donnan potential defined by Eq. 8.31
$\mathcal{F}$	Faraday constant (96,500 C/equiv)
G	Geometric parameter
I	Electric current density $[C/(cm^2)(sec)]$
K	Permeability defined by Eq. 3.17
K_i	Distribution coefficient of molecular species i defined by Eq. 8.4
K_{12}	Distribution coefficient defined by Eqs. 8.85 or 8.88
L_A	Phenomenological coefficient defined by Eqs. 8.19 and 8.20
L_v	Phenomenological coefficient defined by Eq. 8.19
N_i	Molar flux of molecular species i through a membrane
N_v'	Volumetric flux of a solution through a membrane
P	Pressure
$P*$	Standard pressure

$\mathcal{P}$	Net Pressure defined by Eq. 8.10
Q_A'	Solute permeability defined by Eq. 8.14
Q_B	Solvent permeability defined by Eq. 8.11
Q_v	Solvent permeability defined by Eq. 8.17
R	Gas constant
SR	Solute rejection defined by Eq. 8.15
T	Temperature
W	Molar solute flow rate
a_i	Activity of molecular species i
f	Fractional recovery of a metal ion defined by Eq. 8.78
l	Thickness of a membrane
m	Constant
q	Volumetric flow Rate
r	Pore radius
t	Time
v_i	Molar volume of molecular species i
v_{AY}	Average molar volume of molecular species A and Y, defined by Eq. 8.29
x,z	Distance
z_i	Electrochemical valence of ionic species i
α	Diffusivity ratio
β	Valence ratio
γ_i	Activity coefficient defined by Eq. 8.5
$\gamma_\pm$	Mean activity coefficient defined by Eq. 8.35
δ	Thickness of the boundary layer
ϵ	Porosity
ϕ	Electric potential
λ'	Scale parameter characterizing homogeneity
μ	Viscosity
μ_i	Ordinary chemical potential of molecular species i
μ_i^*	Standard chemical potential
μ_i^e	Electrochemical potential defined by Eq. 8.26
ν	Total number of ionic species dissociated from one electrolyte solute in a solution
ν_i	Number of ionic species i dissociated from one electrolyte solute in a solution
π	Osmotic pressure defined by Eq. 8.7
π_s	Swelling pressure defined by Eq. 8.2
σ	Phenomenological constant defined by Eq. 8.20
τ	Constant
ω	Sign parameter defined by Eq. 8.50

ζ Zeta potential

Superscripts

$^-$(overbar)   Referes to the membrane phase

I, II   Refer to solutions

Subscripts

i, j Refer to a molecular or ionic species

A   Refers to a solute

B   Refers to solvent

E   Refers to the diffusate stream

F   Refers to the feed stream

S   Refers to the stripping stream

Y   Refers to a co-ion

1,2   Refer to counterions

References

1. R. P. Durbin, *J. Gen. Physiol.*, **44**, 315 (1960).
2. J. D. Ferry, *Chem. Rev.*, **18**, 373 (1936).
3. H. Faxén, *Ann. Phys.*, **68**, 89 (1922).
4. K. S. Spiegler, *Trans. Faraday Soc.*, **54**, 1408 (1958).
5. U. Merten, "Transport Properties of Osmotic Membranes," in *Desalination by Reverse Osmosis*, U. Merten, Ed., MIT Press, Cambridge, Mass., 1966.
6. A. Katchalsky and P. F. Curran, *Nonequilibrium Thermodynamics in Biophysics*, Harvard University Press, Cambridge, Mass., 1967.
7. G. S. Manning, *J. Phys. Chem.*, **76**, 393 (1972).
8. F. G. Helfferich, *Ion Exchange*, McGraw-Hill, New York, 1962.
9. S. B. Tuwiner, *Duffusion and Membrane Technology*, Reinhold, New York, 1962.
10. H. M. Kelley, D. Randall, and R. M. Wallace, *DuPont Innovation*, **4**, 4 (1973).
11. B. H. Vromen, *Ind. Eng. Chem.*, **54**, (6) 20 (1962).
12. B. E. Conway, in *Physical Chemistry, an Advanced Treatise*, Vol. IXA/-Electrochemistry, H. Eyring, Ed., Academic, New York, 1970, Chapt. 1.
13. J. Newman, in *Advances in Electrochemistry and Electrochemical Engineering*, Vol. 5, C. W. Tobias, Ed., Wiley-Interscience, New York, 1967.
14. R. Schlögl, *Z. Physik. Chem. (Frankfurt)*, **1**, 305 (1954).
15. G. Schmid, *Z. Electrochem.*, **54**, 424 (1950).
16. B. R. Breslau and I. F. Miller, *IEC Fund.*, **10**, 554 (1971).

17. H. Freundlich, *Colloid & Capillary Chemistry*, translated by H.S.Hatfiled, E. P. Dutton, New York, 1922.
18. R. Schlögl, *Z. Physik, Chem. (Frankfurt)*, 3, 73 (1955).
19. E. J. Breton and C. E. Reid, *J. Appl. Polymer Sci.*, 1, 133 (1959).
20. S. Loeb and S. Sourirajan, "Sea Water Demineralization by means of a Semipermeable Membrane," Univ. Calif. (Los Angeles), *Dept. Eng. Rept.*, No. 60-60 (1961).
21. K. Kammermeyer and D. H. Hagerbaumer, *A.I.Ch.E.J.*, 1, 215 (1955).
22. H. K. Lonsdale, "Theory and Practice of Reverse Osmosis and Ultrafiltration," in *Industrial Processing with Membranes*, R. E. Lacey and S. Loeb, Eds., Wiley-Interscience, New York, 1972, Chap. 8.
23. S. Sourirajan, *Reverse Osmosis*, Academic, New York, 1970.
24. A. S. Michaels, H. J. Bixler, and R. M. Hodges, Jr., *J. Colloid Sci.*, 20, 1034 (1965).
25. K. A. Kraus, A. J. Shor, and J. S. Johnson, *Desalination*, 2, 243 (1967).
26. L. C. Flowers, D. E. Sestrich, and D. Berg, U.S. Office of Saline Water, *Res. Dev. Prog. Rep.*, No. 224 (1966).
27. K. A. Kraus, A. E. Marcinkowsky, J. S. Johnson, and A. J. Shor, *Science*, 151, 194 (1966).
28. A. Walch, *CZ-Chem. Tech.*, 2, 7 (1973).
29. U. Merten, H. K. Lonsdale, R. L. Riley, and K. D. Vos, U.S. Office of Saline Water, *Res. Dev. Prog. Rep.*, No. 208 (1966).
30. K. D. Vos, F. O. Burris, Jr., and R. L. Riley, *J. Appl. Polymer Sci.*, 10, 825 (1966).
31. S. Loeb, "Preparation and Performance of High-flux Cellulose Acetate Desalination Membrane," in *Desalination by Reverse Osmosis*, U. Merten, Ed., MIT Press, Cambridge, Mass., 1966.
32. E. Glueckauf, *Proc. First International Symp. on Water Desalination, Washington, D.C.*, U.S. Office of Saline Water, Washington, D.C., 1965, Vol. 1, p. 143.
33. R. Blunk, "A Study of Criteria for the Semipermeability of Cellulose Acetate Membranes to Aqueous Solutions," *UCLA Water Resources Center Rep.*, WRCC-88 (1964).
34. H. K. Lonsdale, C. E. Milstead, B. P. Cross, and F. M. Graber, "Study of Rejection of Various Solutes by Reverse Osmosis Membranes," U.S. Office of Saline Water, *Res. Dev. Prog. Rep.*, No. 447, (1969).
35. Abcor, Inc., Bulletin MS-71-10, 1971.
36. M. C. Porter and A. S. Michaels (a) *Chem. Tech.*, 1, 56 (1971); (b) *ibid.*, 1, 248 (1971); (c) *ibid.*, 1, 440 (1971); (d) *ibid.*, 1, 633 (1971); (e) *ibid.*, 2, 56 (1972). Literature references, Amicon Corp., 21 Hartwell Ave, Lexington, Mass. 02173.
37. R. W. Okey, "The Treatment of Industrial Waste by Pressure-Driven Membrane Processes," in *Industrial Processing with Membranes*, R. E. Lacey and S. Loeb, Eds., Wiley-Interscience, New York, 1972.
38. A. S. Michaels, *Chem. Eng. Progr.*, 64, 31 (1968).

39. R. N. Rickles, *Membrane Technology and Economics,* Noyes Development Corp., Park Ridge, N. J., 1967.

40. J. A. Lane and J. W. Riggle, *Chem. Eng. Prog. Symp. Series,* 24, 127 (1959).

41. L. Cerini, U.S. Patent 1,719,754 (1929).

42. F. G. Donnan, *Chem. Rev.,* 1, 73 (1924).

43. R. M. Wallace, *IEC Proc. Des. Dev.,* 6, 423 (1967).

44. S. S. Melsheimer, H. M. Kelley, L. F. Landon, and R. M. Wallace, Preprint paper No. 51b presented in 74th National AIChE Meeting, March 12, 1973, New Orleans, Louisiana.

45. F. Martin, private communication to R. Kesting, March 1963.

46. R. E. Kesting, A. Vincent, and J. Eberlin, *OSW Research and Development Report* 117, August (1964).

47. R. E. Kesting, OSW Patent Application SAL-830 (1965).

48. R. E. Kesting, W. J. Subcasky, and J. D. Paton, *J. Colloid Interfac. Sci.,* 28, 156 (1968).

49. N. N. Li, U.S. Patent 3,410,794 (1968).

50. N. N. Li, *A.I.Ch.E. J.,* 17, 459 (1971).

51. N. N. Li, *Ind. Eng. Chem. Proc. Des. Develop.,* 10, 215 (1971).

52. C. J. McAuliffe, *Phys. Chem.,* 70, 1267 (1966).

53. N. D. Shah and T. C. Owens, *Ind. Eng. Chem. Prod. Res. Develop.,* 11, 58 (1972).

54. N. N. Li and A. L. Shrier, in *Recent Developments in Separation Science,* Vol. 1, N. N. Li, Ed., Chemical Rubber Co., Cleveland, Ohio, 1972.

55. R. L. Cook, "Gas Separation by Permeation through Liquid Membranes", Ph.D. thesis, University of Iowa, Iowa City, Iowa, 1973.; R. L. Cook and R. W. Tock, *Sepn. Sci.,* 9 (3), 185 (1974).

56. W. D. Stein, *The Movement of Molecules across Cell Membranes,* Academic, New York, 1967.

57. R. Höber and S. L. Ørskov, *Arch. Ges. Physiol.,* 231, 599 (1933).

58. J. B. Wittenberg, *Biol. Bull.,* 1171, 402 (1959).

59. P. F. Scholander, *Science,* 131, 585 (1960).

60. W. L. Robb and D. L. Reinhard, U.S. Patent 3,335,545 (1967).

61. W. J. Ward III and W. L. Robb, *Science,* 156, 1481 (1967).

62. W. J. Ward III, *A.I.Ch.E. J.,* 16, 405 (1970).

63. W. J. Ward III and C. K. Neulander, Final Report, Contract No. Ph-86-68-76. HEW, U.S.P.H.S., NAPCA 510 Wooster Pike, Cincinnati, Ohio 45227.

64. F. Sauer, *Chem. Ing. Tech.,* 43, 880 (1971).

65. R. G. Carbonell and M. D. Kostin, *A.I.Ch.E. J.,* 18, 1 (1972).

66. D. A. Self, G. Kay, M. D. Lilly, and P. Dunnill, *A.I.Ch.E. J.,* 11, 337 (1969).

67. R. J. H. Wilson and M. D. Lilly, *Biotech. Bioeng.,* 11, 349 (1969).

68. A. K. Sharp, G. Kay, and M. D. Lilly, *Biotech. Bioeng.,* 11, 363 (1969).

69. S. P. O'Neill, P. Dunnill, and M. D. Lilly, *Biotech. Bioeng.,* 13, 337 (1971).

70. G. Kay and M. D. Lilly, *Biochim. Biophys. Acta,* 198, 276 (1970).

71. R. Axen and S. Ernback, *Europ. J. Biochem.,* 18, 351 (1971).

72. D. Gabel and B. V. Hofsten, *Europ. J. Biochem.*, **15**, 410 (1970).

73. K. Mosbach, *Acta Chem. Scand.*, **24**, 2084 (1970).

74. K. Mosbach and B. Mattiasson, *Acta Chem. Scand.*, **24**, 2084 (1970).

75. B. Mattiasson and K. Mosbach, *Biochim. Biophys. Acta*, **235**, 253 (1971).

76. J. C. Lee, *Biochim. Biophys. Acta*, **235**, 435 (1971).

77. M. L. Green and G. Crutchfield, *Biochem. J.*, **115**, 183 (1969).

78. J. K. Inman and H. M. Dintzis, *Biochem.*, **8**, 4074 (1969).

79. L. Goldstein, M. Pecht, S. Blumberg, D. Atlas, and Y. Levin, *Biochem.*, **9**, 2322 (1970).

80. L. Goldstein, in *Fermentation Advances*, D. Perlman, Ed., Academic Press, New York, 1969, p. 391.

81. C. K. Glassmeyer and J. D. Ogle, *Biochem.*, **10**, 786 (1970).

82. H. Filippusson and W. E. Hornby, *Biochem. J.*, **120**, 215 (1970).

83. A. B. Patel, S. N. Pennington, and H. D. Brown, *Biochim. Biophys. Acta*, **178**, 626 (1969).

84. N. Weliky, F. S. Brown, and E. C. Dale, *Archives Biochem. Biophys.*, **131**, 1 (1969).

85. H. H. Weetall and L. S. Hersh, *Biochim. Biophys. Acta*, **185**, 464 (1969).

86. H. H. Weetall, *Nature*, **223**, 595 (1969).

87. H. H. Weetall, *Science*, **166**, 615 (1969).

88. H. H. Weetall and G. Baum, *Biotech. Bioeng.*, **12**, 339 (1970).

89. W. F. Line, A. Kwong, and H. H. Weetall, *Biochim. Biophys. Acta*, **242**, 194 (1971).

90. R. Goldman, H. I. Silman, S. R. Caplan, O. Kedem, and E. Katchalski, *Science*, **150**, 758 (1965).

91. R. Goldman, O. Kedem, I. H. Silman, S. R. Caplan, and E. Katchalski, *Biochem.*, **7**, 486 (1968).

92. R. Haynes and K. A. Walsh, *Biochem. Biophys. Res. Commun.*, **36**, 235 (1969).

93. G. Broun, E. Sélégny, S. Arrameas, and D. Thomas, *Biochim. Biophys. Acta*, **185**, 260 (1969).

94. R. A. Messing, *J. Amer. Chem. Soc.*, **91**, 2370 (1969).

95. R. A. Messing, *Enzymol.*, **38**, 370 (1970).

96. R. A. Messing, *Enzymol.*, **39**, 12 (1970).

97. C. Schwabe, *Biochem.*, **8**, 795 (1969).

98. M. J. Bachler, G. W. Strandberg, and K. L. Smiley, *Biotech. Bioeng.*, **12**, 85 (1970).

99. T. Tosa, T. Mori, and I. Chibata, *Agr. Biol. Chem.*, **33**, 1053 (1969).

100. T. Tosa, T. Mori, and I. Chibata, *Enzymol.*, **40**, 49 (1971).

101. P. Bernfeld and R. E. Bieber, *Arch. Biochem. Biophys.*, **131**, 587 (1967).

102. P. Bernfeld, R. E. Bieber, and D. M. Watson, *Biochim. Biophys. Acta*, **191**, 570 (1969).

103. K. Mosbach and P. O. Larson, *Biotech. Bioeng.*, **12**, 19 (1970).

104. T. M. S. Chang, *Science*, **146**, 524 (1964).

105. P. R. Rony, *Biotech. Bioeng.*, **13**, 431 (1971).

106. É. Sélégny, S. Arrameas, G. Broun, and D. Thomas, *Compt. Rend.,* **266,** 1431 (1968).

107. É. Séléngy, G. Broun, J. Geffory, and D. Thomas, *J. Chim. Phys.,* **66,** 391 (1969).

108. R. Goldman, O. Kedem, and E. Katchalski, *Biochem.,* **7,** 4518 (1968).

109. R. Goldman, O. Kedem, and E. Katchalski, *Biochem.,* **10,** 165 (1971).

110. R. Goldman and E. Katchalski, *J. Theor. Biol.,* **32,** 243 (1971).

111. D. F. Ollis, J. B. Thompson, and E. T. Wolynic, *A.I.Ch.E. J.,* **18,** 457 (1972).

112. J. F. Jennings, and R. C. Binning, U.S. Patent 2,956,070 (1960).

113. R. C. Binning J. T. Kelly, U.S. Patent 2,913,507 (1959).

114. L. W. Gosser, W. H. Knoth and G. W. Parshall, *J. Amer. Chem. Soc.,* **95,** 3436 (1973).

115. A. S. Michaels, "Ultrafiltration," in *Progress in Separation and Purification,* Vol. 1, E. S. Perry, Ed., Interscience, New York, 1968.

116. C. J. van Oss, "Ultrafiltration Membranes," in *Progress in Separation and Purification,* Vol. 3, E. S. Perry and C. J. van Oss, Eds., Wiley-Interscience, New York, 1970.

117. J. S. Schultz, J. D. Goddard, and S. R. Suchdeo, *A.I.Ch. E. J.,* **20,** (3) 417 (1974).

Chapter IX

ELECTROMEMBRANE PROCESSES

Processes that proceed under the influence of an electric potential with the flow of a direct current are described by a variety of terms. One of the more common notations refers to them as "electrically driven membrane processes." Although this expression is quite descriptive, we prefer the term *electromembrane processes* as used in Lacy and Loeb's book [1].

The most important of these processes is electrodialysis as far as large-scale use is concerned. The other processes such as electro-osmosis and forced-flow electrophoresis have been used extensively in laboratory investigations, and have occasionally been proposed for larger-scale installation. Brief discussions of these processes will be presented.

1 GENERAL CONSIDERATIONS

Electromembrane processes belong to the category of liquid-phase separations. The fundamental aspects of membrane separation of ionic species in the absence of an external electric potential have been treated extensively in Chapter VIII. There, it was shown that by proper utilization of interionic interaction a driving force for ion separation can be attained that is appreciably higher than that due to the concentration gradient only. Also, pumping of ionic species from a solution of lower concentration to one of higher concentration is possible. However, such separations are always restricted by the Donnan equilibrium principle, and ionic isolation beyond that equilibrium concentration is not possible. Furthermore, the attainable ionic flux is usually too small for an economic process.

These limitations can be overcome by applying an external electric potential. With an appropriate stack of membranes and a suitably high electric potential, even a trace of an ionic species can be removed beyond the concentration of the Donnan equilibrium. The ionic flux in this case is mainly due to the electric transference. Therefore, an arbitrarily high flux within a certain limit can be achieved by maintaining a high electric current (or a high external electric potential) across the membrane. Consequently, an extremely thin membrane is not so essential in this case as in other membrane processes.

The principle of electrically driven membrane processes has been known since the beginning of this century. However, the recognition of the process, particularly of electrodialysis, as an effective industrial operation should be credited to the invention of a multicell electrodialyzer by Meyer and Strauss in 1940 [2]. With such a multicell dialyzer, electrical energy requirements could be reduced compared to single-cell operation. Also, troublesome gases and other corrosive substances produced at the electrodes could be restricted to the end sections of the multicell. The present advance of the process as an efficient commercial unit process is largely due to investigations supported by the U. S. Office of Saline Water in conjunction with the desalination program.

At the present time, the desalination of sea water on a large scale by electro-dialysis is not so economical as by the evaporation process. However, electrodialysis is more versatile than evaporation in small- or moderate-scale installations. Even in large-scale desalination, the process can be effectively employed in preconcentration of feed stock prior to evaporation. In desalination of brackish waters, electrodialysis is definitely more effective than evaporation, and even competes with the reverse-osmosis process. But the electromembrane process is uniquely suited to demineralization of substances such as food and beverage products that are sensitive to heat and high pressure. Current and possible uses of electromembrane processes are summarized in Table 9.1, which is based largely on information published by Lacey [1] and Rickles [3].

Table 9.1   Summary of Potential Applications of Electromembrane Processes

1. *Demineralization and Deionization*
   Desalination of sea and brackish water
   Removal of inorganics, for examples, nitrate, phosphate from sewage effluent
   Demineralization of whey and other protein solutions
   Deashing of chemicals, for examples, glycerol, polymers, and drugs
   Treatment of industrial wastes, for examples, plating waste, mining waste, and radioactive waste
   Pulp and paper applications

2. *Concentration of Electrolytes*
   Production of salt from sea water
   Recovery of high purity uranium tetrafluoride from uranium solutions

3. *Ion-Replacement Reactions*
   Anion replacement, for example, sweetening of citrus juice
   Cation replacement, for example, removal of radioactive strontium from milk
   Controlled pH adjustment of solutions without direct addition of acids or bases

4. *Metathesis Reaction* (double decomposition reaction)

5. *Separation of Electrolysis Products*

6. *Fractionation of Electrolytes*

2 DESCRIPTION OF ELECTROMEMBRANE PROCESSES

There are many operating parameters that will affect the design of a process based on the principle of the electric transference of a charged species in a solution under an external electric potential. The processes may differ in the array of the membrane stack, in utilizing either forced or natural convection, and also in employing either dialytic or osmotic flux of a solution through the membrane. The following description covers processes that have been relevant to practical applications.

Electrodialysis

The most common membrane stack is an alternate array of cation- and anion-exchange membranes between two electrodes as shown in Fig. 9.1. When a sufficiently high external electric potential is applied, the electric current

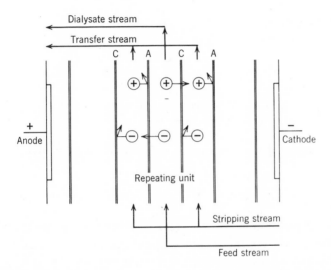

Fig. 9.1 Electrodialysis diagram: C = cation-exchange membrane; A = anion-exchange membrane.

carries cations of the feed stream into the transfer stream through the cation-exchange membrane facing the cathode. But anions are electrically transferred into the transfer stream on the opposite side through the anion-exchange membrane. On the other hand, cations in the transfer stream are rejected by the anion-exchange membrane on the cathode side, and anions by the cation-exchange membrane on the opposite side. Thus, the feed stream is stripped of electrolyte solutes into the two transfer streams around the feed compartment, and the transferred ions remain in the transfer compartment. This electrically driven membrane process is generally called electrodialysis in the narrow sence.

Electrosorption

This process is quite similar to the above electrodialysis, but is not continuous. The membrane stack of this process consists of a number of flattened membrane bags where one face of the bag structure exhibits cation-exchange properties and the other face possesses anion-exchange properties, as shown in Fig. 9.2. When an electric current passes through the entire membrane stack, the cations of the

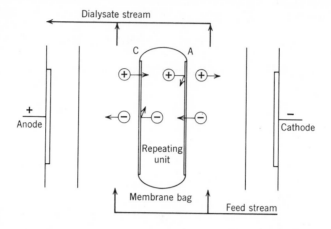

Fig. 9.2 Electrosorption schematic [1]: C = cation-exchange membrane; A = anion-exchange membrane.

solution outside the membrane bag are transferred into the bag through its face that exhibits cation-exchange properties, but the anions are carried in through the other face of the membrane bag. The solution outside the bags is thus depleted, whereas the solutions within the bags are concentrated. The concentrated solution can subsequently be removed from the bags by reversing the direction of the direct electric current. No gaskets or solution manifolds are required. This simplicity and a high available membrane area are its main advantages. The process has been extensively studied by Lacey and Lang [4, 5].

Transport Depletion

As of now, most synthetic polymeric anion-exchange membranes are not so chemically stable as cation-exchange membranes. When the feed streams contain high-molecular-weight anions, they usually adsorb on the anion-exchange membrane and thus foul the surface. At a sufficiently high electric current, the total ionic concentration near the membrane-dialysate stream interface becomes severely depleted. The concentration is often polarized to such an extent that the pH value near the interface changes by the so-called "water splitting" as explained later. In this case, the anion-exchange membrane, which is unstable at a high pH value, generally deteriorates, and the effective membrane life is shortened. To eliminate such membrane deterioration, Lacey [6] studied an array of cation-exchange and neutral membranes rather than the usual anion-cation-membrane combination. With this arrangement, the concentration polarization and the adsorption of large anions could be minimized. Thus, operating parameters such as electric current and stream-flow rates can be chosen with greater flexibility. However, the degree of demineralization for a given electric current is generally reduced compared to the process of ordinary electrodialysis. This process is often called transport depletion and is particularly useful in demineralization of whey. There, the troublesome denaturation of proteins due to a pH change can also be minimized, and large electrically charged proteins can be simultaneously separated by choosing appropriate neutral membranes. Figure 9.3 illustrates the case.

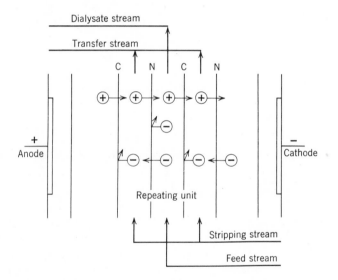

Fig. 9.3 Transport depletion array: C = cation-exchange membrane; N = neutral membrane.

Continuous Ion Exchange

The membrane stack in this process usually consists of all cation- or all anion-exchange membranes. For example, the sweetening of citrus juice is usually accomplished by continuous ion replacement of the citrate ion of the untreated juice with the hydroxyl ion of the stripping stream in a stack of all anion-exchange membranes, as shown in Fig. 9.4. An example in Chapter XV (p. 492) illustrates the process in detail.

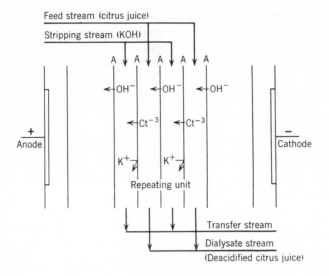

Fig. 9.4 Arrangement for continuous ion exchange: A = anion-exchange membrane; Ct^{-3} = citrate ion.

Metathesis Reaction

This reaction is a double decomposition reaction of two electrolytes as, for example,

$$NaBr + AgNO_3 \rightleftharpoons AgBr + NaNO_3 .$$

This type of reaction can be continuously achieved in an electrodialysis cell that comprises anion- and cation-exchange membranes in an alternate sequence, as shown in Fig. 9.5. The bromide ion of the NaBr compartment is electrically transferred into the gelatin emulsion stream through the anion-exchange membrane. The silver ion is carried into the emulsion stream through the cation-exchange membrane from the neighbor AgBr compartment. Thus, a photographic emulsion of AgBr, which is free of extraneous electrolytes, can be continuously produced. The electrolyte $NaNO_3$ is formed in the $NaNO_3$ compartment. In a similar way other metathesis reactions can be achieved, such as the production of NaOH and $CaCl_2$ from NaCl and $Ca(OH)_2$; KNO_3 and NaCl from KCl and $NaNO_3$, and so on.

Forced-Flow Electrophoresis

Most colloids are negatively charged under normal conditions in biological systems and in polluted waters. Such colloids can be removed or concentrated by this process, which has been investigated by Bier [7]. The electrophoresis

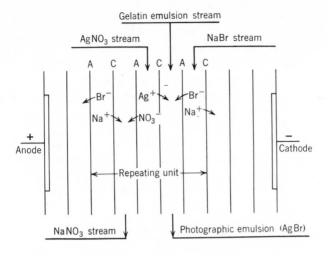

Fig. 9.5 Preparation of photographic emulsion by the metathesis reaction.

cell consists of an alternate array of membranes and filters as shown in Fig. 9.6. The membranes allow the passage of electric current and small ions, if any, but retain large colloids. On the other hand, most colloids and water freely pass through the filters. When a sufficiently high external electric potential is applied on the membrane stack, the negatively charged colloids are concentrated in the compartment on the anode side of the filter. By applying an external pressure on the feed stream, the solvent water is squeezed out through the filter into the opposite compartment. On the other hand, the small anions, which may be present in the feed compartment, are transferred through the membrane into

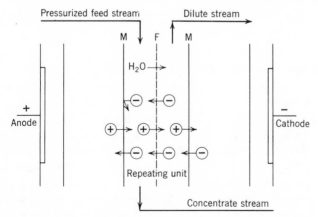

Fig. 9.6 Forced-flow electrophoresis scheme [7] : M = dense membrane; F = filter.

the compartment on the anode side of the membrane, but the cation is carried out through the filter into the compartment on the cathode side of the filter. Bier's proposal for large-scale use is covered in Chapter XV.

If the feed stream were to contain small ions to be simultaneously removed, an ion-exchange membrane could thus be effectively used rather than a neutral membrane. With an anion membrane, the small cations transferred through the filter can be prevented from further electric transference through the membrane on the cathode side.

Desalination of saline water can also be achieved by this process using an appropriate array of the membrane stack. A possible membrane stack would be an alternate array of cation-exchange and neutral membranes as shown in Fig. 9.7. The cation-exchange membrane is of such a high capacity and so

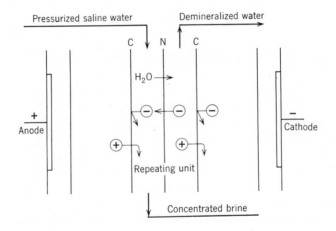

Fig. 9.7 Desalination by forced electrophoresis: C = cation-exchange membrane; N = neutral membrane.

dense that its anion rejection is almost complete, but its permeability to the solvent water is very small. However, the neutral membrane is highly permeable to the solvent water, yet its salt rejection is not necessarily great. With this array, the external electric potential forces the salt to remain in the feed compartment, although the external pressure squeezes out the solvent water into the neighbor compartment through the neutral membrane on the anode side. In effect, this process is a hybrid process, that is, a combination of electrically driven and pressure-driven membrane operations. The main advantages of such a desalination process would be that only relatively little electric energy and low external pressure are needed. In addition, the neutral membranes need not be so thin as the active skin of reverse-osmosis membranes.

BALC Process

BALC is the abbreviation of *barrier-anolyte-liquor-catholyte* combination. This process was recently proposed for the processing of pulping spent liquors to renovate cooking liquors and pulping chemicals for reuse, and also to recover lignin, carbohydrate, and other wood-derived chemicals [8]. The water reclamation is an additional but important objective. The BALC cell consists of a liquid barrier stream as well as anolyte, liquor, and catholyte streams. These four streams are separated by a cation-exchange, neutral, cation-exchange, and anion-exchange membrane in the respective order as shown in Fig. 9.8. The cation of the feed liquor is electrically transferred through the cation exchange

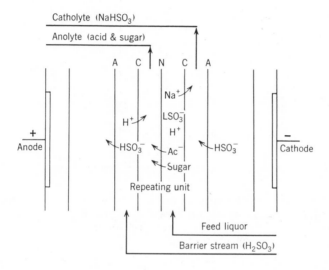

Fig. 9.8 BALC process scheme: A = anion-exchange membrane; C = cation-exchange membrane; N = neutral membrane.

membrane into the catholyte compartment, and the anion is carried through the neutral membrane into the anolyte compartment. The barrier stream supplies the required ions to both the catholyte and the anolyte streams through the cation- and the anion-exchange membranes, respectively. Aqueous sulfurous acid solution is usually used as the barrier stream. Thus, the feed liquor is treated to recover the base as cooking acid in the catholyte stream. By choosing suitable neutral membranes it is possible to further separate high-molecular-weight, base-free lignosulfonic acid from low-molecular-weight, inorganic and organic acids along with sugar complexes in the anolyte stream.

Electrogravitation

In the processes discussed so far, all stream-flow rates are relatively high, such that the effect of natural convection due to gravity is negligible. However, some high-molecular-weight ionic species can effectively be separated by the process of electrogravitation or electrodecantation. The membranes used in this process are also permselective. When the membrane stack consists of only cation-exchange membranes, for example, as shown in Fig. 9.9, the anion in one compartment, for instance, compartment I, is rejected by the membrane

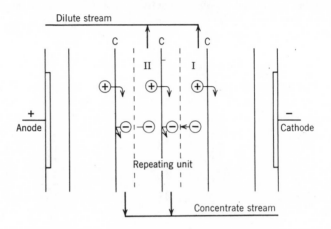

Fig. 9.9 Cell arrangement for electrogravitation: C = cation-exchange membrane. Note that dashed vertical lines do not signify any membrane structure. They represent an arbitrary phase boundary where the concentrated stream, moving downward, and the dilute stream, moving upward, come together.

on the anode side. But the cation of the neighbor compartment on the anode side, that is, compartment II, is electrically transferred into compartment I. Therefore, the total ionic concentration is increased near the membrane-solution interface on the anode side of compartment I. Since the density of the solution on the anode side is greater than that on the cathode side, the solution is slowly circulated in each compartment due to gravity. The solution at the upper side of the compartment is thus demineralized, but the lower part of the solution is concentrated. The feed solution is usually introduced very slowly at the middle of each compartment; in Fig. 9.9, this would be a stream perpendicular to the plane of the diagram. As the electric resistance in a solution is generally inversely proportional to its concentration of ionic solutes, the density of the depleted layer on the cathode side of each compartment is further reduced by inherent electric heating, and the natural convective separation is augmented by such heating. Compared to other electromembrane processes, electric gravitation presents advantages in its extreme simplicity, low pumping energy, and easy maintenance. However, this process is not competitive with other processes in desalination because of its inherent slowness in separation [9].

Electro-osmosis

An electrically driven osmotic, rather than dialytic, process can also be devised. As shown in Chapter VIII, even the neutral water in the pores of an electrically charged membrane behaves like an electrically charged species as a whole and is transferred by an electric current. When a feed stream contains large solutes, electrically charged or neutral, the stream can be concentrated by electrically transferring out the water into the transfer stream. The small ions in the feed stream, if any, can also be removed. In this osmotic process, the pore size, the relative content of the hydrophilic substance to the hydrophobic, and also the thermal stability of the electrically charged membrane are important. However, this process has not been competitive with other processes. This is due to difficulty in preparing an appropriate membrane that possesses a high water permeability and a high capacity or zeta potential as well. If the total ionic concentration in the feed stream is small, the electric resistance of the solutions is so great that a large amount of electric energy is required.

The foregoing presentations involve the use of neutral and charged membranes. It is worthwhile to indicate the schemes of membrane and membrane combinations that may be employed: Table 9.2 typifies actual practice.

Table 9.2 Membrane Usage in Electromembrane Processes

Arrangement	Process
Cation membrane only	Continuous cation exchange Electrogravitation
Anion membrane only	Continuous anion exchange
Cation and anion membrane	Electrodialysis Metathesis reaction Electrosorption
Cation and neutral membrane	Transport depletion
Cation plus anion plus neutral membrane	BALC systems
Neutral membrane plus filter	Forced-flow electrophoresis

3 ELECTROCHEMICAL KINETICS IN ELECTROMEMBRANE PROCESSES

In an electromembrane process, a sufficiently high external electric potential is commonly applied to a membrane stack so as to allow a considerable electric current across the stack. In this case, the transport of an ionic species from one compartment to another is largely due to its transference by the electric current. In the design and operation of such an electromembrane module, the following items of information are essential: relationships of *external electric potential* at a given concentration distribution across the membrane stack and

> the current density,
> the ionic flux,
> the osmotic flow.

These relationships need to be discussed for each of the existing phases, that is, two or more solution phases and the membrane as a phase by itself.

Subsequently, the discussion will deal with complete membrane stacks. The specific cases will be for cation plus anion combination (electrodialysis), and anion-anion operation (continuous anion exchange).

Qualitative Considerations

An electric potential across a membrane stack essentially consists of the concentration potential and the ohmic potential drop. The concentration potential is generated by the concentration distribution of ionic species across the stack. In an electromembrane process, the electric transference of an ionic species to be separated often proceeds in the direction from its dilute stream to its concentrate stream. In this case, the external electric potential is partly consumed to overcome the concentration potential. In addition, the external potential is partly dissipated into irreversible heat by the ohmic resistance across the stack. Thus, the overall external electric potential drop across the membrane stack consists of

1. concentration potentials in membrane phases;
2. concentration potentials at the membrane-solution interfaces;
3. concentration potentials in solution phases;
4. ohmic potential drop in solution phases;
5. ohmic potential drop in membrane phases;
6. electrode potential.

The concentration potential 2 at the phase boundary is simply the Donnan potential, which has already been discussed in Chapter VIII. The sum of the concentration potentials 1 and 2 is often called the membrane potential, and potential 3 is termed the junction potential. The electrode potential arises from the various electrode reactions and their subsequent concentration polarization, as well as the ohmic potential drop in the electrode compartment. This electrode potential can usually be minimized by a high flow rate of the electrode rinse stream, which contains an appropriate electrolyte solute. Furthermore, under suitable conditions [10] the electrode potential of one type (e.g., anode potential) tends to cancel out against the potential of the other type (e.g., cathode potential). Therefore, the electrode potential will not be treated in this section. The reader is referred to the book by Vetter [11] for a detailed discussion.

As in other membrane processes involving liquid phases, the concentration polarization is also very important. When a counterion transports through an ion-exchange membrane from a dilute stream to a coucentrate stream, the counterion is depleted at the membrane-solution interface in the dilute stream, but the co-ion is momentarily accumulated. In order to meet the electro-neutrality condition, the accumulated co-ion migrates away from the interface into the bulk solution of the dilute stream. Thus, the total ionic concentration in the dilute compartment is depleted at both interfaces, as shown in Fig. 9.10. On the other hand, the total ionic concentration in the concentrate compartment

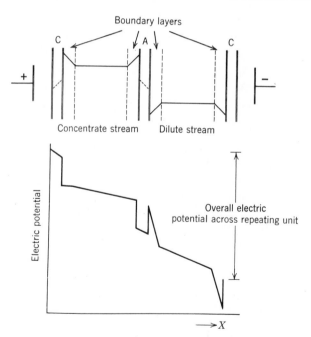

Fig. 9.10 Schematic illustration of the distribution of the electric potential in a repeating unit of electrodialysis.

is accumulated at both interfaces for the analogous reason. As such concentration polarization occurs, the concentration potential across the ion-exchange membrane is further enhanced, and the diffusional leakage of ionic species through the membrane increases in the opposite direction to that of the electric transference. Therefore, the net ionic transport is severely retarded for a given electric potential. In the presence of such concentration polarization, the ohmic potential drop in the dilute stream also increases to a considerable magnitude. This occurs because the ohmic electric resistance of an electrolyte solution is, in general, inversely proportional to its ionic concentration. As a higher external electric potential is applied to obtain a higher electric current, the ionic concentration of the dilute stream is more polarized, often to such an extent that the total ionic concentration at the interface is virtually nil. In this case, the solvent water is electrolyzed, and the electrolyzed hydrogen or hydroxide ions compete with the original counterion in their electric transference across the ion-exchange membrane. The phenomenon of such electrolysis is often called "water splitting." Water splitting thus retards the transport of the ionic species to be separated for a given electric current. The splitting also induces a change in the pH value at the interface. The anion-exchange membrane, which is inherently unstable at a high pH value, consequently deteriorates. Therefore, it is a general rule to take special care to reduce concentration polarization around the anion-exchange membrane. The asymptotic electric current at which the total ionic concentration at the interface would reach zero is called the limiting electric current. This current is an important parameter in determining the degree of concentration polarization. The distribution of the overall electric potential across a cell pair is schematically illustrated in Fig. 9.10.

Quantitative Treatment

An electromembrane module is comprised of the membrane and the solution phases. The ionic concentration in each phase is so small in its most practical applications that the activity coefficient and diffusivity are almost constant and uniform throughout the phase. In this case, the following Nernst-Planck equation is valid for transport of ionic species i in the absence of the convective transport (see Chapter VIII):

$$N_i = -D_i \left(\frac{dC_i}{dx} + z_i C_i \frac{\mathcal{F}}{RT} \frac{d\phi}{dx} \right). \tag{9.1}$$

Equation 9.1 is subject to the electroneutrality condition:

$$\sum_i z_i C_i = \omega \overline{C}_R$$

and the charge conservation:

$$\frac{I}{\mathcal{F}} = \sum_i z_i N_i \tag{9.3}$$

Here, the summation is to be carried out for all ionic species present in the phase. The sign indicator ω is defined as

$$\omega = \begin{array}{ll} +1 & \text{for the cation-exchange membrane phase} \\ 0 & \text{for the neutral membrane and solution phase} \\ -1 & \text{for the anion-exchange membrane phase.} \end{array}$$

Combining the flux equation, Eq. 9.1, and the charge conservation, Eq. 9.3, the gradient of the electric potential can be explicitly expressed as

$$\frac{d\phi}{dx} = - \frac{RT}{\mathcal{F}} \frac{\sum_i D_i z_i \dfrac{dC_i}{dx}}{\sum_i D_i z_i^2 C_i} - \frac{RT}{\mathcal{F}^2} \frac{I}{\sum_i D_i z_i^2 C_i} \quad . \tag{9.4}$$

The first term on the right-hand side of Eq. 9.4 represents the concentration potential, and the ohmic potential drop corresponds to the second term. It is customary in electrochemistry to define the transference number t_j as

$$t_j = \frac{D_j z_j C_j}{\sum_i D_i z_i C_i} \tag{9.5}$$

and the specific electric conductance κ as

$$\kappa = \frac{\mathcal{F}^2}{RT} \sum_i D_i z_i^2 C_i \quad . \tag{9.6}$$

Substituting these quantities into Eq. 9.4 gives

$$\frac{d\phi}{dx} = - \frac{RT}{\mathcal{F}} \sum_i \frac{t_i}{C_i} \frac{dC_i}{dx} - \frac{I}{\kappa} \quad . \tag{9.7}$$

Eliminating the gradient of the electric potential in Eq. 9.1 by substituting Eq. 9.7, the ionic flux N_j can also be expressed as

$$N_j = - D_j \left(\frac{dC_j}{dx} - z_j C_j \sum_i \frac{t_i}{C_i} \frac{dC_i}{dx} \right) + t_j \frac{I}{\mathcal{F}} \quad . \tag{9.8}$$

Thus, the ionic flux N_j consists of the self-diffusion, the migration by the concentration potential, and the electric transference by the electric current.

The physical meaning of the transference number t_j of an ionic species j is self-evident from Eq. 9.8. It is the number of moles of ionic species j transferred by one faraday of electricity in the absence of the concentration gradient and electro-osmosis, if the phase is an ion-exchange membrane. It should not be confused with the so-called transport number, which is defined as the number of equivalents of the ionic species transferred by one faraday of electricity. Thus, the transport number is just the product of the transference number and the electrochemical valence of the ionic species. The following relationship for transference numbers is valid by definition:

$$\sum_i z_i t_i = 1 .$$
(9.9)

In the presence of a considerable electro-osmotic flux in a membrane phase, the above-mentioned physical meaning of the transference number is valid only in the frame of the moving solution. With respect to the frame of the stationary membrane phase, the apparent transference t_j' can be defined as

$$t_j' = \frac{D_j C_j z_j}{\sum_i D_i z_i^2 C_i - \overline{D}_v \dfrac{RT}{\epsilon} \overline{C}_R^2} .$$
(9.10)

Here the validity of the Schmid equation, Eq. 8.72, for the electro-osmotic flux is assumed. If a considerable pressure gradient exists across the electro-osmotic membrane, the streaming potential, which was discussed in Chapter III, constitutes the overall electric potential in addition to the concentration potential and the ohmic potential drop. In most practical electromembrane processes, the ion-exchange membrane is so dense and its capacity is not so large that the following condition is usually satisfied:

$$\sum_i D_i z_i^2 C_i \gg \overline{D}_v \frac{RT \overline{C}_R^2}{\epsilon} .$$

In this case, electro-osmosis does not significantly influence the individual ionic fluxes but only the solvent flux. Also, the external pressure gradient is not so large that the contribution of the streaming potential is negligible and the ionic flux is not affected by the pressure gradient.

The distribution of the electric potential over a membrane stack can be obtained by integrating Eq. 9.4 for a given electric current and the concentration distribution. Integration of Eq. 9.8 for a given ionic flux and electric current density leads, in turn, to the concentration distribution. The exact integration of these differential equations for an arbitrary number of ionic species is rather difficult. Only approximate integration in a closed form is available.

The subsequent discussions cover the cases of demineralization and continuous ion exchange. Demineralization is discussed only for the presence of a single electrolyte. Continuous ion exchange is treated for the case where two ions of the same type and a common ion of the other type are present. The more general cases where multiple electrolytes are present are discussed only briefly.

Transport in Membrane Phase

Electromembrane processes call for the use of ion-exchange membranes and neutral membranes, either by themselves or in combination. The overall transport phenomena involve transport in solutions and within the membrane. This section deals with a detailed discussion of membrane transport.

Transport in Neutral Membranes

Here, the discussion covers demineralization with neutral membranes, which represents the transport depletion at neutral membranes in a membrane stack. This involves both current-density and ionic-flux phenomena in the membrane phase only. If the contiguous solutions in contact with a neutral membrane contain only one *single electrolyte solute,* the electroneutrality condition becomes

$$z_+ \overline{C}_+ = |z_-|\ \overline{C}_- \ . \tag{9.11}$$

Here, the bar refers to the membrane phase and subscripts + and − refer to cation and anion, respectively. In this case, the transference number and the electric conductance can be expressed simply as

$$\overline{t}_\pm = \frac{\pm\ \overline{D}_\pm}{\overline{D}_+ z_+ + \overline{D}_-\ |z_-|} \tag{9.12}$$

and

$$\overline{\kappa} = \frac{\mathcal{F}^2}{RT}\ z_+ \overline{C}_+\ (\overline{D}_+ z_+ + \overline{D}_-\ |z_-|) \ . \tag{9.13}$$

Thus, the transference number is independent of ionic concentrations. The flux equation for the cation, Eq. 9.8, can then be integrated to

$$N_+ = \overline{D}_e \left(\frac{\overline{C}_+^{\mathrm{I}} - \overline{C}_+^{\mathrm{II}}}{l} \right) + \overline{t}_+ \frac{I}{\mathcal{F}} \ . \tag{9.14}$$

The effective diffusivity $\overline{D}_e$ is defined as

$$\overline{D}_e = \frac{\overline{D}_+ \overline{D}_- (z_+ + |z_-|)}{\overline{D}_+ z_+ + \overline{D}_-\ |z_-|} \ . \tag{9.15}$$

and superscripts I and II refer to the two contiguous solutions. The distribution of the electric potential can be obtained by integrating Eq. 9.7 and using Eq. 9.14 as

$$\overline{\phi}^{II} - \overline{\phi}^{I} = \frac{RT}{\mathcal{F}} (\overline{t}_+ - |\overline{t}_-|) \ln \frac{\overline{C}^{I}}{\overline{C}^{II}} - \frac{l \, I}{\overline{\kappa}_{lm}} . \qquad (9.16)$$

The log-mean average of the electric conductance $\overline{\kappa}_{lm}$ is defined by

$$\overline{\kappa}_{lm} = \frac{\overline{\kappa}(\overline{C}^{I}_+) - \overline{\kappa}(\overline{C}^{II}_+)}{\ln \dfrac{\overline{\kappa}(\overline{C}^{I}_-)}{\overline{\kappa}(\overline{C}^{II}_-)}} . \qquad (9.17)$$

Thus, the electric potential across the membrane consists of the concentration potential and the ohmic potential drop, represented by the first and second terms on the right-hand side of Eq. 9.16, respectively. If the diffusivities of both ions are the same in particular, that is, $\overline{t}_+ = |\overline{t}_-|$, the concentration potential becomes zero, and only the ohmic potential drop constitutes the electric potential.

In most practical applications, the distribution coefficient K of the neutral membrane, which is defined as

$$K = \frac{\overline{\gamma}_\pm}{\gamma_\pm}$$
$$= \frac{\overline{C}_+}{C_+} = \frac{\overline{C}_-}{C_-} , \qquad (9.18)$$

is almost constant and uniform throughout the membrane phase. In this case, the phase-boundary potentials at both interfaces between the membrane and the solution phases tend to cancel each other in the absence of the specific adsorption of one of the ionic species. The membrane potential E_m can then be expressed in terms of solution concentrations as

$$E_m = \phi^{II} - \phi^{I}$$
$$= (\phi^{II} - \overline{\phi}^{II}) - (\phi^{I} - \overline{\phi}^{I}) + (\overline{\phi}^{II} - \overline{\phi}^{I})$$
$$= \overline{\phi}^{II} - \overline{\phi}^{I}$$
$$= \frac{RT}{\mathcal{F}} (\overline{t}_+ - |\overline{t}_-|) \ln \frac{C^{I}_+}{C^{II}_+} - R_m I . \qquad (9.19)$$

The membrane area resistance R_m is defined by

$$R_m = \frac{l}{\bar{\kappa}_{lm}} = \frac{l}{K} \frac{\ln\left[\bar{\kappa}(C_+^{I})/\bar{\kappa}(C_+^{II})\right]}{\bar{\kappa}(C_+^{I}) - \bar{\kappa}(C_+^{II})} . \tag{9.20}$$

The quantities without the bar refer to the solution phase. Also, the integrated flux equation, Eq. 9.14, becomes

$$N_+ = \bar{D}_e K \left(\frac{C_+^{I} - C_+^{II}}{l}\right) + \bar{t}_+ \left(\frac{I}{\mathcal{F}}\right). \tag{9.21}$$

As indicated by Eq. 9.21, the troublesome diffusional leakage of the ion, which is represented by the first term of the right-hand side, can be reduced by employing a thick membrane. Then, however, the membrane resistance increases correspondingly. The opposing two effects must, therefore, be suitably balanced. It is important to note that a thin membrane is not so essential in an electromembrane process as in other membrane processes. In electrochemistry it is customary to define the current efficiency of one ion η as the number of equivalents of the ion transported per one faraday of electricity, that is, for the cation in the neutral membrane:

$$\eta_+ = \frac{z_+ N_+ \mathcal{F}}{I}$$

$$= \bar{t}_+ z_+ - \left(\frac{\bar{D}_e K z_+ \mathcal{F}}{I}\right) \left(\frac{C_+^{II} - C_+^{I}}{l}\right) . \tag{9.22}$$

In the case where *multiple* ionic species are present in the membrane phase, the transference numbers are not constant, and consequently the integrations of Eqs. 9.7 and 9.8 are quite complicated. The sophisticated mathematical manipulation, which was used by Schlögl [12] in the absence of the electric transference, may be directly used in this case. However, the following two approximations are worthy of consideration. In the first instance, it may be assumed that an ionic species present at every point in the membrane phase could be formed by mixing ionic species present at both phase boundaries. This assumption was originally made by Henderson [13] in the absence of electric transference. Then, the integration of Eq. 9.7 yields

$$\bar{\phi}^{II} - \bar{\phi}^{I} = \frac{\sum\limits_{i} \bar{D}_i z_i (\bar{C}_i^{II} - \bar{C}_i^{I})}{\sum\limits_{i} \bar{D}_i z_i^2 (\bar{C}_i^{II} - \bar{C}_i^{I})} \frac{RT}{\mathcal{F}} \ln \frac{\sum\limits_{i} \bar{D}_i z_i^2 \bar{C}_i^{I}}{\sum\limits_{i} \bar{D}_i z_i^2 \bar{C}_i^{II}} - R_m I , \tag{9.23}$$

where the membrane area resistance R_m is given by

$$R_m = \int_0^l \frac{dx}{\overline{\kappa}} = \int_0^l \frac{dx}{\sum_i \overline{D_i z_i^2 \, C_i}} \,. \qquad (9.24)$$

The second approximation is valid if the electric transference dominates over the diffusional leakage, and also if the concentration difference across the membrane is not large. Then the transference numbers may be approximated to be constant and uniform throughout the membrane phase. In this case, the overall electric potential across the membrane can be obtained by integrating Eq. 9.7 into

$$\phi^{II} - \phi^{I} = - \frac{RT}{\mathcal{F}} \sum_i \overline{t_i} \ln \frac{C_i^{II}}{C_i^{I}} - RI. \qquad (9.25)$$

It should be noted that the area resistance in this case generally depends on the electric current density for a given concentration distribution. Ionic fluxes may also be expressed approximately, similarly to Eq. 9.21, by defining an appropriate effective diffusivity.

Transport in Ion-Exchange Membranes

The following treatment concerns all electromembrane processes where ion-exchange membranes are used, either by themselves or in combination with neutral membranes. Again the discussion deals only with the membrane phase, and is limited to *one electrolyte solute.*

Consider an electrolyte solution in which an ion-exchange membrane of capacity $\omega \overline{C}_R$ is immersed. The solution contains one 1,1-valent electrolyte solute. If the system is in thermodynamic equilibrium, the concentration potential between the solution and the membrane phases is the Donnan potential as expressed in Eq. 8.31. Also, the concentration distribution of ions between the two phases is given by Eq. 8.34. Substituting Eq. 8.34 into Eq. 8.31, the Donnan potential E_{Don} can be expressed in terms of solution concentrations to be

$$E_{Don} = \overline{\phi} - \phi$$

$$= \frac{\omega RT}{\mathcal{F}} \ln \left(\frac{2C_A K}{\overline{C}_R + (C_R^2 + 4K^2 C_A^2)^{1/2}} \right), \qquad (9.26)$$

$$\overline{C}_A = \frac{\overline{C}_R + (C_R^2 + 4K^2 C_A^2)^{1/2}}{2}; \quad \overline{C}_Y = \overline{C}_A - \overline{C}_R \qquad (9.27)$$

$$K = \frac{\gamma_\pm}{\overline{\gamma}_\pm} \,. \qquad (9.28)$$

Subscripts A and Y refer to the counterion and co-ion, respectively. Here, the effect of the swelling pressure is assumed to be negligible. If the effect is significant, it can be included in the mean-activity coefficient $\overline{\gamma}_{\pm}$. Then, Eq. 9.26 is still approximately valid, particularly in the case where the ionic sizes of the counterion and co-ion are similar. For a z_A, z_Y- or z_Y, z_A-valent electrolyte solute, the following generalization is possible:

$$E_{Don} = \frac{RT}{z_A \mathcal{F}} \ln \left(\frac{KC_A}{\overline{C}_A} \right),$$ (9.29)

$$\left(\frac{\overline{C}_A}{C_A} \right)^{z_Y} \left(\frac{\overline{C}_A}{C_A} - \frac{\overline{C}_R}{|z_A|C_A} \right)^{z_A} = K^{|z_A|+|z_Y|};$$

$$|z_Y|\overline{C}_Y = |z_A|\overline{C}_A - \overline{C}_R .$$ (9.30)

Here K is the ratio of the mean-activity coefficients of the two phases defined by Eq. 8.35. If, in particular, $\overline{C}_R \gg KC_A$, Eq. 9.30 reduces to

$$|z_A|\overline{C}_A = \overline{C}_R \left[1 + \left(\frac{K |z_A|C_A}{\overline{C}_R} \right)^{1+|z_Y/z_A|} \right].$$ (9.31)

Suppose that the ion-exchange membrane separates two solutions that contain the same electrolyte, but possess different concentrations. Also, a sufficiently high external electric potential is applied on the system such that an electric current flows across the membrane. Then, the concentration profile of the counterion and the distribution of the electric potential can be obtained for a given ionic flux and electric current density by integrating Eqs. 9.8 and 9.7, respectively. It should be noted that the transference number in the ion-exchange membrane is not independent of its ionic concentrations as in neutral membranes. In this case, Eqs. 9.7 and 9.8 can be rearranged into more convenient forms:

$$\frac{\mathcal{F}}{RT} \frac{d\overline{\phi}}{dx} = - \frac{(\overline{D}_A - \overline{D}_Y)z_A}{\overline{D}_A z_A^2 \overline{C}_A + \overline{D}_Y z_Y^2 \overline{C}_Y} \left(\frac{d\overline{C}_A}{dx} \right) - \frac{1}{\overline{D}_A z_A^2 \overline{C}_A + \overline{D}_Y z_Y^2 \overline{C}_Y} \left(\frac{I}{\mathcal{F}} \right)$$ (9.32)

and

$$N_A = - \frac{\overline{D}_A \overline{D}_Y (z_A^2 \overline{C}_A - z_Y^2 \overline{C}_Y)}{\overline{D}_A z_A^2 \overline{C}_A + \overline{D}_Y z_Y^2 \overline{C}_Y} \left(\frac{d\overline{C}_A}{dx} \right) + \frac{\overline{D}_A z_A \overline{C}_A}{\overline{D}_A z_A^2 \overline{C}_A + \overline{D}_Y z_Y^2 \overline{C}_Y} \left(\frac{I}{\mathcal{F}} \right).$$ (9.33)

Using the electroneutrality condition, Eq. 9.2, integrations of Eqs. 9.31 and 9.32 yield

$$\phi^{II} - \phi^{I} = \frac{RT}{z_A \mathcal{F}} \left(\frac{1-(1-a)\eta_A}{1-(1-a\beta)\eta_A}\right) \ln \left(\frac{|z_A|\overline{C}_A^{I} \; [1-\eta_A(1-a\beta)] \; - \; \eta_A a\beta\overline{C}_R}{|z_A|\overline{C}_A^{II} \; [1-\eta_A(1-a\beta)] \; - \; \eta_A a\beta\overline{C}_R}\right) \quad (9.34)$$

and

$$\frac{I}{\mathcal{F}} = \frac{\overline{D}_A z_A a(1-\beta)}{l \; [1-\eta_A(1-a\beta)]} \left[(\overline{C}_A^{II} - \overline{C}_A^{I}) + \left(\frac{\overline{C}_R \beta \; [1-(1-a)\eta_A]}{|z_A|(1-\beta) \; [1-(1-a)\eta_A]}\right) \cdot \right.$$
$$\left. \ln \left(\frac{|z_A|\overline{C}_A^{I} \; [1-\eta_A(1-a\beta)] \; - \; \eta_A a\beta\overline{C}_R}{|z_A|\overline{C}_A^{II} \; [1-\eta_A(1-a\beta)] \; - \; \eta_A a\beta\overline{C}_R}\right)\right], \quad (9.35)$$

where

$$a = \frac{\overline{D}_Y}{\overline{D}_A}, \qquad \beta = \frac{z_Y}{z_A}, \quad (9.36)$$

and the current efficiency of the counterion A is defined as

$$\eta_A = \frac{|z_A|N_A \mathcal{F}}{I}. \quad (9.37)$$

Then Eqs. 9.34 and 9.35 can be expressed in terms of solution concentrations by combining Eq. 9.30 with these equations. The overall electric potential across the membrane is the potential given by Eq. 9.34 plus the Donnan potentials at both phase boundaries, Eq. 9.29. Thus the current efficiency of the counterion η_A for given concentration distribution and the external electric potential can be computed from Eqs. 9.34 and 9.29, and the electric current density is obtained from Eq. 9.35.

In order to characterize an ion-exchange membrane, the concentration potential has generally been measured in the absence of the external electric potential. A 1,1-valent electrolyte solute is usually used. In this case, that is, $\eta_A \to \infty$, the membrane potential can be expressed from Eqs. 9.34, 9.26 and 9.27 as

$$E_m = \frac{\omega RT}{\mathcal{F}} \left[\ln \left(\frac{\overline{C}_R + [\overline{C}_R^2 + 4 \; K^2(C_A^{II})^2]^{1/2} \; C_A^{I}}{\overline{C}_R + [\overline{C}_R^2 + 4 \; K^2(C_A^{I})^2]^{1/2} \; C_A^{II}}\right) + \right.$$
$$\left. \left(\frac{1-a}{1-a}\right) \ln \left(\frac{[\overline{C}_R^2 + 4 \; K^2(C_A^{I})^2]^{1/2} \; - \; \frac{(1-a)}{(1+a)} \; \overline{C}_R}{[\overline{C}_R^2 + 4 \; K^2(C_A^{II})^2]^{1/2} \; - \; \frac{(1-a)}{(1+a)} \; \overline{C}_R}\right)\right] \quad (9.38)$$

This equation was originally derived by Teorell [14] and also by Meyer and his co-workers [15]. If the capacity of the membrane is very high compared to ionic concentration of the solutions, that is, $\overline{C}_R \gg KC_A$, the membrane potential is reduced to

$$E_m = \frac{\omega R T}{\mathcal{F}} \left\{ \ln\left(\frac{C_A^{I}}{C_A^{II}}\right) + \frac{K^2}{a} \left[\left(\frac{C_A^{II}}{\overline{C}_R}\right)^2 - \left(\frac{C_A^{I}}{\overline{C}_R}\right)^2 \right] \right\}. \quad (9.39)$$

The first term in the bracket of this equation represents the classical Nernst equation. Figure 9.11 shows the membrane potential for a given concentration distribution, which was published by Meyer [15]. The figure can be used to determine the capacity of the membrane and the mobility ratio a from a series of measurements with different concentration distributions. However, the

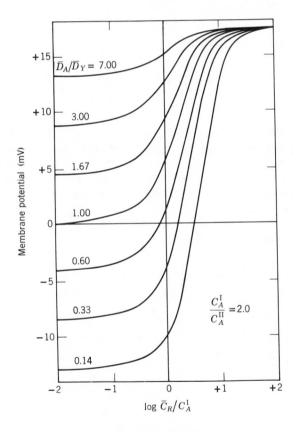

Fig. 9.11   Membrane potential of an ion-exchange membrane [15].

procedure should not be extended to high concentrations of the solution phase, since activity coefficients and diffusivities of ionic species in the membrane phase strongly depend on ionic concentrations. This case was treated by Helfferich [16].

In the cases discussed it was assumed that the diffusional leakage of the counterion across the ion-exchange membrane is comparable to its electric transference. In most practical operations a sufficiently high electric current density is, however, commonly used, and also the difference of ionic concentrations across the membrane is often small. Consequently, the ionic transport through the membrane is primarily by its electric transference. In this case, the transference number $\overline{t}_A$ and the effective diffusivity $\overline{D}_e$, which are defined as

$$\overline{t}_A = \frac{\overline{D}_A z_A \overline{C}_A}{\overline{D}_A z_A^2 \overline{C}_A + \overline{D}_Y z_Y^2 \overline{C}_Y} \qquad (9.40)$$

and

$$\overline{D}_e = \frac{\overline{D}_A \overline{D}_Y (z_A^2 \overline{C}_A + z_Y^2 \overline{C}_Y)}{\overline{D}_A z_A^2 \overline{C}_A + \overline{D}_Y z_Y^2 \overline{C}_Y}, \qquad (9.41)$$

can be approximated as being constant and uniform throughout the membrane. This approximation is particularly valid for a high-capacity membrane and low ionic concentration of the solution phases. Then, the Donnan exclusion with respect to the co-ion Y is very effective and thus the transference number approaches asympototically to $1/z_A$, and the effective diffusivity to $\overline{D}_Y$. The integrations of Eqs. 9.7 and 9.8 with constant $\overline{t}_A$ and $\overline{D}_e$ are rather simple and very similar to the case of the neutral membranes. The electric potential drop within the membrane becomes, from Eq. 9.7,

$$\overline{\phi}^{II} - \overline{\phi}^{I} = -\frac{RT}{\mathscr{F}} \left[\overline{t}_A \ln \left(\frac{\overline{C}_A^{II}}{\overline{C}_A^{I}} \right) + \overline{t}_Y \ln \left(\frac{\overline{C}_Y^{II}}{\overline{C}_Y^{I}} \right) \right] - \frac{l}{\kappa_{lm}} I. \quad (9.42)$$

Then the overall electric potential drop across the membrane can be obtained from Eqs. 9.29, 9.30, and 9.42 as

$$\phi^{II} - \phi^{I} = (\phi^{II} - \overline{\phi}^{II}) - (\phi^{I} - \overline{\phi}^{I}) + (\overline{\phi}^{II} - \overline{\phi}^{I})$$

$$= -\frac{RT}{\mathscr{F}} (\overline{t}_A + \overline{t}_Y) \ln \frac{C_A^{II}}{C_A^{I}} - R_m I. \qquad (9.43)$$

The membrane area resistance R_m is defined as

$$R_m = \frac{l}{\kappa_{lm}} - \frac{l \, \ln[\bar{\kappa}(\overline{C}_A^I)/\bar{\kappa}(\overline{C}_A^{II})]}{\bar{\kappa}(\overline{C}_A^I) - \bar{\kappa}(\overline{C}_A^{II})} . \tag{9.44}$$

Thus, the overall potential can be separated into the membrane potential and the ohmic potential drop, as in the case of the neutral membrane. The area resistance R_m in this case is, however, not an explicit function of the current efficiency as in Eq. 9.34, which indicates that R_m depends on the electric current density I for a given concentration distribution.

Also, the integrated flux equation from Eq. 9.8 becomes

$$N_A = -\overline{D}_e \left(\frac{\overline{C}_A^{II} - \overline{C}_A^I}{l} \right) + \bar{t}_A \left(\frac{I}{\mathcal{F}} . \right) \tag{9.45}$$

This equation can be expressed in terms of solution concentrations by using the boundary condition, Eq. 9.30. For a 1,1-valent electrolyte solute, for example, Eq. 9.27 yields the flux N_A:

$$N_A = -\frac{\overline{D}_e}{2l} \, [\overline{C}_R^2 + 4K^2(C_A^{II})^2]^{\frac{1}{2}} - [\overline{C}_R^2 + 4K^2(C_A^I)^2]^{\frac{1}{2}} + \bar{t}_A \frac{I}{\mathcal{F}}. \tag{9.46}$$

If, in particular, $\overline{C}_R \gg 2KC_A$, the equation simplifies to

$$N_A = -\left(\frac{\overline{D}_e K^2}{l\overline{C}_R} \right) [(C_A^{II})^2 - (C_A^I)^2] + \bar{t}_A \left(\frac{I}{\mathcal{F}} \right). \tag{9.47}$$

Because of their relative simplicity, Eqs. 9.42 to 9.47 are most commonly used in place of Eqs. 9.34 and 9.35 even though the latter equations more realistically represent actual conditions. However, Eq. 9.38, which is derived from Eqs. 9.34 and 9.35, should be used for characterization of the membrane.

Ion-Exchange Membranes with More than One Electrolyte Solute

Suppose that the electromembrane process involves two counterions, 1 and 2, and a common co-ion Y, as in a continuous ion-exchange process. The total ionic concentrations in the solution phases are relatively small compared to the capacity of the ion-exchange membrane. Then, the Donnan exclusion with respect to the co-ion is so effective that the concentration of the co-ion in the membrane phase is not appreciable compared to those of the counterions. In this case, the transport mechanism is very similar to the previous case, Since thermodynamic equilibrium is usually maintained at the solution-membrane interface, the Donnan potential and the concentration distribution can be expressed, similarly to Eqs. 9.29 and 9.30, as

$$E_{\text{Don}} = \frac{RT}{z_1 \mathcal{F}} \ln \frac{\gamma_1 C_1}{\overline{\gamma}_1 \overline{C}_1} , \tag{9.48}$$

$$\frac{\overline{C_1}}{C_1} = \left(\frac{\gamma_1}{\overline{\gamma_1}}\right)\left(\frac{\overline{\gamma_2}}{\gamma_2}\right)^{\frac{z_1}{z_2}} \left(\frac{\overline{C_R} - |z_1|\ \overline{C_1}}{|z_Y|\ C_Y - |z_1|\ C_1}\right)^{\frac{z_1}{z_2}} , \qquad (4.49)$$

$$\overline{C_2} = \frac{1}{|z_2|} \left(\overline{C_R} - |z_1|\ \overline{C_1}\right) . \qquad (9.50)$$

If, for example, the electrochemical valences of the two counterions are the same, that is, $|z_1| = |z_2| = z$, the concentration distribution of the counterion 1 between the two phases yields

$$\overline{C_1} = \left(\frac{K\overline{C_R}}{z}\right)\left(\frac{C_1}{C_2 + KC_1}\right) ,$$

$$K = \left(\frac{\gamma_1}{\overline{\gamma_1}}\right)\left(\frac{\overline{\gamma_2}}{\gamma_2}\right) . \qquad (9.51)$$

In the case where the diffusional leakage is comparable to or larger than the electric transference, the electric potential drop within the membrane and the electric current dinsity for a fixed concentration distribution are given by Eqs. 9.34 and 9.35, respectively. All that needs to be done is to substitute quantities subscribed by 1 and 2 for those subscribed by A and Y. If the electrochemical valences of the two counterions are the same, that is, $\beta = 1$, the electric current density is expressed only by the logarithmic term in the bracket.

When the electric transference is much larger than the diffusional leakage, and also the concentration gradient is not large, the transference numbers and the effective diffusivities of the two counterions are almost constant. In this case, the overall membrane potential and the ionic flux are given by Eqs. 9.43 and 9.45, respectively. If $|z_1| = |z_2| = z$, the ionic flux can be obtained from Eqs. 9.45 and 9.51 as

$$N_A = - \frac{\overline{D_e}K\overline{C_R}}{z} \left[\frac{C_1^{II}}{C_2^{II} + KC_1^{II}} - \frac{C_1^{I}}{C_2^{I} + KC_1^{I}}\right] + \overline{t_1}\left(\frac{I}{\mathcal{F}}\right) . \qquad (9.52)$$

For a system in which there exist multiple ionic species, the electric potential and the fluxes can be rigorously obtained by the Schlögl method, which was discussed in the section on neutral membranes (Section 3, P. 209). Also, the Henderson approximation, Eq. 9.23, or the approximation of the constant transference numbers, Eq. 9.25, should be suitable for multispecies treatment.

Osmotic Flow

This flow phenomenon is primarily restricted to the membrane itself, because its resistance in the liquid-phase boundary layers is very small in all dialytic electromembrane processes. As discussed in Chapter VIII, the solvent water in an ion-exchange membrane behaves like an ionic species as a whole and is transferred by the electric potential gradient as well as the osmotic and the hydraulic pressure gradients. In most dialytic electromembrane processes, the osmotic flux generally has an adverse effect. A significant degree of osmosis dilutes the diffusate (transfer) stream and results in an extra burden to that stream.

As already mentioned, the ion-exchange membranes, which are commonly used in dialytic electromembrane processes, are quite dense, and their capacity is not large. Thus, the individual ionic flux is not greatly influenced by electro-osmosis. Also, the concentration potential is usually small compared to the total external electric potential. In this case, an integrated form of the osmotic flux can be obtained from Eq. 8,72 as

$$N_v' = - \frac{\overline{D}_v}{l} \left[\wp^{II} - \wp^{I} + \frac{\omega \overline{C}_R \mathscr{F}}{\epsilon} (\overline{\phi}^{I} - \overline{\phi}^{II}) \right] .$$ (9.53)

When the ohmic potential drop is mainly responsible for the total electric potential, Eq. 9.53 further simplifies to

$$N_v' = - \frac{\overline{D}_v}{l} (\wp^{II} - \wp^{I}) + \overline{t}_v \frac{I}{\mathscr{F}} .$$ (9.54)

Here, the so-called electro-osmotic transport number $\overline{t}_v$ is conveniently introduced as

$$\overline{t}_v = - \frac{\omega \overline{D}_v \overline{C}_R \mathscr{F}^2 R_m}{\epsilon} .$$ (9.55)

Physically, the transport number is the volume of the solvent water transferred by one faraday of electricity in the absence of the hydraulic and the osmotic pressure gradients. However, when the concentration potential has a significant effect, Eq. 9.54 is not valid because the volume of water transferred by one faraday also depends on the ionic concentration distribution and the current efficiency.

Most ionic species are transported through an ion-exchange membrane in their hydrated form. The corresponding transport of the solvent water in the hydrated form is commonly of the same order of magnitude as by pure electro-osmosis. Thus the apparent electro-osmotic transport number should be considered, including the effect of hydration. If the diffusional leakage is very small compared to electric transference in ionic transport through the membrane, the apparent electro-osmotic transport number can be approximated by

$$\vec{i}_v' = -\frac{\omega \overline{D}_v \overline{C}_R \mathcal{F}^2 R_m}{\epsilon} + \sum_i \vec{t}_i \, n_{hi} \,, \tag{9.56}$$

where n_{hi} is the hydrate number of ionic species i and the summation is to be carried over all ionic species participating in transport across the membrane. Figure 9.12 illustrates the typical relationship between the apparent electro-osmotic flux and the electric current. This figure was prepared by Breslau and Miller [17] for $0.1N$ lithium aqueous solution with the ACI(DK-1) cation-exchange membrane.

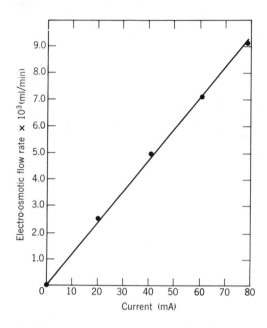

Fig 9.12 Electro-osmotic flow through the ACI(DK-1) cation-exchange membrane for $0.1N$ lithium aqueous solution. Reprinted by permission from Ref. 17. Copyright 1971 by the American Chemical Society.

Transport in Solution Phases

The previous section discussed the electrokinetic processes in the membrane phase proper. The subsequent treatise deals primarily with the phenomenon of concentration polarization in the solution phases.

The ionic concentration of the solution phase is always polarized at the solution-membrane interface because of the permselectivity of the membrane. Polarization is consequently more severe in an ion-exchange membrane than in a neutral membrane. With an ion-exchange membrane, the concentration at the interface is further accumulated in the concentrate compartment but depleted in the dilute compartment, as shown in Fig. 9.13. In an electro-membrane process involving demineralization, the concentration depletion is more of a troublesome problem than the concentration buildup, because the extremely low concentration at the interface creates a considerable ohmic potential drop and also induces water splitting. The deterioration of an anion-exchange membrane, if any, by the pH change due to water splitting is an additional problem.

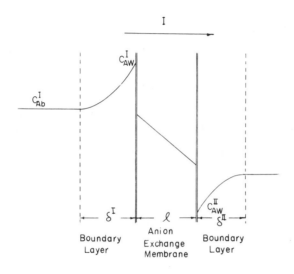

Fig. 9.13 Concentration polarization across an anion-exchange membrane.

The so-called Nernst film concept as discussed in Chapter VIII will be used extensively in this section. The solution phase in a membrane stack can be divided into two boundary films near the solution-membrane interface and a bulk phase between them, as schematically illustrated in Fig. 9.13. The ionic concentration in the bulk phase is constant and uniform, whereas the concentration in the boundary film is distributed over a uniform thickness δ. In the bulk phase the ohmic potential drop is entirely responsible for the electric potential drop. The ionic transport in the bulk phase in the direction parallel to the membrane surface is via convective flow and bulk backdiffusion. In the stagnant boundary film, on the other hand, the ionic transport occurs only in the direction normal to the membrane surface, and its mechanism is both the molecular diffusion and the electric transference. The electric potential drop in both phases is almost absent in the direction parallel to the membrane surface. The validity of this film concept has been established in turbulent flow, where the boundary layer is very thin and uniform along the membrane surface. Even in laminar flow, this concept can be effectively applied if suitably averaged quantities are employed. This approximation can also be extended to the case where natural convection is significant. The more rigorous theory will be treated in Chapter XIII. There it is shown that the boundary-layer thickness δ depends in a rather complicated manner on the kinematic viscosity, effective diffusivity, the ratio of ionic fluxes, the stream-flow rates, and the dimensional ratio of the channel width to the longitudinal dimension of the stack. When natural convection dominates over forced convection, the density coefficient with respect to a concentration variation is also important. The boundary-layer thickness generally increases as the flux ratio and the ratio of the effective diffusivity to the kinematic viscosity increase. On the other hand, a high stream-flow rate and a low dimensional ratio reduce the thickness. Since natural convection is enhanced by the density coefficient, a high value of the coefficient also keeps the film thin.

Solution Phases in Demineralization

When a solution contains only one single electrolyte, the transference numbers and the effective diffusivities are constant, as shown in the section on neutral membranes (Section 3, p. 207). Consequently, integration of Eqs. 9.7 and 9.8 to obtain the ionic flux and the electric potential drop is simple and essentially the same as in a neutral membrane. If the solution is associated with an anion-exchange membrane, the ionic flux of the counterion A across the boundary layer is thus given by

$$N_A = - D_e \left(\frac{C_{Ab}^{II} - C_{AW}^{II}}{\delta^{II}} \right) + \bar{t}_A \frac{I}{\mathcal{F}} . \tag{9.57}$$

where the subscripts b and W refer to the bulk phase and the membrane-solution interface, respectively (see Fig. 9.13). Similarly, the electric potential drop across the film becomes

$$\phi_b^{II} - \phi_W^{II} = - \frac{RT}{\mathcal{F}} (t_A + t_Y) \ln \frac{C_{Ab}^{II}}{C_{AW}^{II}} - \frac{I \, \delta^{II}}{\kappa_b^{II} - \kappa_W^{II}} \ln \frac{C_{Ab}^{II}}{C_{AW}^{II}} . \tag{9.58}$$

and the electric conductance in the solution is defined as

$$\kappa_i^{II} = \frac{\mathcal{F}^2}{RT} (D_A z_A - D_Y z_Y) C_{Ai} z_A, \quad i = b, W. \tag{9.59}$$

Equation 9.58 thus shows clearly that the depletion of the concentration C_{AW} of the counterion at the interface results in a rather substantial increase of the electric potential across the boundary film. An additional increase will occur when the electric current density I is raised. When the concentration C_{AW} approaches zero, the electric potential increases rapidly for a small increment of the current density, and the asymptotic value is called the limiting current density. In practice, however, the electric potential does not diverge to infinity at this limiting current density. As the potential increases to a certain value, the solvent water is electrolyzed or "split," and the hydrogen or the hydroxide ion competes with the original counterion in electric transference. Therefore, the operation of an electromembrane module must be restricted to a current density smaller than the limiting value. The limitation is particularly serious when the process involves demineralization.

The limiting current density can be expressed by combining Eqs. 9.57 and 9.48. If, for example, the electrochemical valences of the counterion and the co-ion are the same, that is, $|z_A| = |z_Y| = z$, the current density becomes

$$(\overline{t_A} - t_A) \frac{I}{\mathcal{F}} = - D_e \left(\frac{C_{Ab}^{II} - C_{AW}^{II}}{\delta^{II}} \right) + \frac{\overline{D}_e z K^2}{\overline{C}_R} [(C_{AW}^{I})^2 - (C_{AW}^{II})^2] . \quad (9.60)$$

Here it is assumed that the electric transference dominates over the diffusional leakage in the membrane phase and also that the capacity of the ion-exchange membrane is very high compared to the interfacial concentration in the boundary film. When the concentration C_{AW} at the interface is depleted to zero, but the solvent water is not split, the limiting current density, $I_l^{II} > 0$, in compartment II (Fig. 9.13) can be obtained as

$$|\overline{t_A} - t_A| \left(\frac{I_l^{II}}{\mathcal{F}} \right) = \frac{D_e}{\delta^{II}} C_{Ab}^{II} + \frac{\overline{D}_e z K^2}{\overline{C}_R} (C_{AW}^{I})^2 . \quad (9.61)$$

Obviously, the I_l^{II} increases with the bulk concentration C_{Ab} and the effective diffusivity D_e. But the value decreases as the difference of the two transference numbers, $|\overline{t_A} - t_A|$, and the boundary-layer thickness δ^{II} increase. In particular, the effect of the thickness δ^{II} is very important. A high limiting current can be achieved by controlling the thickness with a high stream-flow rate in a thin channel of the dilute compartment. The diffusional leakage also enhances the limiting value, but leakage is, of course, not desirable.

Since the diffusional leakage term in Eqs. 9.60 and 9.61 is often small compared to the diffusion term across the boundary film, the limiting current density is more frequently defined as

$$I_l^{II} = \frac{\mathcal{F} D_e C_{Ab}^{II}}{\delta^{II} |\overline{t_A} - t_A|} . \quad (9.62)$$

This then implies that the ionic transport through the membrane is only via electric transference, and thus the current efficiency η_A of the counterion A becomes just the transport number $z_A \bar{t}_A$ of the ion in the membrane. In this case, the concentration distribution across the boundary film is, from Eqs. 9.60 and 9.62, simplified to

$$\frac{C_{AW}^{II}}{C_{Ab}^{II}} = 1 - \frac{I}{I_l^{II}} , \tag{9.63}$$

and the electric potential, Eq. 9.58, can be reduced to

$$\phi_b^{II} - \phi_W^{II} = \frac{RT}{\mathcal{F}} \left[(t_A + t_Y) + \frac{|t_A t_Y|}{\bar{t}_A - t_A} \left(1 - \frac{z_Y}{z_A} \right) \right] \ln \left(1 - \frac{I}{I_l^{II}} \right). \tag{9.64}$$

Similarly, the ionic flux and the electric potential in the concentrate compartment I become

$$N_A = D_e \frac{C_{AW}^{II} - C_{Ab}^{I}}{\delta^I} + t_A \frac{I}{\mathcal{F}} , \tag{9.65}$$

and

$$\phi_b^I - \phi_W^I = \left[\frac{RT}{\mathcal{F}} (t_A - t_Y) + \frac{I \delta^I}{\kappa_b^I - \kappa_W^I} \right] \ln \left(1 + \frac{C_{AW}^I}{C_{Ab}^I} \right). \tag{9.66}$$

If the diffusional leakage through the membrane is very small compared to diffusion across the boundary film, the electric potential, Eq. 9.66, becomes

$$\phi_b^I - \phi_W^I = \frac{RT}{\mathcal{F}} \left[(t_A + t_Y) + \frac{|t_A t_Y|}{\bar{t}_A - t_A} \left(1 - \frac{z_Y}{z_A} \right) \right] \ln \left(1 + \frac{I}{I_l^I} \right). \tag{9.67}$$

Here, the limiting current density I_l^I is defined, in analogy to Eq. 9.62, as

$$I_l^I = \frac{\mathcal{F} D_e C_{Ab}^I}{\delta^I |\bar{t}_A - t_A|} . \tag{9.68}$$

In contrast to the conditions in the dilute compartment, the electric potential in the concentrate compartment never diverges to infinity at any concentration $C_{AW}^I (C_{AW}^I > C_{AW}^{II})$. Therefore the physical meaning of the limiting parameter I_l^I is not the same as I_l^{II}.

In the bulk phase in the middle of a compartment, the concentration distribution is uniform in the direction normal to the membrane surface. Consequently, the electric potential drop arises only from the ohmic potential drop.

Solution Phases in Continuous Ion Exchange

Here the solution phase contains two counterions, 1 and 2, a common co-ion Y. The solution is continuously fed into a compartment between two ion-exchange membranes. The ionic concentration is so small compared to the capacity of the membranes that the permselectivity of the membranes is almost perfect with respect to the counterions. In other words, the ionic flux of the co-ion across these ion-exchange membranes is almost zero. This is the typical situation in a continuous ion-exchange process. In order to obtain the distribution of the electric potential and the ionic concentrations, one could attempt to solve a system of simultaneous differential equations of the Nernst-Planck type with the electroneutrality condition and the charge conservation. However, such a solution has not been achieved in a closed form, except for the case where the electrochemical valences of the counterions are the same. Of course, treatment by the iteration method should be possible for the general case. Nevertheless, the following qualitative statements are valid. When an external electric potential is applied to this system, the co-ion is immediately shifted at the beginning and distributed in the compartment such that the following relationship is satisfied at every point:

$$\left(\frac{\mathcal{F}}{RT}\right)\left(\frac{d\phi}{dx}\right) = -\left(\frac{1}{z_Y C_Y}\right)\left(\frac{dC_Y}{dx}\right). \tag{9.69}$$

Thus, the external potential is completely compensated with the concentration potential created by the concentration distribution of the co-ion. Also, one of the counterions is enriched near the surface of the ion-exchange membrane through which the counterion is removed from the compartment. The other counterion, on the other hand, accumulates at the surface of the other membrane through which the counterion is introduced into the compartment. Consequently, the total ion concentration is higher near the surface of the membrane across which a counterion is introduced, but the concentration of the original counterion to be stripped is lower near this surface. In this region, the effect of the electric potential on the migration of the original counterion is therefore "quenched," as would be the case of an ion in the presence of a large amount of supporting ions. The ionic transport of the ion is almost entirely by its self-diffusion. The transport of the counterion to be introduced, on the other hand, is primarily by its electric transference. Similarly, the opposite situation can be expected to occur near the surface of the other membrane. Therefore, the concentration polarization in this case of a continuous ion-exchange process is not as serious as in demineralization. When a high electric potential is used, the boundary layer near a membrane plays a rather favorable role to prevent the counterion, which was introduced through the membrane on the opposite side, from transferring further through the membrane.

Total Electric Potential Drop Across an Electromembrane

This section discusses the behavior of one membrane and its solution phases. It is confined to the case where the external electric potential is very high and the difference of ionic concentrations across the membrane is very small. Consequently, the ionic transport through the membrane is mainly due to electric transference. This case closely corresponds to most practical membrane processes, particularly electrodialysis in desalination.

Consider an electromembrane associated with solutions containing *one single electrolyte,* as shown in Fig. 9.15. The total electric potential drop across such a membrane, $E = \phi_6 - \phi_1$, is then composed of the ohmic potential drop in the bulk phase of compartment I ($\phi_2 - \phi_1$); the concentration potential plus the ohmic potential drop across the boundary film in compartment I ($\phi_3 - \phi_2$); the membrane potential plus the ohmic potential drop across the membrane ($\phi_4 - \phi_3$); the concentration potential plus the ohmic potential drop across the boundary film in compartment II ($\phi_5 - \phi_4$); and finally, the ohmic potential drop in the bulk phase of compartment II ($\phi_6 - \phi_5$). Thus, the total electric potential drop can be expressed as

$$
\begin{aligned}
E &= (\phi_2 - \phi_1) + (\phi_3 - \phi_2) + (\phi_4 - \phi_3) + (\phi_5 - \phi_4) + (\phi_6 - \phi_5) \\[2mm]
&= - I \left[\frac{b^{I}/2 - \delta^{I}}{C_b^{I} \, \Lambda^{I}} + \frac{b^{II}/2 - \delta^{II}}{C_b^{II} \, \Lambda^{II}} + R_m \right] \\[2mm]
&\quad + \frac{RT}{\mathcal{F}} \left[(t_+ + t_-) \ln \frac{C_b^{I}}{C_W^{I}} + (\bar{t}_+ + \bar{t}_-) \ln \frac{C_W^{I}}{C_W^{II}} + (t_+ + t_-) \ln \frac{C_W^{II}}{C_b^{II}} \right] \\[2mm]
&\quad - I \left[\frac{\delta^{I}}{\Lambda^{I} \, (C_b^{I} - C_W^{I})} \ln \frac{C_b^{I}}{C_W^{I}} + \frac{\delta^{II}}{\Lambda^{II} \, (C_b^{II} - C_W^{II})} \ln \frac{C_b^{II}}{C_W^{II}} \right] .
\end{aligned}
\tag{9.70}
$$

Here, the equivalent electric conductivity Λ in the solution phase was introduced such that

$$
\Lambda = \frac{\mathcal{F}^2}{RT} \, (D_+ z_+ - D_- z_-) .
\tag{9.71}
$$

Equation 9.70 holds exactly for a neutral membrane. Also, the equation is approximately valid for an ion-exchange membrane when the electric transference is controlling in the membrane phase and the concentration difference across the membrane is relatively small.

Combination of Eq. 9.70 with flux equations across all existing phases leads to a relationship between the electric potential and current density. The relationship then depends on bulk concentrations, C_b^I and C_b^{II}, and also flow conditions in the two compartments, that is, boundary-layer thicknesses δ^I and δ^{II}, If the diffusional leakage through the membrane is small compared to the diffusion term across the boundary layer, the relationship across the anion exchange membrane in Fig. 9.15 simplifies to

$$E = - I \left[\frac{b^I/2 - \delta^I}{C_b^I \Lambda^I} + \frac{b^{II}/2 - \delta^{II}}{C_b^{II} \Lambda^{II}} + R_m \right]$$

$$- \frac{RT}{\mathcal{F}} \left[\frac{z_+ - z_-)|t_+ t_-|}{z_-(\bar{t}_- - t_-)} + (t_+ + t_-) - (\bar{t}_+ + \bar{t}_-) \ln \left(\frac{1 + \frac{I}{I_l^I}}{1 - \frac{I}{I_l^{II}}} \right) \right]$$

$$- \frac{RT}{\mathcal{F}} (\bar{t}_+ + \bar{t}_-) \ln \frac{C_b^I}{C_w^{II}} . \tag{9.72}$$

Here, it is assumed that transference numbers in compartment I are the same as in compartment II. Thus, the relationship, Eq. 9.72, has as parameters the limiting current densities, I_l^I and I_l^{II}, and bulk concentrations, C_b^I and C_b^{II}. When the current density, $I > 0$, approaches the limiting value I_l^{II} (or I_l^I if $I < 0$), the electric potential diverges to infinity. Figure 9.14 illustrates the relationship for the system of an anion-exchange membrane [$\bar{t}_- = -0.98$, $R_m = 11$ (Ω) (cm^2)] with a 0.03N KCl solution. This figure was originally prepared by Spiegler [18] using an equation that is essentially the same as Eq. 9.72.

Transport through the Membrane Stack

The overall electrokinetic phenomenon in a membrane stack can be understood by combining all individual phenomena in existing membrane and solution phases, which have so far been discussed separately in previous sections. The membrane stack differs from one electromembrane process to another. The discussion will, however, be confined to the process of electrodialysis involving demineralization of one single electrolyte. This case corresponds to the desalination operation. Its extension to other electromembrane processes would be analogous

Repeating Unit of a Membrane Stack

The membrane stack generally consists of a number of repeating units. In electrodialysis, for example, an anion- and a cation-exchange membrane constitute the repeating unit, as shown in Fig. 9.15. Thus, the total electric

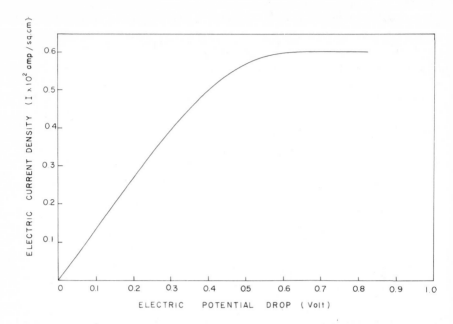

Fig. 9.14   Electric current density versus electric potential drop across an anion-exchange membrane, calculated from Eq. 9.72. [$\bar{t}_- = -0.98$; Rm = 11 $(\Omega)(cm^2)$; 0.03N KCl solution] [18].

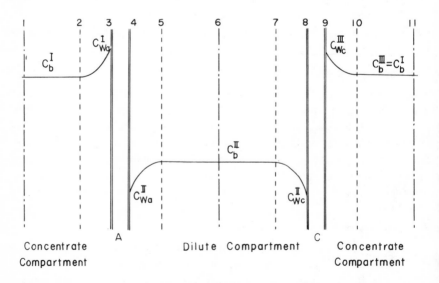

Fig. 9.15   Concentration polarization in a repeating unit of electrodialysis. (A = anion-exchange membrane; C = cation-exchange membrane.)

226

potential drop E_u across the repeating unit is the potential across the anion-exchange membrane, $E_a = \phi_6 - \phi_1$, plus the potential across the cation-exchange $E_c = \phi_{11} - \phi_6$, so that

$$E_u = \phi_{11} - \phi_1$$

$$= E_a + E_c \ . \tag{9.73}$$

Then, the overall potential drop E_T across the membrane stack becomes

$$E_T = n \, E_u + E_e \ , \tag{9.74}$$

where n is the number of the repeating units in the stack, and E_e is the electrode potential. The potential drops E_a and E_c are given by Eqs. 9.70 or 9.72. Eliminating the concentrations at the membrane-solution interfaces in Eq. 9.73 by using the flux equations, an explicit relationship between the electric potential and current density can be obtained for a given bulk concentration. Then the corresponding current efficiency can be calculated.

Consider a repeating unit of electrodialysis, as in Fig. 9.15. A solution containing one single electrolyte is to be dialyzed. In such a repeating unit, the net transport rate of the cation out of the dilute compartment will be

$$N_+ = N_{+c} - N_{+a} \ , \tag{9.75}$$

$$N_{+a} = \frac{k^{\mathrm{I}}}{z_+} (C_b^{\mathrm{I}} - C_{Wa}^{\mathrm{I}}) + t_+ \left(\frac{I}{\mathcal{F}} \right) \tag{9.76}$$

$$= \frac{U_a}{z_+} \left[(C_{Wa}^{\mathrm{I}})^{1-\frac{1}{\beta}} - (C_{Wa}^{\mathrm{II}})^{1-\frac{1}{\beta}} \right] + \bar{t}_{+a} \left(\frac{I}{\mathcal{F}} \right) \tag{9.77}$$

$$= \frac{k^{\mathrm{II}}}{z_+} (C_{Wa}^{\mathrm{II}} - C_b^{\mathrm{II}}) + t_+ \left(\frac{I}{\mathcal{F}} \right), \tag{9.78}$$

$$N_{+c} = \frac{k^{\mathrm{II}}}{z_+} (C_b^{\mathrm{II}} - C_{Wc}^{\mathrm{II}}) + t_+ \left(\frac{I}{\mathcal{F}} \right) \tag{9.79}$$

$$= \frac{U_c}{z_+} \left[(C_{Wc}^{\mathrm{II}})^{1-\frac{1}{\beta}} - (C_{Wc}^{\mathrm{I}})^{1-\frac{1}{\beta}} \right] + \bar{t}_{+c} \left(\frac{I}{\mathcal{F}} \right) \tag{9.80}$$

$$= \frac{k^{\mathrm{I}}}{z_+} (C_{Wc}^{\mathrm{I}} - C_b^{\mathrm{I}}) + t_+ \left(\frac{I}{\mathcal{F}} \right), \tag{9.81}$$

where subscripts c and a refer to the cation- and the anion-exchange membranes, respectively. Also, the mass-transfer coefficients k^{I} and k^{II} across the boundary layers and the dialysis coefficients U_c and U_a across the membranes are conveniently introduced such that

$$k^{\mathrm{I}} = \frac{D_e^{\mathrm{I}}}{\delta^{\mathrm{I}}}; \quad k^{\mathrm{II}} = \frac{D_e^{\mathrm{II}}}{\delta^{\mathrm{II}}} \qquad (9.82)$$

and

$$U_a = \frac{\overline{D}_{ea} \, K_-^{1-\frac{1}{\beta}}}{\overline{C}_{Ra}^{\left|\frac{1}{\beta}\right|} \, l_a} \; ; \quad U_c = \frac{\overline{D}_{ec} \, K_c^{1-\frac{1}{\beta}}}{\overline{C}_{Rc}^{\left|\frac{1}{\beta}\right|} \, l_c} \; ;$$

$$\beta = \frac{z_-}{z_+} < 0 \quad . \qquad (9.83)$$

Here the solution concentrations are assumed to be so small that D_e and t_+ are constant and uniform for all compartments, and also the Donnan exclusion is very efficient. Furthermore, the stream-flow conditions in the concentrate compartments I and III are the same, and therefore the mass-transfer coefficients are identical. Consequently, $C_{Wc}^{\mathrm{III}} = C_{Wc}^{\mathrm{I}}$.

Thus, the net current efficiency η_+ becomes, from Eqs. 9.75, 9.77, and 9.80,

$$\eta_+ \equiv \frac{z_+ N_+ \mathcal{F}}{I}$$

$$= z_+ (\overline{t}_{+c} - \overline{t}_{+a}) - \frac{\mathcal{F}}{I} \left[U_c \left((C_{Wc})^{1-\frac{1}{\beta}} - (C_{Wc})^{1-\frac{1}{\beta}} \right) + U_a \left((C_{Wa})^{1-\frac{1}{\beta}} \right. \right.$$

$$\left. \left. - (C_{Wa})^{1-\frac{1}{\beta}} \right) \right] \quad . \qquad (9.84)$$

As indicated by this equation, a high current efficiency can be obtained by employing cation- and anion-exchange membranes that possess high transference numbers, $\overline{t}_{+c}$ and $\overline{t}_{-a}$, but small dialysis coefficients, U_c and U_a. Also, operation of such a module with a high current density reduces the contribution of the diffusional leakages and therefore enhances the current efficiency. However, an extremely high current density always greatly polarizes the solution concentrations, leading to a considerable ohmic potential drop and also a high diffusional leakage across the membrane as stated earlier. Therefore, the current density should be less than the limiting current density.

When the diffusional leakage is very small compared to electric transference, and current efficiency reduces to

$$\eta_+ = z_+(\overline{t}_{+c} - \overline{t}_{+a})$$

$$= z_+\overline{t}_{+c} + z_-\overline{t}_a - 1. \tag{9.85}$$

In this case, the electric potential E_u, from Eq. 9.70, across the repeating unit reduces to

$$E_u = -\left(\frac{b^{\mathrm{I}} - 2\delta^{\mathrm{I}}}{C_b^{\mathrm{I}}\,\Lambda^{\mathrm{I}}} + \frac{b^{\mathrm{II}} - 2\delta^{\mathrm{II}}}{C_b^{\mathrm{II}}\,\Lambda^{\mathrm{II}}} + R_{ma} + R_{ma}\right).I$$

$$- \frac{RT}{\mathcal{F}}\left[\left(1 - \frac{z_-}{z_+}\right)\frac{|t_+t_-|}{t_+ - t_{+a}} + (t_+ + t_-) - (\overline{t}_{+a} - \overline{t}_{-a})\right]\ln\left(\frac{1 + \dfrac{I}{I_{la}^{\mathrm{I}}}}{1 + \dfrac{I}{I_{la}^{\mathrm{II}}}}\right)$$

$$- \frac{RT}{\mathcal{F}}\left[\left(1 - \frac{z_-}{z_+}\right)\frac{|t_+t_-|}{\overline{t}_{+c} - t_+} + (t_+ + t_-) + (\overline{t}_{+c} + \overline{t}_{-c})\right]\ln\left(\frac{1 + \dfrac{I}{I_{lc}^{\mathrm{I}}}}{1 - \dfrac{I}{I_{lc}^{\mathrm{II}}}}\right)$$

$$- \frac{RT}{\mathcal{F}}(\overline{t}_{+c} - \overline{t}_{-a} + \overline{t}_{-c} - \overline{t}_{-a})\ln\frac{C_b^{\mathrm{I}}}{C_b^{\mathrm{II}}}, \tag{9.86}$$

where limiting current densities are defined as

$$I_{la}^{\mathrm{I}} = \frac{\mathcal{F}k^{\mathrm{I}}(C_b^{\mathrm{I}})}{z_+(t_+ - \overline{t}_{+a})}; \quad I_{lc}^{\mathrm{I}} = \frac{\mathcal{F}k^{\mathrm{I}}(C_b^{\mathrm{I}})}{z_+(\overline{t}_{+c} - t_+)};$$

$$I_{la}^{\mathrm{II}} = \frac{\mathcal{F}k^{\mathrm{II}}(C_b^{\mathrm{II}})}{z_+(t_+ - \overline{t}_{+a})}; \quad I_{lc}^{\mathrm{II}} = \frac{\mathcal{F}k^{\mathrm{II}}(C_b^{\mathrm{II}})}{z_+(\overline{t}_{+c} - t_+)}. \tag{9.87}$$

When the diffusional leakage is considerable, an appropriate iteration scheme may be used to solve Eqs. 9.73 through 9.81 with the solution of Eqs. 9.85 and 9.86 as an initial estimate.

Experimental Data

The electric potential across the repeating unit in electrodialysis with one single electrolyte solute can be approximately expressed by Eq. 9.86. The first bracket term on the right-hand side of the equation represents the ohmic potential drops across the membrane and the bulk solution phases; and the last term corresponds to the concentration potential in the absence of concentration polarization. On the other hand, the second and third bracket terms are essentially the potential drop arising from concentration polarization around the anion and the cation exchange membranes, respectively. If the current density is small compared to the limiting value, I_{Id}^{II} or I_{Ic}^{II}, the polarization potentials are also almost linear with respect to the current density as in the ohmic potential drop of the first bracket term. The linearity persists often until the current density closely approaches the limiting value. In other words, the polarization potentials also satisfy the simple Ohm's law in this range of the current density. Hence, it is convenient to approximate the electric potential across the repeating unit as

$$E_u = - R_u I - V_u \qquad (9.88)$$

or the overall potential across the membrane stack by

$$E_T = - nR_u I - nV_u - E_e \quad . \qquad (9.89)$$

Here R_u is the apparent electric resistance, and V_u is the concentration potential across the repeating unit in the absence of concentration polarization, that is, the last term on the right-hand side of Eq. 9.86. Thus, the total apparent electric resistance across the membrane stack R_T is given by

$$R_T \equiv - \frac{E_u}{I}$$

$$= nR_u + \frac{nV_u + E_e}{I} \quad . \qquad (9.90)$$

When the apparent resistance R_T is plotted against the reciprocal current density, a straight line with a slope of $nV_u + E_e$ is thus obtained for small values of current density, as shown by the right-rising branch of the curve in Fig. 9.16. This figure was experimentally prepared for an aqueous sodium sulfate solution by Cowan and Brown [19]. As the current density approaches close to the limiting value, the concentrations near the membrane-solution interface in the dilute compartment are greatly depleted. The contribution of the polarization potential becomes increasingly significant, and the electric potential is sharply enhanced according to Eq. 9.86. Therefore, the curve of the apparent resistance against the reciprocal current density would pass through a sharp minimum. However, the minimum point is not so sharp in practice as is indicated in Eq. 9.86 (see the second minimum in Fig. 9.16). This is becuase the diffusional leakage through the membrane is correspondingly enhanced. Therefore, the interfacial concentrations are never depleted to zero in practice. Also, hydrogen or hydroxyl ions that are produced by water splitting participate

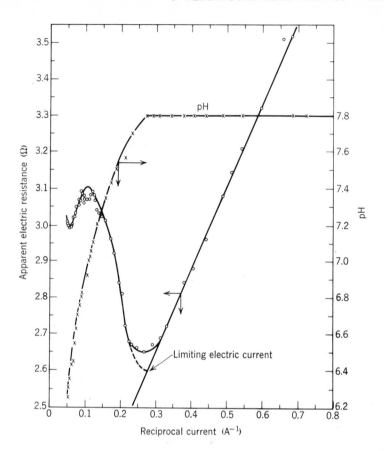

Fig. 9.16 Apparent electric resistant across a membrane stack. (Two diluting channels 0.32 cm thick with 718 μmho/cm conductivity, flowing at 94 cm/sec.). Reprinted by by permission from Ref. 19. Copyright 1959 by the American Chemical Society.

increasingly in electric transference. This phenomenon is clearly indicated by the sharp drop of the pH value around the minimum point in Fig. 9.16. Therefore, Eq. 9.86, which was derived by neglecting the diffusional leakage and water splitting, is not valid in this region. The deviation may also be attributed to the fact that the bulk concentration in the dilute compartment is not uniform but distributed in the longitudinal direction parallel to the membrane surface. The concentration at the inlet of the dilute compartment is always smaller than at the outlet.

When the electric potential increases further, more solvent water is split. Beyond a certain value of the potential drop. the electric current is primarily due to electric transference of hydrogen or hydroxyl ions. Again, referring to Fig. 9.16, it is evident that the potential drop across the polarized boundary

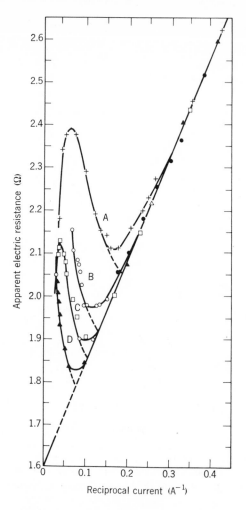

Fig. 9.17 Effect of stream velocity in dilute compartment on the apparent electric resistance (3000 μmho/cm, 0.64-cm-thick channels, A:29; B:43; C:58; D:87 cm/sec). Reprinted by permission from Ref. 19. Copyright 1959 by the American Chemical Society.

layer becomes almost proportional to the current density. This behavior is shown by the rising branch on the left side of the curve. The curve of the apparent resistance against the reciprocal current density recovers a positive slope. The existence of the maximum point in Fig. 9.16 indicates this situation. In the practical operation of electrodialysis, an internally staged system is often employed as in the experiment of Cowan and Brown. A dialysate stream from one dilute compartment is fed back into another dilute compartment in the membrane stack. In this case, the above-mentioned behavior would

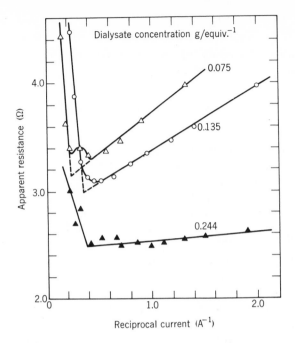

Fig. 9.18 Effect of dialysate concentration on the apparent electric resistance (11 concentrate compartments, 10 dilute compartments, 1/32 in. thick; AMF 3129-B-56 and CSIR-TA membranes; C_b^I = 1.0N Nacl in the concentrate compartment; 10 cm/sec stream velocities). Reprinted by permisstion from Ref. 21. Copyright 1965 by the American Chemical Society.

be repeated again in the next dilute compartment after considerable water splitting occurs in the previous dilute compartment. The first minimum point may be interpreted this way. It should be noted that even a trace of ionic impurity in the feed stream, if any, would also significantly affect the behavior around the limiting value of the current density.

Experimental determination of the limiting current density is, therefore, impossible unless a complete theoretical analysis is made including the effect of the diffusional leakage, water splitting, longitudinal concentration distribution and effect of ionic impurity, if present. However, somewhat arbitrary and approximate methods for the experimental determination of the limiting value have been proposed. The method employed by Rosenberg and Tirrell [20] consists of a series of observations of the pH value with a membrane stack that is comprised of only anion-exchange or cation-exchange membranes. On the other hand, Cowan and Brown [19] arbitrarily set the limiting value by extrapolating the curve of the negative slope to the straight line as shown by the dotted line in Fig. 9.16. In most practical cases, the limiting values obtained by the two methods are in good agreement.

Figure 9.17 illustrates the relationships between the apparent electric resistance and the reciprocal current density for different stream velocities in the dilute compartment. Figure 9.18 shows the effect of the concentration of the dilute compartment on the relationship.

4 DESIGN AND OPERATION OF ELECTROMEMBRANE MODULES

The material concerning design and operation falls under the heading of engineering aspects, and its discussion might be more properly covered in Chapter XIII. However, the uniqueness of the electrophenomena and the development of the design equations in the preceding section make it more appropriate to present the qualitative design aspects at this point.

Electromembrane Module

Most electromembrane modules are commercially available in a plate-and frame type with or without tortuous paths. As in other membrane processes, simplicity, easy maintenance, and low pressure loss are essential requirements. The materials of construction must also be corrosion resistant and electrically nonconducting. Plastics or plastic-lined steel is often suitable for this purpose. It is important to take special care with pipe lines connecting internal or external stages so as to minimize electrical leakage through solutions in these lines. When a module is used for food and beverage processing, crevices or pockets on the inside surface of the module must be avoided as bacterial growth will readily occur in such places.

A basic solution is generally produced at the cathode, whereas an acidic solution is at the anode. Therefore, any electrically conducting material can be used as a cathode. However, selection of a suitable anode material requires special care, since most conducting materials tend to deteriorate by severe oxidation in an acidic solution. Platinum is very durable but expensive. Platinum-coated metals, some heavy-metal oxides, and magnetite are satisfactory in some cases as anode materials. Graphite is often suitable despite its tendency to deteriorate gradually. It is not expensive and is easy to handle [22]. The anode deterioration can be minimized by employing a high flow rate of the anode rinse stream with appropriate pH adjustment. The use of anode and cathode rinse streams at a high flow rate with proper pH is also important to prevent the produced hydrogen or hydroxyl ions from being introduced into the process streams.

Electromembranes

A summary of preparation of various electromembranes, neutral or electrically charged, is presented in Chapter XIV. Most commercial ion-exchange membranes possess an area resistance of less than 20 $(\Omega)(cm^2)$; the apparent transference number for a 1,1-valent electrolyte is in the range of 0.85 to 0.90 at a concentration of $0.5/1.0N$; membrane thickness is in the range of 10 to 20 mils. The concentration and size of the ionic species to be separated greatly influence the membrane performance. Donnan exclusion is very efficient when the capacity of the membrane is high compared to the solution concentration and its pore size is large compared to the ionic size. Consequently, a membrane of suitably large pore sizes is generally required for a large ionic species in a hydrated form.

The membrane life depends largely on the nature of the feed solution and the operating temperature. Multivalent or large counterions are specifically adsorbed (often irreversibly) on the pore wall of the ion-exchange membrane, and so neutralize the fixed ionic group of the membrane. Consequently, its transference number is reduced, and the area resistance increases adversely. In demineralization, where such ions are secondary, the membrane life is, therefore, considerably prolonged by prior removal of these ions. Oxidation, pH adjustment, or other chemical and physical treatments are commonly employed for pretreatment of the feed solution. Removal of particulate and colloidal materials, if any, from the feed solution improves the membrane performance. When the above-mentioned multivalent or large ionic species are not secondary, or cannot be effectively removed in prior treatment, a membrane of a suitably large pore size should be used. If these ions are inorganic, some pretreatment normally is required to prevent membrane fouling by their precipitates. Addition of a reducing agent or a precipitation inhibitor and pH adjustment are typical examples. Also, reversing of electrode polarity and adding a detergent or an acid are often used to clean the fouled membrane. The pH adjustment in the feed solution is also very important to improve the performance of polymeric anion-exchange membranes, which are inherently unstable at high pH values. Most polymeric ion-exchange membranes, particularly anion-exchange membranes, easily deteriorate at a high temperature. Therefore, high-temperature operation is undesirable, even if the electric resistance of a solution is greatly reduced at a higher temperature.

Operation of Electromembrane Modules

In order to achieve a high degree of enrichment or stripping of a solute in an electromembrane process, a large membrane area and a high external electric potential are generally required, within certain limitations. The limitations are usually imposed by the nature of the feed solution, concentration polarization, economic consideration, and electro-osmosis. Obviously, separation can not be achieved beyond the solubility of an electrolyte solute to be separated.

When the ionic concentration of the solution phase is high compared to the capacity of the ion-exchange membrane, the membrane performance is greatly reduced, as mentioned earlier. On the other hand, separation of an extremely low concentration of the electrolyte solute is also prohibitive, since the electric resistance of such a solution is very high.

As the solute concentration in the solution phase, particularly in the dilute compartment, is polarized, the current efficiency is reduced, and the electric energy is increasingly dissipated into heat generation by ohmic resistance. Polarization occurs often to such an extent that the electric energy is also consumed by undesirable water splitting. Therefore, the external electric *
potential must be restricted such that the electric current density may not exceed the limiting current density. For a given electric potential, the concentration polarization can be minimized by employing high stream-flow rates and thin channels. Also, an appropriate turbulence promoter or tortuous paths will reduce the concentration polarization a great deal so that a higher limiting current density can be obtained.

The overall electric potential drop across a membrane stack is largely due to the ohmic potential drop for a high value of current density. The corresponding electric energy per unit membrane area, which is a product of the electric potential drop and the current density, increases almost with the square of the current density. Since the flux of an electrolyte solute to be separated is approximately proportional to the current density, the electric energy required for a unit amount of the solute to be separated increases directly with the current density. On the other hand, the fixed cost of an electromembrane module decreases often with the current density for a given amount of the solute separated. Hence, there always exists an optimum current density, at which the total cost, the sum of the energy cost and the fixed cost, is minimized. However, this optimum value often exceeds the maximum feasible limiting value that can be practically obtained for a dilute stream at a high flow rate in a thin channel with a turbulent promoter [23]. Thus, the operating current density is restricted primarily by the concentration polarization rather than by economic optimization.

The solvent flux is another major factor that reduces the overall performance of the electromembrane module. Electro-osmosis is primarily responsible for the solvent flux when the solution contains a hydrophobic organic electrolyte. If the electrolyte solute is inorganic, transport of the solvent water in a hydrated form is also important. Since the apparent electro-osmotic water-transport number is usually greater than the transference number of the ionic species to be separated, separation is not possible beyond a certain concentration. For example, concentration of saline water cannot generally be achieved with most commercial electromembranes beyond about 27%, since the water-transport number of saline water is about 12 [24].

Nomenclature

Symbol	*Meaning*
C	Ionic concentration or equivalent concentration of an electrolyte
C_A	Concentration of counterion A
C_Y	Concentration of co-ion Y
C_b	Equivalent concentration in the bulk phase
C_W	Equivalent concentration at the membrane-solution interface
C_1, C_2	Concentrations of counterions 1 and 2
C_{Ab}	Concentration of counterion A in the bulk phase
C_{AW}	Concentration of counterion A at the membrane-solution interface
C_{Wa}	Equivalent concentration at the surface of an anion-exchange membrane
C_{Wc}	Equivalent concentration at the surface of a cation-exchange membrane
$\overline{C}_R$	Capacity of an ion-exchange membrane
$\overline{C}_{Ra}$	Capacity of an anion-exchange membrane
$\overline{C}_{Rc}$	Capacity of a cation-exchange membrane
D_i	Diffusivity of ionic species i
D_e	Effective diffusivity defined by Eqs. 9.15 or 9.41
$\overline{D}_v$	Effective diffusivity of the solvent water defined by Eq. 8.72
E	Electric potential drop
E_a, E_c	Total electric potential drops across an anion- and a cation-exchange membrane
E_e	Electric potential drop across the two electrode compartments
E_{Don}	Donnan potential
E_m	Membrane potential
E_T	Total electric potential drop across a membrane stack
E_u	Total electric potential drop across a repeating unit
$\mathcal{F}$	Faraday constant
I	Electric current density
I_l	Limiting current density

I_{la}, I_{lc}	Limiting current densities for an anion- and a cation-exchange membrane
K	Distribution coefficient, or ratio of activity coefficients between the membrane and the solution phases, defined by Eqs. 9.18, 9.28, and 9.51
N_i	Molar flux of ionic species i
N_+	Molar flux of the cation
N_{+a}, N_{+c}	Molar fluxes of the cation through an anion- and cation-exchange membrane
N_v'	Volumetric water flux
$\mathscr{P}$	Total pressure, or sum of the hydraulic and the osmotic pressures
R	Gas constant
R_m	Membrane area resistance
R_T	Total area resistance
T	Temperature
U_a, U_c	Dialysis coefficients of an anion- and a cation-exchange membrane, defined by Eq. 9.83.
b	Channel width
k	Mass-transfer coefficient across the boundary layer, defined by Eq. 9.82
l	Thickness of a membrane
n	Number of repeating units in a membrane stack
n_{hi}	Hydration number of ionic species i
t_i	Transference number of ionic species i, defined by Eq. 9.5
t_i'	Apparent transference number of ionic species i, defined by Eq. 9.10
$\overline{t_v}$	Electro-osmotic transport number
$\overline{t_v}'$	Apparent electro-osmotic trnasport number, defined by Eq. 9.56
x	Space coordinate
z_i	Electrochemical valence of ionic species i
z	Electrochemical valence of an electrolyte when $z_+ = \lfloor z \rfloor$
a	Diffusivity ratio (see Eq. 9.36)
β	Electrochemical valence ratio (see Eq. 9.36)
γ_i	Activity coefficient of ionic species i
$\gamma_\pm$	Mean activity coefficient of an electrolyte, defined by Eq. 8.35
δ	Thickness of the boundary layer

ϵ	Porosity of a membrane
ϕ	Electrical potential
ϕ_i	Electric potential at position i, $i=1,2\cdots,11$ (see Fig. 9.15)
κ	Electric conductivity
κ_{lm}	Log-mean electric conductivity defined by Eq. 9.17
Λ	Equivalent electric conductivity of a solution
ω	Sign indicator
η	Current efficiency

Superscript

$^-$(overbar)	Refers to a membrane phase
I, II, III	Refer to solution phases

Subscript

A	Refers to the counterion
W	Refers to the membrane-solution interface
Y	Refers to the co-ion
a	Refers to an anion-exchange membrane
b	Refers to the bulk solution phase
c	Refers to a cation-exchange membrane
i,j	Refers to ionic species
1,2	Refers to counterions
+,-	Refers to the cation and the anion
$+a,+c$	Refers to the cations associated with an anion- and a cation-exchange membrane
$-a,-c$	Refers to the anions associated with an anion- and a cation-exchange membrane

References

1. R. E. Lacey, "Basis of Electromembrane Processes," in *Industrial Processing with Membranes*, R. E. Lacey and S. Loeb, Eds., Wiley-Interscience, New York, 1972.
2. K. H. Meyer and W. Strauss, *Helv. Chim. Acta.*, **23**, 795 (1940).
3. R. N. Rickles, *Membranes, Technology and Economics*, Noyes Developemnt Co., Park Ridge, N. J., 1967.
4. R. E. Lacey and E. W. Lang, *U.S. Off. Saline Water Res. Dev. Rep.*, **106** (1964).
5. R. E. Lacey and E. W. Lang, *U.S. Off. Saline Water Res. Dev. Rep.*, **398** (1969).

6. R. E. Lacey, *U.S. Off. Saline Water Res. Dev. Rep.*, **80** (1963).

7. M. Bier, "Forced-Flow Electrophoresis and Its Biomedical Applications," in *Membrane Porcesses for Industry*, Proceedings of Symposium sponsored by Southern REsearch Institute, Birmingham, Alabama, May 19-20, 1966.

8. G. A. Dubey, Can. Patent 677,654 (1964); A. J. Wiley and J. M. Holderby, *Pulp Paper Mag. Can.*, **61**, T-212 (1960); G. A. Dubey, T. R. McElhinney, and A. J. Wiley, *Tappi*, **48**, 95 (1965).

9. E. W. Lang and E. L. Huffman, *U.S. Off. Saline Water Res. Dev. Rep.*, **439** (1969).

10. K. Spiegler, *U.S. Off. Saline Water Res. Dev. Rep.*, **353** (1968).

11. K. J. Vetter, *Electrochemical Kinetics, Theoretical and Experimental Aspects*, translated by Scripta Technica, Academic, New York, 1967.

12. R. Schlögl, *Z. Physik. Chem. (Frankfurt)*, **1**, 305 (1954).

13. P. Henderson, *Z. Phsik Chem.*, **59**, 118 (1907); *ibid.*, **63**, 325 (1908).

14. T. Teorell, *Proc. Soc. Exptl. Biol.*, **33**, 282 (1935).

15. K. H. Meyer and J. F. Sievers, *Helv. Chim Acta*, **19**, 649, 665 (1936); K. H. Meyer and H. Mark, *Makromolekulare Chemie*, 3rd ed., Akad. Verlagsanstalt Gest und Portig, Leipzig, 1953.

16. F. Helfferich, *Ion Exchange*, McGraw-Hill, New York, 1962.

17. B. R. Breslau and I. F. Miller, *I. E. C. Fund.*, **10**, 554 (1971).

18. K. S. Spiegler, *U.S. Off. Saline Water Res. Dev. Rep.*, **353** (1968).

19. D. A. Cowan and J. H. Brown, *Ind. Eng. Chem.*, **51**, 1445 (1959).

20. N. W. Rosenberg and C. E. Tirrell, *Ind. Eng. Chem.*, **49**, 781 (1957).

21. W. G. B. Manderslott and R. E. Hicks, *I. E. C. Proc. Des. Dev.*, **4**, 304 (1965).

22. T. A. Davis and G. F. Brockman, "Physicochemical Aspects of Electromembrane Processes," in *Industrial Processing with Membranes*, R. E. Lacey and S. Loeb, Eds., Wiley, New York, 1972.

23. L. H. Schaffer and M. S. Mintz, in *Principles of Desalination*, K. S. Speigler, Ed., Academic, New York, 1966.

24. S. B. Tuwiner, *Diffusion and Membrane Technology*, Reinhold, New York, 1962.

Chapter **X**

BIOMEDICAL APPLICATIONS

The example of the use of membranes in biomedical applications is an indication of technological progress to a point where we can now assist the human body in its constant fight against disease and aging. Surely much of our chemical and technical competence was derived from observations of the animal organism, which represents an unbelievably sophisticated physicochemical system. It is thus gratifying that we have come full circle and are able to repay some of our debt.

Biological membranes, that is, living tissue, will be touched upon only briefly; the main subject of this chapter concerns the use of synthetic membranes in biological applications. Some such uses are: membranes in hemodialysis, organ-preservation devices, controlled release of drugs and medication from implants, and related applications of biomaterials.

1 BIOLOGICAL MEMBRANES

Biological membranes encompass a great variety of structures from eggshells to highly sophisticated colloid systems. The function of every such creation is that of a barrier, usually permeable to some constituents, while excluding others. The truly biological membrane requires the spark of life for its proper functioning and as such is as yet simply out of reach of synthetic imitation.

In living biological membranes the reaction kinetics are extremely complicated and represent the conditions of nonlinear multivariable systems. Additionally, electrochemical phenomena are usually involved in the functioning of such membranes. Actually, the area of living membranes is outside of the scope of a treatise on "membranes in separation" directed toward physical and chemical sciences. However, membrane technology, especially in its recognition of transport phenomena, owes much to the study of biological processes, and thus a limited account of pertinent references is desirable.

A recent paper by Singer and Nicolson [1] views cell membranes as two-dimensional solutions of oriented globular proteins and lipids. An extensive bibliography is of great help to the investigators in this field. An informative generalized discussion of the structure of cell membranes was presented by Fox [2]. The earlier studies of drug transport across membranes by Hogben and co-workers [3-5] contain some fundamental findings. In 1967, the November issue of *Berichte der Bunsegesellschaft* was devoted entirely to "mass transport through membranes in chemistry and biology" with 30 contributions covering the whole spectrum of membranology.

A considerable number of books have been published on the subject of biological membranes and the transport phenomena associated with them. Some of the more recent books are those of Stein [6], who dealt with the movement of molecules across cell membranes, a collaborative volume edited by Chapman [7], which treated structures and functional interpretations, and a modern treatise by Kotyk and Janacek [8], which covered theoretical and experimental aspects of transport in cell membranes.

A fascinating and up-to-date paper by Heldt [9] covers energy exchange in mitochondria. It treats membrane transport of substrates, phosphate, and adeninnucleotides. The 71 citations constitute valuable reference material.

2 MEMBRANES IN BIOMEDICAL APPLICATIONS

The most widely known use of synthetic membranes in medicine today is undoubtedly that of the artificial kidney. This represents a liquid-phase transfer process with constituent flow in both directions. Another example of this type of exchange is the use of plastics encapsulation of drugs and medications.

Many of the other applications of membranes, however, are in gas transport, usually both ways, through membranes. This use is likely to grow in magnitude as organ preservation becomes of increasing importance. Even today the membrane gas exchanger in the heart-lung machine is of great importance and may well displace most other procedures now used in blood oxygenation.

Artificial Kidneys—Hemodialysis

The human kidney has a number of important functions: excretion of waste products such as urea, creatinine, and uric acid from the blood; regulation of water and electrolyte balance; maintenance of acid-base balance; regulation of blood pressure; and formation of red blood cells. The first three functions can be closely imitated by an artificial kidney. Such a kidney consists essentially of a hemodialyzer and its supporting equipment. A hemodialyzer is a membrane-containing apparatus to diayze blood. It is a one-step operation, whereas the natural kidney works in a two-step operation to eliminate the waste materials and keep the balance of electrolytes and water.

In the natural kidney, the first step of regulating the composition of the blood is ultrafiltration through the glomerular capillaries. The glomerular filtrate contains essentially the same concentration of small molecular constituents, but less of the larger molecular substances than that present in the plasma. A tremendous amount of water, 180 liters per day, and dissolved substances are filtered through the nephrons of the two kidneys, although only about 1.5 liters of urine are produced daily. Therefore, most of the material must be reabsorbed into the blood. Over 99% of water, glucose, amino acids, and the ions that are filtered into the nephrons should be reabsorbed; on the other hand, only 40% of the urea in the filtrate is returned to the blood.

In an artificial kidney, the removal of the waste materials is achieved by hemodialysis, which denotes the dialysis of blood through a semipermeable membrane into a dialyzing fluid called dialysate. There are basically four types of hemodialyzers that have been used in clinical application.

The first one was developed by Kolff [10-14] during World War II. It was a rotating drum covered by cellophane sausage tubing. The blood flows through the inside of the tubing, which is in contact with the dialysate so that the waste material can permeate through the membrane from the blood side to the dialysate side according to its own concentration gradient.

The second type is a stationary disposable coil developed by Kolff originally and modified by Alwall [15, 16] to reduce blood priming volume, and it is now available from Travenol Laboratories (Morton Grove, Illinois). Cellophane tubing is wound between two cylinders and compressed to allow both dialysis and ultrafiltration.

The third type of artifical kidney is the parallel horizontal-flow dialyzer, or the plate-and-frame type. Skeggs and Leonards [17, 18] developed a counter-current-flow dialyzer using flat cellophane sheets and rubber mats. Among many other variations of this model, the Kiil dialyzer is the most widely used. It consists of flat cellophane sheets and plastic spacers [19].

Gambro Lundia Dialyzer

A disposable unit was recently developed by a Swedish firm, AB Gambro [20]. The Gambro Lundia dialyzer is a disposable system constructed of 17 layers of plates and membranes sandwiched together just like a filter press, as shown in Fig. 10.1. Cuprophan, a cuprammonium cellophane, is employed as

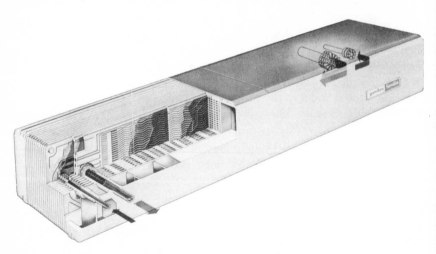

Fig. 10.1   Gambro Lundia dialyzer; flat membrane design [20].

a membrane with a thickness of 18 μ. The total dialyzing area is one square meter per unit. The priming volume is 90 ml at a pressure drop of 30 mm Hg. It costs about $20. The dialyzer can be used at different flow rates in a single-pass system, as well as in a recirculating system. The optimum perfor-mance may be achieved by placing the dialyzer in the vertical position with countercurrent dialysate flow. A typical performance characteristic is shown in

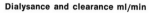

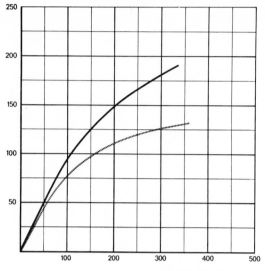

Fig. 10.2 Clearance (dotted curve) and dialysance (solid curve), for Gambro Lundia dialyzer, calculated from Eqs. 10.1 and 10.2; same test at different flow rates, with dialysate flow of 500 ml/min [20].

Fig. 10.2. Here, the clearance and dialysance are calculated by the following equations:

$$C' = \frac{C_{bi} - C_{bo}}{C_{bi}} \, Q_{bo} \qquad (10.1)$$

$$D' = \frac{C_{bi} - C_{bo}}{C_{bi} - C_{do}} \, Q_{bo} \qquad (10.2)$$

where C_{bi}, C_{bo}, C_{do} are the concentrations of the test substance in the blood inlet, in the blood outlet, and in the dialysate outlet, respectively, and Q_{bo} is the blood the flow rate.

Extracorporeal Dialyzer

A disposable hemodialyzer cartridge is commercially available from Extracorporeal Medical Specialties, Inc. A plastic mesh is used as a backing and spacer of a spiral-wound cuprophan membrane. Each unit has an effective membrane surface area of 0.31 to 0.84 m^2.

Ultra-Flow II

A coil hemodialyzer developed by Travenol Laboratories is similar in construction to the Extracorporeal Dialyzer. The triangular screen support reduces the masking effects by minimizing the contact area between membrane and screen. This unit is also disposable.

The fourth type of artificial kidney is the most recent one developed by utilizing hollow fibers of various polymeric materials. The advantage of using hollow fibers is in the fact that a large surface area can be contained in a small volume device. This permits a small blood priming volume and maintains constant blood and dialysate volume regardless of pressure changes.

Cordis Dow Kidney

A commercially available unit is the Cordis Dow hollow-fiber artificial kidney marketed by Cordis Corporation, Miami, Florida. This compact, disposable unit, which also can serve as a hemodialyzer, has been tested clinically [21, 22] and is widely marketed. The cost also is about $20. The overall dimensions are very small as shown in Fig. 10.3. The construction is quite similar to the

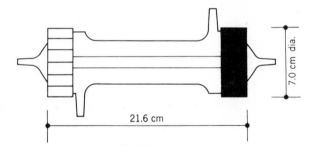

Fig. 10.3 Dimensions of Cordis-Dow kidney; hollow-fiber design [21, 22].

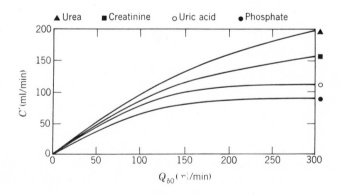

Fig. 10.4 Clearance curves for Cordis Dow kidney.

tube-and-shell type of heat exchanger. Regenerated cellulose is used as fiber material with 225 μ i.d. and 30 μ wall thickness. The total surface area is one square meter per unit. Fourteen thousand such fibers are packed in a rig. transparent plastic container. Typical clearance curves are shown in Fig. 10.4 at a dialysate flow rate of 500 ml/min.

Monsanto Kidney

There are many other companies that are trying to come up with an inexpensive, small, disposable artificial kidney. Figure 10.5 shows the parts of the Monsanto kidney [23]. Monsanto Research Corporation selected modified polyacrylonitrile hollow fibers among many other polymers they tested. Each fiber measures 400 μ o.d. and 300 μ i.d. The amine-modified polyacrylonitrile can be ionically coupled with heparin to provide a nonthrombogenic membrane.

So far, the Monsanto Kidney represents a laboratory development that has been found satisfactory in clinical trials. The main advantage of this apparatus

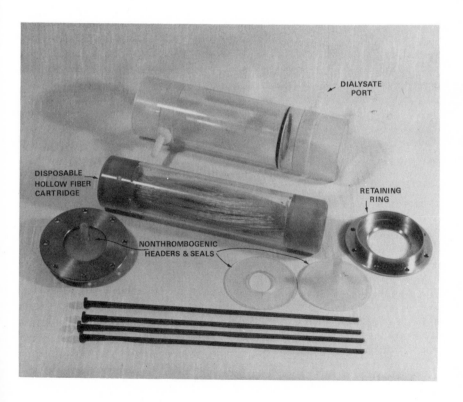

Fig. 10.5 Components of Monsanto kidney. Courtesy of the Monsanto Research Corp.

is the nonthrombogenic nature of all surfaces that are in contact with the patient's blood. A cost estimate is not yet available. The construction of the prototype is such that it can be disassembled for cleaning and sterilization.

Table 10.1   Typical Data for Hemodialysis Treatment on Kiil Artifical Kidney

One suare meter of area using Cuprophane PT-150 membrane
Total dialysis time: 8 hr

Blood flow rate	200 ml/min
Dialysate flow rate	500 ml/min
Pressure of blood side	50 mm Hg
Pressure on dialysate side	-50 mm Hg

Composition of dialysate fluid – all concentrations in mequiv/liter. Total solution used per treatment: 100 gal

Sodium	135	Magnesium	1
Chloride	103	Potassium	1.2
Acetate	35	Calcium	3

Concentrations from patient's blood samples

mequiv/liter Sodium	pre	143	Hematocrit (vol %)		14
	post	139	Hemoglobin (g %)		4.9
mequiv/liter Potassium		6.9	White blood cells		9900
		4.6			
mequiv/liter Calcium		9.7	Serum iron (μg %)		55
		10.0			
mequiv/liter Phosphate		7.7	Total iron binding capacity		
		5.8	(μg %)		340
			Clotting time		2 min
mequiv/liter Magnesium		2.1	Supine blood pressure	pre	138/84
		1.9	(mm Hg)	post	122/80
mequiv/liter Chloride-		93	Standing blood pressure		118/64
bicarbonate		19	(mm Hg)		
mg % Blood urea nitrogen		100	Pulse (per min)	pre	92
		43		post	84
mg % Creatinine		15.6	Temperature (°C)	pre	35.5
		8.0		post	36.9
mg % Uric acid		10.1	Weight (kg)	pre	84.9
		4.8		post	82.0
mg % Total protein-albumin		7.2			
		3.9			
mg % Chloesterol		198			
mg %Glucose		111			

General Considerations

In the development of the artificial kidney, the primary importance is to search for a new membrane that will selectively separate the waste materials from blood. The equipment design is also quite important. How can one improve the efficiency of the hemodialyzer by reducing the stagnation and turbulence of blood flow? How do we increase the area of dialysis, keeping the total volume at a minimum? In addition, the following features are desirable: minimum blood trauma, high permeability, disposability after use, ease of sterilization before use, thromboresistant biocompactibility.

The development of heparin as an anticoagulant was instrumental for the wide application of the present artificial kidney. Equally important was the production of modified regenerated cellulose membranes. Another indispensable factor in the promotion of the artificial kidney is the development of the shunted plastic tubes that will connect a patient to the dialyzer. This device made frequent dialyses possible without surgery at each dialysis treatment.

Finally, it should be mentioned that the artificial kidney has also been used to remove drugs, poisons, and Strontium 90 from the blood and to treat schizophrenia, in addition to the normal application for chronic kidney disorders.

Clinical Record

Routine hemodialysis is today an accepted fact. Patients in all stages of kidney disease, or without kidneys, are being treated in numerous hospital facilities. Also, the use of home dialysis units is growing. An example of an artificial kidney treatment is presented in Table 10.1. The patient was a white male, age 50, with chronic renal disease, being dialyzed three times per week for 8 hr at a session. This particular patient had an implanted external shunt and also an internal arterial-venous shunt. The latter represents a direct connection of an artery with a vein in one of the extermities. Because of the higher pressure in the artery, a bulging out of the vein (called a fistula) forms at the shunt location. The patient is connected to the kidney by insertion of a hypodermic needle into the fistula, and blood return in accomplished by another needle into the same fistula at locations that can easily be several centimeters apart.

3 LIQUID PHASE – GAS TRANSPORT

The artificial kidney process involves transport in liquid phase, and the moving constituents are primarily dissolved solids such as urea, proteins, and salts. Coincidentally some gas exchange may take place, but this is not of great importance.

The exchange in the other devices, that is, blood oxygenators and the like, however, is in essence dissolved gas transport.

Dissolved Gas Permeation

When a liquid is divided into two parts by a membrane and each side contains a different concentration of dissolved gas, there will be permeation of the dissolved gas from the high-concentration side to the low-concentration side. The gas permeation via the liquid-membrane-liquid system is especially interesting on two accounts: first, it gives the information on the boundary resistances and possible permeation mechanisms through the membrane; second, it has wide applications in bioengineering such as in blood oxygenators.

As dissolved gas permeates through a membrane in the liquid phase, the general phenomenological treatment discussed in Chapter II should apply. The final equation in the thickness analysis, Eq. 2.15, is given as

$$\frac{1}{Q} = \frac{1}{DS_m} + (r_1 + r_2)\frac{1}{l} .$$
(10.3)

A system of dissolved oxygen in water [24] exhibits this type of permeability variation with the thickness of silicone-rubber membrane as shown in Fig. 10.6.

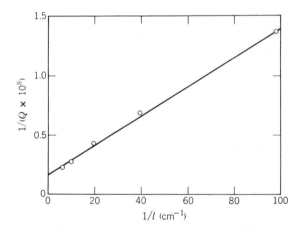

Fig. 10.6 Transport of dissolved oxygen; Analysis of effect of membrane thickness (silicone rubber) on surface resistance [26]; Q = permeability, l = thickness.

The intercept of the straight line gives the limiting value of permeability at infinite thickness. The value is 6.0×10^{-8} in standard units, which is very close to the gas-phase permeability, suggesting that the transport mechanism of oxygen through silicone rubber is about the same as in the case of gas-phase permeation. The slope of the straight line represents the resistances of the boundary layers. It is surprising that the boundary resistances for a 4-mils-thick silicone-rubber membrane amounted to 89% of the total resistance.

Yasuda [25] reported the permeabilities of dissolved oxygen through varieties of polymer membranes. However, the data do not permit the analysis of boundary resistances, because they are taken only for one thickness of the membrane.

A similar study was done for a carbon dioxide system [26]. The result is shown in Fig. 10.7. From the intercept of the straight line, the limiting value of permeability at infinite thickness was obtained as 302×10^{-9}, which is again the same as the gas-phase permeability. The resistances of the boundary layers are not so great as those for the dissolved oxygen system.

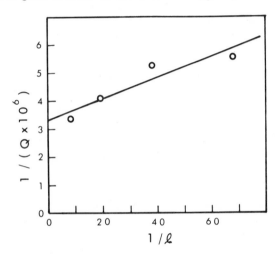

Fig. 10.7   Same as Fig. 10.6, but for dissolved CO_2 [26].

Blood Oxygenators

There is a growing need in clinical medicine for blood oxygenation. Some of these applications are: patients with reversible pulmonary disease (posttransplant lung) in respiratory failure needing temporary supplemental oxygenation; improvement of blood oxygenation for cardiac support as a temporary life-support system; oxygenation of a dynamic system for organ preservation; and partial or total implantable surrogate lung.

Four different types of oxygenators have been used clinically.

Bubble Oxygenator

The simplest kind of oxygenator is one where air is bubbled through blood. The gas exchange is very efficient, but considerable foaming occurs and the small bubbles are difficult to remove from blood after oxygenation. Also, protein denaturation is high, and blood cells are damaged.

Screen Oxygenator

When blood is allowed to flow over a vertical wire-mesh screen, a mild turbulence is produced in the descending blood film, and it causes increased oxygen transfer to blood. The main disadvantages of this arrangement are its large priming volume and difficulty of cleaning.

Disk Oxygenator

When vertical disks revolving on a horizontal axis dip into a trough of blood, thin films of blood will coat the disks, and they are exposed to the air to be oxygenated. This blood film is rapidly and constantly renewed. The disk oxygenator also requires large priming volume, and it tends to produce foaming at the higher rates of disk rotation.

Membrane Oxygenator

The direct contact of blood and oxygen in the above oxygenators produces the following undesirable side effects [27-29]: blood-cell damage, protien denaturation, and microscopic gas emboli. These can be prevented if the blood is separated from the gas by means of a membrane that has high permeabilities to oxygen and carbon dioxide.

There are basically two types of membrane oxygenators: one with sheet membranes and the other with capillary membranes. As far as the design and construction are concerned, the membrane oxygenators are much like the artificial kidneys. The flat-plate-and-sheet membrane oxygenator has been constructed and used as a Kiil dialyzer. The spiral-coil membrane oxygenator is made just as a spiral-coil artificial kidney. Also, capillary membrane oxygenators have been constructed similar to the capillary dialyzers of the Cordis Dow or Monsanto kidneys.

A major difference is in the selection of the membrane. In the case of a dialyzer, one should choose a membrane that can selectively pass the waste material from the blood: most frequently the regenerated cellulose membrane is used for this purpose. However, in the case of an oxygenator, the membrane should possess high gas permeabilities. Silicone rubber is the best material found so far [30, 31] compared to the most permeable nonsilicone plastics. It is also highly biocompatible [32-35], can be autoclaved, and if necessary, the surface can be heparinized [32, 33]. The only shortcoming is that its tensile strength is not very high.

The earlier work by Kolff [36] and Clowes [37, 38] led to more sophisticated recent developments [39-46] of membrane oxygenators. Considerable improvement has already been made on the membrane-supporting systems, which is one of the major difficulties with membrane oxygenators. In order to make a compact unit, spiral-coil membrane oxygenators [47] and capillary bundle oxygenators [48-51] were fabricated.

Liquid-Membrane Oxygenator

Finally, this newest type of oxygenator should be mentioned. This technique is one of the applications of the new separation method utilizing liquid surfactant membranes developed by Li (see Chapter VIII). It is still in the experimental stage. The idea is to produce fluorocarbon liquid membranes around oxygen bubbles in blood and have the gaseous exchange take place through these liquid membranes. The advantages are high rates of gas permeation and elimination of blood hemolysis and protein denaturation. After the oxygenation is done, the encapsulated bubbles are collected in a separatory funnel to be broken down and separated.

4 CONTROLLED RELEASE OF DRUGS BY ENCAPSULATION

The technique of drug encapsulation has received much attention in recent years. It provides a uniform and long-lasting dosage form for many nutrients and drugs. The more conventional ways of drug administration, injection or oral tablets, frequently cause undesirable side effects, as the concentration of drugs in the body increases sharply yielding a peak, which is considered to be responsible for the side effects. In other situations, it is desired to have large quantities of drugs in the body and be released slowly over a long period of time. Examples are contraceptive devices containing progesterone and hormone implants to make up the deficiency caused by endocrine failure. The other advantages of drug implants are effective local administration where the drug is needed and in some cases when it is needed by having a builtin trigger mechanism. For instance, body chemistry changes in the presence of disease. Thus, if this change can trigger the release of a drug, it can be used very efficiently to control the disease without undesirable side effects.

Numerous polymeric materials have been studied for implants. The basic requirements are

1. Materials should be biocompatible.
2. They must have high permeabilities of drugs.

It is well known that silicone rubber is highly biocompatible [32-35] and also possesses very high permeabilities [53-58] for many drugs and hormones. Therefore, it is logical to investigate its usefulness in drug encapsulation.

A recent study of testosterone encapsulation by silicone-rubber membrane [59] shows that the release rate can be effectively controlled by changing the thickness of the membrane and also by modifying the membrane structure. Pellets of crystalline testosterone were made using a press and set of dies. These pellets were then coated by a thin layer of silicone-rubber membrane. By measuring the permeation rate of testosterone, the thickness effect upon

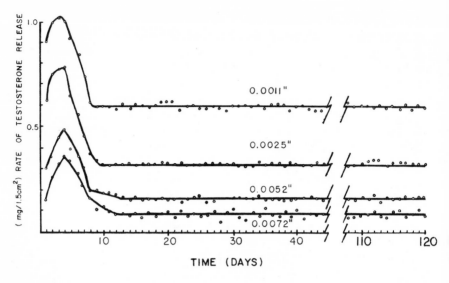

Fig. 10.8 Testosterone release from encapsulated pellets; effect of silicone-rubber membrane [59].

the hormone release was established as shown in Fig. 10.8. After the initial unsteady-state permeation, the release rate remains essentially constant over a

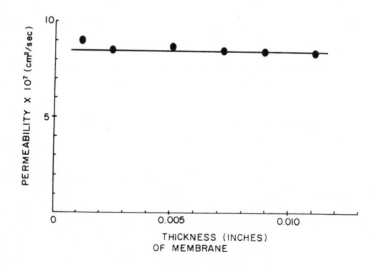

Fig. 10.9 Permeability of testosterone through silicone rubber is independent of membrane thickness.

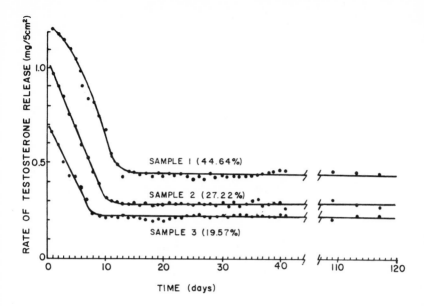

Fig. 10.10 Release of testosterone from mixture with silicone rubber as a function of original percent in membrane mixture.

long period of time. These steady-state levels of release rates confirm the permeation equation, which states that the permeation rate is inversley proportional to the membrane thickness:

$$F = -QS_m \frac{\Delta C}{l} . \tag{10.4}$$

However, the permeability constant is invariant with thickness, as shown in Fig. 10.9

Instead of using pure silicone rubber as the membrane material, mixtures of testosterone and silicone rubber may be employed to increase the permeation rate. The idea is to provide a porous membrane so that testosterone can leach out through the holes more readily than through the solid membrane. Results are summarized in Fig. 10.10 for testosterone contents up to 45%. The figure illustrates the steady-state release rate after the initial unsteady-state dies out. The ultimate levels of steady release show a definite increase in the testosterone flux with increasing percentage of testosterone; see Fig. 10.11.

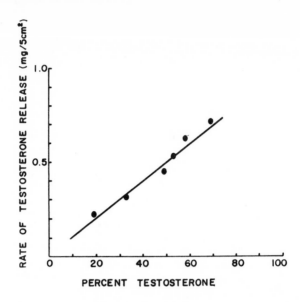

Fig. 10.11   Rate of testosterone release as a function of testosterone content from mixture with silicone rubber.

5 DIALYSIS BY SEMIPERMEABLE MICROCAPSULES

The microencapsulation technique used in microdialysis represents a novel development. Chang [60, 61] has devised a procedure in which semipermeable microcapsules are prepared with an ultrathin semipermeable membrane to permit rapid mass transfer. In ordinary hemodialysis, an important concern is to attain a large membrane area in a small bulk volume. This is easily achieved by microencapsulation.

For example, 10 ml of 20-μ diameter microcapsules can have a total surface area of 25,000 cm^2; this area is larger than that contained in an ordinary artificial kidney. In addition, the membrane can be made as thin as 500Å, whereas the thinnest membrane employed in an artificial kidney is at least 50,000Å.

Also, Chang suggests that microdialysis could be readily combined with other separation processes such as adsorption or catalysis, or both. When the microcapsules contain adsorbent, microdialysis and selective adsorption will take place. In the place of adsorbents, enzymes can be encapsulated to give combined microdialysis and selective catalysis. If both enzyme and adsorbent are included in the microcapsules, all three processes (dialysis, catalysis, and adsorption) will take place simultaneously.

In order to remove waste materials from the blood stream, Chang [62, 63] tested activated charcoal, ion-exchange resins, activated zirconium charcoal, and zirconium phosphate as possible adsorbents. When these are coated with a thin membrane, which permits only the permeation of toxic materials, the microcapsules of adsorbents selectively separate and remove the toxic materials without interacting with plasma proteins and platelets. The removal of creatinine and uric acid was very efficient by the microcapsules of activated charcoal; however, the removal of urea was not significant. The reason is that urea is not adsorbed by activated charcoal even though urea can permeate the coating membrane.

Combined action of microdialysis and catalysis was employed to remove urea. Microencapsulated urease was found to be effective in adsorbing urea and converting it into ammonia [61, 62], thus removing it from the blood stream. Such formed ammonia then should be removed from blood. This was accomplished by adding selective adsorbents such as ion-exchange resins or zirconium phosphate.

This microencapsulation technique demonstrates its usefulness and flexibility not only in hemodialysis but also in any other separation processes. By choosing appropriate membranes, adsorbents, and catalysts, one can achieve various degrees of separation for many different systems.

Nomenclature

Symbol	Meaning
C'	Clearence, as defined by Eq. 10.1
C	Concentration
C_{bi}	Concentration in blood inlet
C_{bo}	Concentration in blood outlet
C_{do}	Concentration in dialysate outlet
D'	Dialysance, as defined by Eq. 10.2
D	Diffusivity
F	Flux
Q	Permeability
Q_{bo}	Blood flow rate
S	Solubility
S_m	Solubility in membrane
l	Membrane thickness
r	Boundary-layer resistance

References

1. S. J. Singer and G. L. Nicolson, *Science*, **175**, 720 (1972).
2. C. F. Fox, *Sci. Amer.*, **226**, (2) 31 (1972).
3. B. B. Brodie and C. A. M. Hogben, *J. Pharm. Pharmacol.*, **9**, 345 (1957).
4. C. A. M. Hogben, *Handbuch der experimentellen Pharmakologie*, B. B. Brodie and J. Gillette, Eds., Vol. 28/1, Springer Verlag, Berlin-New York, 1971.
5. C. A. M. Hogben, *Fed. Proc.*, **19**, 864 (1960).
6. W. D. Stein, *The Movement of Molecules across Cell Membranes*, Academic, New York, 1967.
7. D. Chapman, Ed., *Biological Membranes*, Academic, New York, 1968.
8. A. Kotyk and K. Janacek, *Cell Membrane Transport—Principles and Techniques*, Plenum, New York, 1971.
9. H. W. Heldt, *Agnew. Chem.*, **84**, 792 (1972).
10. W. J. Kolff, H. T. J. Berk, M. Ter Welle, J. W. van der Leg, E. C. van Dijk, and J. van Noordwijk, *Acta Med. Scand.*, **117**, 121 (1944).
11. W. J. Kolff, *Arch. Neerl. de Physoil.*, **28**, 166 (1946).
12. W. J. Kolff, *Belg. Tejdschr. Geneesk*, **2**, 449 (1946).
13. W. J. Kolff, *J. Mt. Sinai Hosp.*, **14**, 171 (1947).
14. W. J. Kolff, *Geneesk, gids.*, **26**, 25 (1948).
15. N. Alwall, *Acta Med. Scand.*, **128**, 317 (1947).
16. N. Alwall, *Acta Med. Scand.*, **196**, 250 (1947).
17. L. T. Skeggs, Jr. and J. R. Leonards, *Sci.*, **108**, 212 (1948).
18. L. T. Skeggs, Jr., J. R. Leonards, and C. R. Heisler, *Proc. Soc. Exper. Biol.*, **72**, 539 (1949).
19. F. Kiil, *Acta Chir. Scand.*, **253**, 142 (1960).
20. T. Lindholm, C. Gullberg, and A. Akerlund, *Scand. J. Urology Nephrology*, *Suppl. 13* (1972); also Bulletins of Gambro, Inc., Wheeling, Ill.
21. R. D. Stewart, B. J. Lipps, E. D. Baretta, W. R. Piering, W. R. Roth, and J. A. Sargent, *Trans. Amer. Soc. Artif. Intern. Organs*, **14**, 121 (1968).
22. F. Gotch, B. J. Lipps, J. Weaver, Jr., J. Brandes, J. Rosin, J. Sargent, and P. D. Oja, *Trans. Amer. Soc. Artif. Intern. Organs*, **15**, 87 (1969); also Bulletins of Cordis Corp., Miami, Fla.
23. I. O. Salyer, A. J. Blardinelli, L. Ball, W. E. Weesner, V. L. Gott, M. D. Ramos, and A. Furuse, *J. Biomed. Matl. Res. Symp.*, **1**, 105 (1970).
24. S-T. Hwang, T. S. Tang, and K. Kammermeyer, *J. Macromolecular Sci. – Phys. B*, **5**, 1 (1971).
25. H. Yasuda, *J. Polymer Sci. A-1*, **5**, 2952 (1967).
26. S-T. Hwang and G. Strong, 164th ACS New York Meeting (1972), **32** (2) Organic Coatings and Plastics Chemistry, 125 (1972).
27. W. H. Lee, D. Krumhaar, G. Derry, D. Sachs, S. H. Lawrences, G. H. A. Clowes, and J. F. Maloney, *Surg. Form*, **12**, 200 (1961).
28. M. A. Shea, R. A. Indeglia, F. D. Dorman, J. F. Haleen, P. L. Blackshear, R. L. Varco, and E. F. Bernstein, *Trans. Amer. Soc. Artif. Int. Organs*, **13**, 116 (1967).

29. M. A. Shea, R. A. Indeglia, and E. F. Bernstein, *Surg. Forum*, **19**, 133 (1968).
30. W. L. Robb, General Electric Report, No. 65-C-031, Schenectady, New York, 1965.
31. P. M. Galletti, M. T. Snider, and S. A. Daniele, *Med. Res. Eng.*, **5**, 20 (1966).
32. G. A. Grode, S. Anderson, and R. D. Falb, *USGRDR*, **70**, 33, PB 188 111 (1970).
33. P. B. Halbert, M. Anken, and A. E. Ushakoff, *USGRDR*, **70**, 33, PB 188 117 (1970).
34. V. L. Gott, *USGRDR*, **69**, 45, PB 186 551 (1969).
35. M. C. Musolf, V. Metevia, and V. D. Hulse, *USGRDR*, **70**, 48, PB 190 666 (1970).
36. W. J. Kolff, D. B. Effler, L. J. Groves, G. Perreboom, and P. P. Moraca, *Cleveland Clin Quart.*, **23**, 69 (1956).
37. G. H. A. Clowes, Jr., A. L. Hopkins, and W. E. Neville, *J. Thoracic Cardiovas. Surg.*, **32**, 630 (1956).
38. G. H. A. Clowes, Jr. and A. L. Hopkins, *Trans. Amer. Soc. Artif. Int. Organs*, **1**, 6 (1956).
39. P. M. Galletti and G. A. Brecher, *Heart-lung Bypass*, Grune and Stratton, New York, 1962, p. 50.
40. T. Kolobow, and R. L. Bowman, *Trans. Amer. Soc. Artif. Int. Organs*, **9**, 238 (1963).
41. E. C. Peirce II and G. Peirce, *J. Surg. Res.*, **3**, 67 (1963).
42. R. G. Buckles, "An analysis of gas exchange in a membrane oxygenator," Ph.D. thesis, Massachusetts Institute of Technology, Cambridge, Mass., 1969.
43. A. J. Landé, R. G. Carlson, R. A. Pershau, R. P. Lange, L. J. Sonstegard, and C. W. Lillehei, *Surg. Clin. N. Amer.*, **47**, 1461 (1967).
44. A. J. Landé, B. Parker, V. Subramanian, H. Solomon, A. Edwards, T. Killip, and C. W. Lillehei, *Circulation*, **38**, 121 (1968).
45. A. J. Landé, S. J. Fillmore, V. Subramanian, R. N. Tiedmann, R. G. Carlson, J. A. Bloch, and C. W. Lillehei, *Trans. Amer. Soc. Artif. Int. Organs*, **15**, 181 (1969).
46. F. Gerbode, J. J. Osborn, and M. L. Bramson, *Amer. J. Surg.*, **114**, 16 (1967).
47. T. Kolobow, W. Zapal, and J. Marcus, in *Organ Perfusion and Preservation*, John C. Norman, Ed., Appleton-Century-Crofts, New York, 1968.
48. B. R. Bodell, J. M. Head, L. R. Head, A. J. Formolo, and J. R. Head, *J. Thoracic Cardiovas. Surg.*, **46**, 639 (1963).
49. R. P. DeFilippi, F. C. Tompkins, J. H. Porter, and G. W. Harris, *USGRDR*, **70**, 48, PB 187 291 (1970).
50. P. Dantowitz, A. S. Borsanyi, and M. C. Deibert, *USGRDR*, **70**, 48, PB 187 293 (1970).
51. T. L. Williams, L. H. Bosher, Jr., W. R. Harlan, Jr., and L. McA. Fisher, *USGRDR*, **70**, 48, PB 187 303 (1970).
52. J. P. Merril, *Sci. Amer.*, **205**, 56 (1961).

53. P. Dzuik and B. Cook, *Endocrinology*, **78**, 208 (1966).
54. E. Garrett and P Chemburkar, *J. Pharm. Sci.*, **57**, 949 (1968).
55. B. J. Culliton, *Science News*, **95**, 555 (1969).
56. J. Folkman and V. H. Mark, *Trans. N. Y. Acad. Sci.*, **30**, 1187 (1968).
57. F. Kincl, G. Benagiano, and I. Angee, *Steroids*, **11**, 673 (1968).
58. S-T. Hwang, R. J. Shea, K. H. Moon, and R. G. Bunge, *Invest. Urol.*, **8**, 245 (1970).
59. R. L. Shippy, S-T. Hwang, and R. G. Bunge, *J. Biomed. Mat. Res.*, **7**, 95 (1973).
60. T. M. S. Chang, *Science*, **146**, 524 (1964).
61. T. M. S. Chang, *Artificial Cells*, Charles C. Thomas, Springfield, Ill., 1972.
62. T. M. S. Chang, *Trans. Amer. Soc. Artif. Intern. Organs*, **12**, 13 (1966).
63. T. M. S. Chang, *Can. J. Physiol. Pharmacol.*, **47**, 1043 (1969).

General References

Artificial Kidney Bibliography, published by National Institute of Arthritis and Metabolic Diseases, obtainable from U.S. Government Printing Office, Washington, D.C. 20402

Dialysis and Transplantation, The Journal of Renal Technology, Vol. 1, 1972; published by Dialysis and Transplantation, 15300 Ventura Blvd., Sherman Oaks, California 91403.

Journal of Biomaterials Research, Sumner N. Levine, Ed., Interscience, New York.

Acta Medicotechnica, Helios Verlag, 1 Berlin 52 (Borsigwalde), Germany.

Branton, D., and D. W. Deamer, *Membrane Structure*, Springer Verlag, New York, 1972.

Chapter **XI**

TRANSFER COEFFICIENTS IN MEMBRANES

Transfer coefficients are involved in every membrane process. They express the rate of transfer of a given component through the membrane under a given set of conditions. In effect, when all of the units in a rate coefficient are properly specified, then the transfer equation itself is given. For instance, if the permeabiltiy of a particular compound through a membrane is stated to be as follows:

permeability of liquid or vapor:

$$Q = (\text{a number}) = \frac{(\text{cc})(\text{cm thickness})}{(\text{hr})(\text{cm}^2 \text{ area } S)(\Delta' \text{ driving force})},$$

the equation for the *flux,* that is, quantity transported per unit time and per unit area, is simply

$$\text{flux} = \frac{\text{cc}}{(\text{hr})(\text{cm}^2)} = \frac{(Q)(\Delta')}{(l)},$$

where driving force Δ' is expressed in any desired units, as, for instance, partial pressure difference or concentration difference.

Obviously, a knowledge of transfer coefficients becomes an elementary necessity. If they can be related to properties of the moving substance and the stationary membrane, valuable and convenient correlations may be established, In many membrane processes the components of a mixture move at their own rates, unencumbered by the presence of other species, and thus a knowledge of transfer coefficients of pure components is often enough to permit calculations of a separation process. An example of this behavior is gas permeation through plastic films as treated in Chapter V.

1 BASIC TRANSPORT CONCEPTS

The transport of fluids across membranes can be characterized by the fundamental relationship

$$Q = \overline{D} S_m \qquad (11.1)$$

when there is no significant amount of boundary resistances. As previously stated, Q is the permeability, $\overline{D}$ the diffusivity, and S_m the solubility in the membrane. For the sake of convenience the term *solubility* will be used to encompass amount of sorption per unit volume of a membrane. With this definition the product form of permeability is then quite useful for microporous membranes and the membranes consisting of polymers, glasses, crystals, and metals.

The diffusivity is a measure of how fast the permeating component goes through the membrane, whereas the solubility signifies how much of the permeating component is present in the membrane. The former is a dynamic quantity (transport coefficient), whereas the latter is an equilibrium property.

The overall permeation rate, therefore, depends equally upon both of these quantities. Even though the diffusivity is high, that is, the rate of diffusion through the membrane is fast, if the solubility is low, that is, the amount of permeating component in the membrane is small, the net flow rate may not be great. Likewise, if the solubility is high but the diffusivity is low, the overall permeation rate may also be slow. In order to have a high permeation rate, hence, both the diffusivity and solubility should be large. Here, the general background of these two quantities will be discussed together with some examples.

Applying the theory of absolute rates to the diffusion process, we can obtain the temperature dependency of diffusivity. The Arrhenius-type relationship has long been observed experimentally, but Eyring and co-workers [1] were the first to derive the absolute value of diffusivity based on the model of an activated state. They treated the diffusion process as one of the more general processes in which every molecule has to go through an unstable intermediate state before any change takes place. Their expression for the diffusivity is

$$\overline{D} = \lambda^2 \ \frac{kT}{h} \ \frac{f^{\ddagger}}{f} \ \exp \ \frac{-E_0}{kT} \ , \tag{11.2}$$

where $f^{\ddagger}$ and f are the partition functions of a diffusing molecule in the activated and normal states, respectively, and E_0 is the activation energy per molecule at $0°K$. Changing the units to a mole basis and using thermodynamic relations, the above equation can be re-expressed as

$$\overline{D} = \lambda^2 \ \frac{kT}{h} \ \exp \ \frac{\Delta S^{\ddagger}}{R} \ \exp \ \frac{-\Delta H^{\ddagger}}{RT} \tag{11.3}$$

or

$$\overline{D} = e\lambda^2 \ \frac{kT}{h} \ \exp \ \frac{\Delta S^{\ddagger}}{R} \ \exp \ \frac{-E_D}{RT} \ , \tag{11.4}$$

where $\Delta S^{\ddagger}$ is the entropy of activation, and E_D, is the observed activation energy.

The last equation is the theoretical basis of the Arrhenius relationship:

$$\overline{D} = \overline{D}_0 \ \exp \ \frac{-E_D}{RT} \ , \tag{11.5}$$

where

$$\overline{D}_0 = e\lambda^2 \, \frac{kT}{h} \, \exp \frac{\Delta S^{\ddagger}}{R} \, . \qquad (11.6)$$

Over a range of small temperature changes, the pre-exponential term is much less temperature sensitive than the exponential term, thus $\overline{D}_0$ stays approxi-

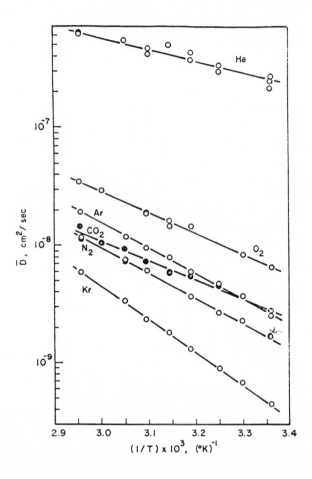

Fig. 11.1 Temperature dependence of diffusivity of helium, oxygen, argon, carbon dioxide, nitrogen, and krypton in PMD. (Modified PVC containing N-methyl dithiocarbamate [2].)

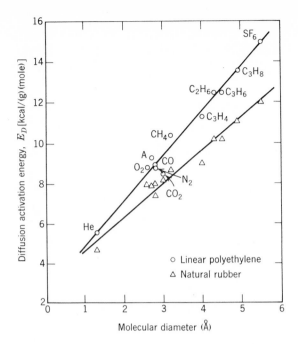

Fig. 11.2 Correlation of activation energies for diffusion of gases in rubber and polyethylene [3].

mately constant as temperature varies. The activation energy is a function of both the diffusing molecule and the membrane. It is also, in general, temperature dependent. However, if the temperature range is small, the variation is so small in many cases that a constant value is often used for the activation energy. This fact is well illustrated in a semilogarithmic plot of the gas diffusivity versus the inverse temperature as shown in Fig. 11.1 [2]. The constant slope of the straight line means that the activation energy is a constant in this temperature range. For a given membrane, the activation energy for diffusion increases as the size of the diffusing molecule increases. An example is shown in Fig. 11.2 for polyethylene and natural-rubber membranes [3].

Analogously, the solutibility expression becomes

$$S_m = S_0 \ \exp \frac{-\Delta H}{RT} , \qquad (11.7)$$

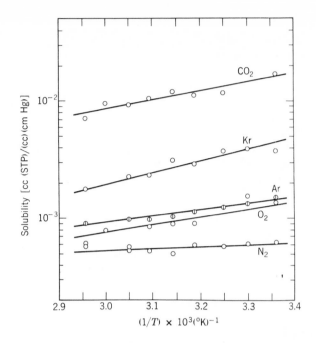

Fig. 11.3 Temperature dependence of solubility of carbon dioxide, nitrogen, krypton, argon, and oxygen in PMD [2].

where ΔH is the heat of solution. In many systems, the solubility changes, but not as pronounced as the diffusivity does with changing temperature. An example is given in Fig. 11.3 from work of Stannett and co-workers [2].

In general, the solubility of a component A in the medium B is determined by the affinity of A and B. The interactions between molecules A and B are caused by the dispersion force. Therefore, it is not difficult to imagine that the greater the force constant is, the higher the solubility becomes. This fact was well illustrated originally by Jolly and Hildebrand [4], and the same type of plot is shown here in Fig. 11.4 from Michaels and Bixler [3]. Similarly, the enthalpy of solution of a gas in the polymer was also related to the Lennard-Jones parameter as shown in Fig. 11.5. Combining the two equations of diffusivity, Eq. 11.5, and of solubility, Eq. 11.7, into the permeability equation, Eq. 11.1, the following expression is obtained:

$$Q = Q_0 \ \exp \frac{-E_Q}{RT} \ , \tag{11.8}$$

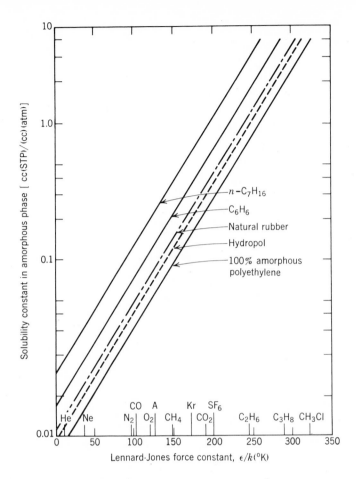

Fig. 11.4 Jolley-Hildebrand correlation of solubility constants (25°C) of gases in amorphous polymers [3].

where E_Q is the activation energy for permeation:

$$E_Q = E_D + \Delta H .\qquad(11.9)$$

The temperature dependence of the overall permeability is illustrated in Fig. 11.6, again from Stannett et al. [2].

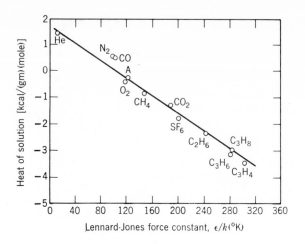

Fig.· 11.5 Jolley-Hildebrand correlation of heats of solution of gases in amorphous polyethylene [3].

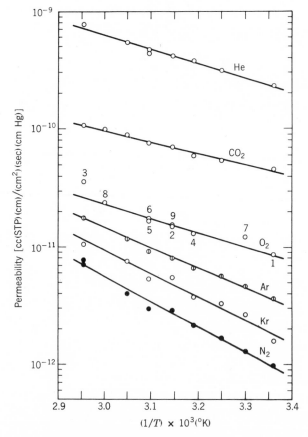

Fig. 11.6 Temperature dependence of permeability of PMD to helium, carbon dioxide, oxygen, argon, krypton, and nitrogen [2].

2 CORRELATIONS, GAS PHASE

It is only natural that attempts have been made to establish correlations of transfer coefficients and system parameters. It also should not be surprising that such attempts have been only marginally successful. The relatively large number of factors that is involved, especially those of membrane structure, introduces an appreciable degree of complexity.

So, as far as generalized correlations are concerned there is a dearth of information. Fairly useful procedures are those of Hwang [5] for surface flow in microporous media and of Li and Henley [6] for gas and vapor permeation through a polymer. But even these correlations tend to be specific rather than general. A fair number of correlations have been developed for particular polymeric membranes: some of the more useful studies are those of Michaels and Bixler [7], Gee [8], Klute [9], Norton [10], and Stern [11]. A rather comprehensive collection of permeability data was correlated by Othmer and Frohlich [12] by relating the permeabilities to the vapor pressures of reference substances. The final arrangement resulted in a convenient nomographic plot. In essence, this method is a collation of experimental data form the literature.

Microporous Membranes

A rather instructive picture of a generalized relationship is shown in Fig. 11.7, where permeability values are plotted as $Q M^{1/2}$, extrapolated to zero

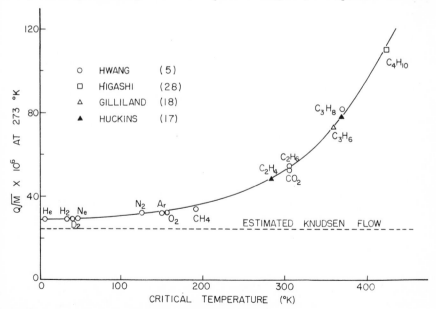

Fig. 11.7 Generalized permeability correlation, microporous Vycor glass system [13]: ○ Hwang [5]; □ Higashi [28]; △ Gilliland [18]; ▲ Huckins [17].

pressure, against a physical property of the gas, that is, critical temperature [13]. Similar curves would be obtained if the parameter were taken as the atmospheric boiling point or the Lennard-Jones potential. Compared with earlier data, the most significant development is the recognition of the existence of surface flow of helium. This places the basic molecular flow line well below the previously accepted basis. As indicated in the figure the base line is considered to be at an estimated value, but it is located reasonably close to what should be a true value. Because the plot is for the quantity $QM^{1/2}$, the base line represents the Knudsen flow for all compounds. Consequently, the difference between the base line and the curve gives an estimate of the surface flow. The curve should be used with some degree of caution. It is too much to expect that such a relatively simple correlation would be exact.

Microporous Vycor glass has an average pore diameter of about 50 Å and a pore volume of about 32%. The quantity Q is expressed in the units: (std cc) (cm thickness)/(sec)(cm^2 area)(cm Hg pressure differential). This definition is the most commonly used one. The solid matrix acts, in principle, as a chemically inert component. It exerts an influence on the surface flow only because it is a capillary system and thus leads to adsorption. Consequently, the behavior of *porous Vycor* glass can be expected to be representative of that of any

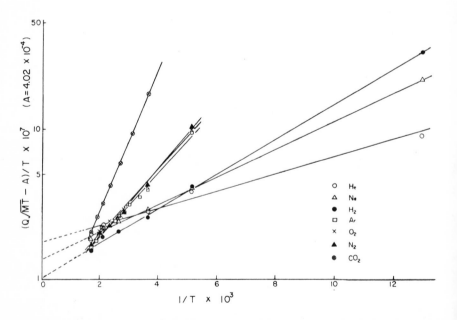

Fig. 11.8 Correlation of surface diffusion in microporous Vycor glass. This figure is reprinted from Ref. 5, through courtesy of The Canadian Journal of Chemical Engineering.

microporous medium. This means that similar correlations should be obtainable for any other microporous solid matrix.

When the surface coverage of the adsorbed molecules is low and the Knudsen regime prevails, it has been shown by Hwang [5] that the best correlation can be achieved for the flow data of many gases through microporous Vycor glass by the following equation:

$$Q(MT)^{\frac{1}{2}} = A + B T \exp \frac{\Delta}{T} . \qquad (11.10)$$

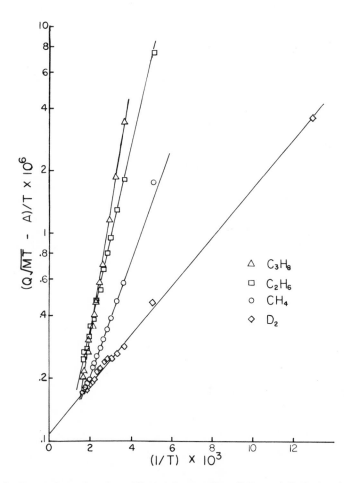

Fig. 11.9 Correlations of surface diffusion for D_2, CH_4, C_2H_6, and C_3H_8 in microporous Vycor glass. Reprinted from Ref. 14, by courtesy of Marcel Dekker, Inc.

Here A represents the Knudsen flow, which thus remains constant for a given microporous medium. The second term represents the surface flow. The values of B and Δ vary from gas to gas. Hwang showed how these values could be estimated. The final result of the surface diffusion correlation is exemplified by Fig. 11.8 and 11.9 for two sets of gases. The slope of the straight line in these figures is the value of Δ, which is defined by

$$\Delta = \frac{\epsilon^* - \epsilon^{\ddagger}}{k} , \tag{11.11}$$

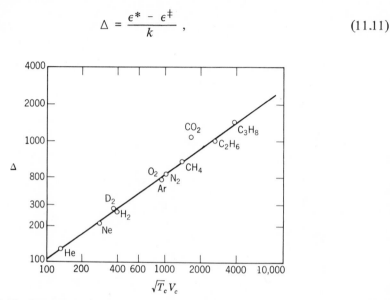

Fig. 11.10 Correlation of Δ with the critical properties [15]. Courtesy of the American Institute of Chemical engineers.

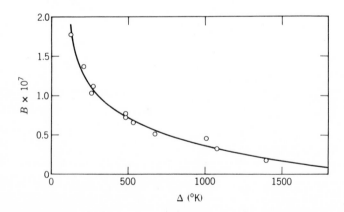

Fig. 11.11 Interrelationship between B and Δ [15]. Courtesy of the American Institute of Chemical Engineers.

where ϵ^* is the gas-solid interaction energy, and $\epsilon^{\ddagger}$ is the activation energy of surface diffusion for an adsorbed molecule. Therefore, the value of Δ is a measure of the interaction energy between the gas molecule and the solid surface. An excellent correlation was obtained between the values of Δ and the critical properties of gases [15] as shown in Fig. 11.10. It was also pointed out that there exists an interrelationship between B and Δ as pictured in Fig. 11.11. Based on these correlations, it was possible to predict the amounts of surface diffusion, and in turn the total permeabilities of several new gases. These estimated values are compared with the experimental data from all different sources in Table 11.1.

Table 11.1 Comparison of the Experimental and Estimated Permeabilities through Vycor Porous Glass[a]

Gas	T (°K)	$Q \times 10^6$ (estimated)	$Q \times 10^6$ (experimental)	Invesitgator
Kr	294	3.55	3.4	Barrer [16]
C_2H_4	304.2	8.0	8.2	Huckins [17]
C_3H_6	298.2	8.47	8.00	Gilliland [18]
n-C_4H_{10}	308.2	12.9	13.0	Higashi [19]

[a]Reference 15.

Nonporous Membranes

A correlation similar to that shown in Fig. 11.7, but for a number of organic compounds in polyethylene, is that of Li and Henley [6] as shown in Fig. 11.12. It is to be expected that similar correlations could be devised for other polymers.

There are two instances of rather broad attempts to relate permeability data with structured parameters. At the outset it must be said that these efforts had to lean rather heavily on empirical consideration. One of these correlations is the well-known *Permachor* characterization by Salame [21], and the other study is that of Malachowski [20], who introduced a *space factor* to allow for temperature effects on the polymer matrix. A considerable amount of effort has already gone into the development of the Permachor analysis, and it has resulted in some rather good correlations especially for homologous series of compounds. This is not the case with the space-factor method, and it remains to be seen if it will prove to be a useful procedure.

Permachor Characterization

The basic idea that underlies the Permachor concept is that all facets of a polymer structure will exert some effect upon its permeability behavior toward gases and vapors. So, Salame [21, 22] developed a scheme of additive (positive or negative) contributions of the structural elements of a polymer to the permeability.

These "structural" contributory values, however, are not the same for all gases or vapors, nor for the polymer. They are different for different permeating

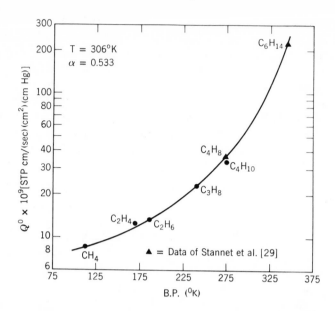

Fig. 11.12 Correlation of Q°, permeability at zero pressure with boiling point [6]; α is volume fraction of amorphous polymer. Courtesy of the American Institute of Chemical Engineers.

species and also vary with the morphology of the polymer. Therefore, a rather comprehensive collection of segmental values is required to utilize Permachor computations and correlations. So far, it is necessary to obtain this information from Salame's versatile publications [21-27].

SAMPLE CALCULATION OF PERMACHOR

From Salame's tables for acrylonitrile structure [22]:

Backbone Segment	n Contribution	Permachor Value
$+(CH_2)+$	1	$\dfrac{40}{N}$
$+(CH)+$	1	$\dfrac{20}{N}$
Branch		
$-C\equiv N$		$\dfrac{180}{N}$

Then

$$\Pi = \frac{40}{N} + \frac{20}{N} + \frac{180}{N} = \frac{240}{N}$$

$$N = \Sigma n = 2; \quad \Pi = \frac{240}{2} = 120$$

Solubility-Space-Factor Method

This method was developed by Malachowski [20] in an attempt to estabilsh a generalized method for *permeability prediction*. The final result was an equation containing only three additive terms:

$$\ln \ Q = a + b \ \frac{T_c}{T} + c \ \frac{d}{T^n} \ , \tag{11.12}$$

where a, b, and c are constants depending on polymer properties; T is the absolute temperature; T_c is the critical absolute temperature of the permeant; d is the molecular diameter of the permeant, which varies with temperature; and n is the space factor.

The T_c/T term introduced by Stern [11] essentially describes the contribution of solubility of the permeating species in the polymer. The molecular diameter d has a bearing on the ease of permeation, and the space factor n considers the effect of the thermal expansion of the polymer membrane. An approximate correlation of the space factor n with the coefficient of cubical expansion of the polymer has been established. The method in application calls for determination of the constants and permeant properties and thus requires a fair amount of data. Malachowski recommends to obtain permeability data for two gases, as, for instance, helium and oxygen or carbon dioxide, at two different temperatures. This information is then sufficient to establish values of the constants for a given polymer. On the whole, the method has some promise, but it still involves a considerable amount of empiricism.

Numerical Data

There is a wealth of information available on permeabilities in the gas phase and on water-vapor transmission. A comprehensive up-to-date collection of data was published recently by Hwang [45]. The discussion in Chapter V contains many examples. A rather good collection of transfer data for nonporous inorganic membranes, that is, metal, glass, and the like, is given in Chapter VI is graphical form.

3 TRANSFER COEFFICIENTS IN LIQUID PHASE

The conditions encountered in liquid flow over and through a membrane introduce the likelihood of extrinsic effects upon the transfer mechanism.

Although gas-phase flow is essentially unhampered by fluid behaviorism, the dependence of the boundary layer upon flow conditions in liquid flow does not readily permit isolating the intrinsic transfer coefficient from experimental observation. This is because the diffusion coefficient of a molecular species in the liquid phase is small compared to that in the gas phase.

Effect of Concentration Polarization

The phenomenon of concentration polarization in the liquid-phase process can have a decisive influence upon the magnitude of transfer coefficients. Figure 11.13 illustrates this influence in a batch process of caustic dialysis. When

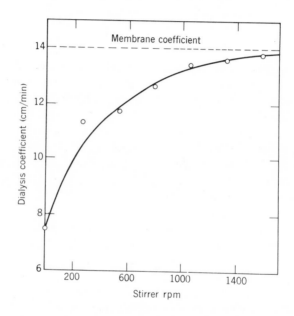

Fig. 11.13 Effect of stirring on batch dialysis coefficient. Reprinted by permission from Ref. 30. Copyright 1951 by the American Chemical Society.

the stirring rate is small and concentration of caustic is, therefore, severely polarized, the overall transfer coefficient is greatly reduced as shown in the figure. With increasing stirring rate, the transfer coefficient is enhanced and then approaches an asymptotic value, which approximately corresponds to the intrinsic coefficient of the membrane.

Strictly speaking, this value is not the true intrinsic coefficient, since complete elimination of concentration polarization with a finite stirring rate is not possible. In order to isolate the intrinsic membrane properties, Hwang et al. [31] proposed a series of experimental observations with various thicknesses of membranes, and a plot of the results as shown in Fig. 11.14 (also see Chapter II). The intercept of the straight line, which corresponds to the value with the infinite thickness of the membrane, is the actual intrinsic value of the membrane. However, this procedure requires several membranes with different thicknesses but of the same uniform structure, which are often difficult to prepare, particularly in the anisotropic type.

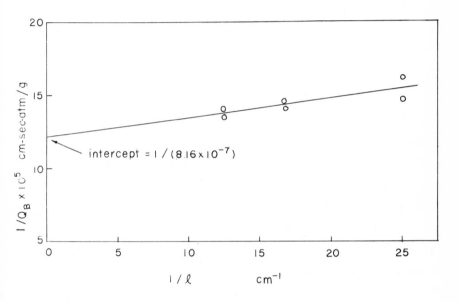

Fig. 11.14 Isolation of intrinsic transfer coefficient (membrane: cellulose acetate thin film composite; solution: 3.5% NaCl aq. solution; data from Riley et al. [32]).

As mentioned earlier, the intrinsic transfer coefficient also depends on the external conditions such as solution concentration, pressure, temperature, and electric current density. When one measures the transfer coefficient of a membrane, it is therefore necessary to specify experimental conditions so that the coefficient is meaningful. Furthermore, when correlations of such coefficients are available, they are likely to be limited to the range of operating conditions used to obtain them. Although such data are then of limited utility, it is nevertheless instructive to present some samples to illustrate the magnitude that can be expected.

Parameters Influencing Transfer Coefficients

A set of transfer coefficients of a membrane in conjunction with a driving force represents a complicated interaction among the membrane, the solvent, and solute molecules. Therefore, the transfer coefficients depend on the nature and structure of the membrane and the molecules to be separated. As discussed in Chapter XIV, the structure of the polymeric membrane is greatly influenced by the method of its preparation. Thus, the nature and composition of the casting solution, gelation, and evaporation, time, and the annealing temperature are important factors that influence the transfer coefficients. Also, the values of the coefficients change gradually as time elapses. This is because of the physical and chemical deterioration of the polymeric membrane, such as compaction of the anisotropic type of membranes, hydrolysis of the cellulose acetate membrane at extremely low and high pH values, and other chemical and enzymatic attacks. The operating parameters such as temperature, the distributions of concentration, pressure, and electric potential across the membrane also influence the coefficients. The effect of concentration polarization has already been discussed; this phenomenon is treated in depth in Chapter XIII.

Pore-Model Correlations

Membranes can be divided into two categories: homogeneous and heterogeneous. As discussed in Chapter VIII, Section 1, the distinction is somewhat arbitrary but should be made in conjunction with the nature and behavior of the molecular or ionic species associated with the membrane. In the general theory of nonequilibrium thermodynamics discussed in Chapter II, the distinction is not necessary. However, the distinction is very convenient and instructive to obtain some insight into the mechanism of dynamic and static behavior of the molecular or ionic species, when such species are associated with the membrane. The understanding, in turn, helps correlate the transfer coefficient in terms of fundamental properties of the species and the membrane. For example, a membrane that is heterogeneous in the λ' scale can be considered to be homogeneous, when the membrane is associated with a molecular or ionic species of either solute or solvent whose size is comparable to λ'. In this case, the transfer coefficient of the species in the membrane can be satisfactorily correlated with the activated diffusion or homogeneous model, presented in Section 1.

Solvent Flow

The heterogeneous pore model, on the other hand, satisfactorily describes the transport mechanism of both solvent and solute molecules, when their sizes are much smaller than the pore size. The solvent permeability can then be approximately correlated with the Poiseuille, Kozeny-Carman, or more generally the Darcy equation (see Chapter II).

Solute Flow

The transport of the solute molecule is mainly via the convective motion with the solvent transport and the molecular diffusion through the solution phase in the pore. The apparent diffusion coefficient of the solute in the membrane is therefore reduced by tortuous paths in the porous channels. Thus, Wheeler [33] related the coefficient of the solute through *large pores* in the membrane phase $\overline{D}_A$ to that in the free solution phase D_A in terms of the porosity ϵ' as

$$\overline{D}_A = D_A \frac{\epsilon'}{2} . \tag{11.13}$$

Here, the tortuosity of the membrane is assumed to be $2/\epsilon'$.

When the solute size is comparable to the pore size, that is, for the *intermediate pore*, it can, however, be expected that the solute transport is hampered by the "hydrodynamic drag," and also influenced by the molecular interactions such as dispersion or electrostatic force. Faxen [34] estimated the hydrodynamic drag factor, $F = \overline{D}_A / D_A$, for a spherical particle of diameter d in a straight cyclindrical pore of radius r as

$$F = 1 - 2.104 \frac{d}{2r} + 2.09 \left(\frac{d}{2r} \right)^3 - 0.95 \left(\frac{d}{2r} \right)^5 \tag{11.14}$$

and, in a parallel-plate pore of width r,

$$F = 1 - 1.004 \frac{d}{2r} + 0.418 \left(\frac{d}{2r} \right)^3 + 0.210 \left(\frac{d}{2r} \right)^4 - 0.169 \left(\frac{d}{2r} \right)^5 ,$$

$$0 < \frac{d}{2r} \leqslant 0.4. \tag{11.15}$$

The more exact drag factor was extensively discussed by Happel and Brenner [35]. These mechanical models often describe the solute transport reasonably well, even if these models do not account for molecular interactions. In particular, the selectivity of one species against another species is better explained by the parallel-plate model than by the cylindrical model [36].

Solute Diffusion Coefficients

As indicated by Eq. 11.14, the diffusion coefficient is generally higher for a smaller-size solute. If the solute is an inorganic ionic species, its transport across the membrane usually occurs in its hydrated form. In this case, diameter

d in Eq. 11.14 corresponds to the hydrated size of the solute species. However, the hydration is hampered in a highly cross-linked polymeric membrane and also in a very fine pore, as in the skin layer of the anisotropic membrane. The diffusion coefficient is then determined largely by the undehydrated size. For example, the diffusion coefficient of lithium ion is smaller than that of potassium ion in the membrane of the *moderate or large* pore size. However, the reverse situation occurs in the membrane of the *very fine pore size*, where there is little space for solvation.

The diffusion coefficient is also greatly influenced by various other factors such as the electrostatic and other molecular interaction between the membrane and the solute species; the interaction among the solute species; pore structure and swelling of the membrane; ion-exchange capacity; and temperature. These effects were discussed in great detail by Helfferich [37]. In general, the transport of the solute that has a greater interaction with the membrane is more retarded by the interaction. For example, the diffusion coefficient of the counterion in an ion-exchange membrane is more reduced than the coefficient of the co-ion, compared to the values in the free solution. Also, the reduction is larger for a polyvalent ion than for a monovalent ion. Such reductions are particularly pronounced in the membrane of high capacity. This is a direct consequence of the electrostatic interaction between the fixed ion of the membrane and the mobile ion. A similar situation can be expected in the system where other physical or chemical interactions, like the dispersion force, prevail. If there exist two mobile ionic species in the membrane phase, the diffusion coefficient of the faster ion is lessened by the presence of the slower ion, but the coefficient of the slower ion is enhanced by the faster ion. Such interionic interaction has already been discussed in Chapter VIII, Section 1.

The effect of the membrane structure is also important. The strong reduction of the diffusion coefficient in the membrane of a high degree of cross-linking is very analogous to the case of gas diffusion or pervaporation. It can also be easily expected that the diffusion coefficient increases with the content of free water, but not necessarily with total water.

In general, the diffusion coefficient increases with the temperature. The activation energy usually assumes a value in the range of 6 to 10 kcal/(g)(mole), which is somewhat higher than in the free solution. This is partial evidence that there exists molecular or ionic interaction between the solute and the membrane matrix.

Reverse Osmosis

In the general theory of irreversible thermodynamics, fluxes of the solvent and solute across an osmotic membrane are given by Eqs. 8.19 and 8.20.

Thus, the three transfer coefficients, L_v, L_A, and σ of these equations characterize a reverse-osmosis membrane. However, two of these coefficients, L_v and L_A are often sufficient, since the reflection coefficient σ assumes an almost constant value close to unity in most cases. Thus, it is more convenient to use the solvent and solute permeabilities, Q_B and Q'_A, defined in the homogeneous diffusion model by Eqs. 8.11 and 8.14. Alternatively, one may employ the solvent flux N_B and the solute rejection SR, defined by Eq. 8.15, to characterize the membrane. In this case, the total pressure drop ΔP should be specified so that the transfer coefficients Q_B and Q'_A can be estimated from Eqs. 8.11 and 8.16. According to these equations, the solvent flux N_B and the solute rejection SR increase with the net pressure drop ΔP. However, the rate of increase of N_B is reduced at a high pressure drop across the anisotropic polymeric membranes, since the solvent permeability Q_B itself decreases slightly with the pressure drop.

As shown in Chapter VIII, both permeabilities Q_B and Q'_A generally increase with the operating temperature. Thus, both fluxes N_B and N_A increase accordingly. But the solute rejection is often slightly reduced at a high temperature, since the activation energy of the water transport is less than the energy of the transport of most electrolyte solutes. The typical activation energy is in the order of 5 to 6 kcal/mole for the solvent water, and 7 to 8 kcal/mole for alkali halides, in the cellulose acetate membranes [38].

The physical and chemical deterioration of the polymeric membranes, and rejection of various solutes have been covered adequately in Chapter VIII.

Numerical Data

Walch [39, 40] presented data for a number of cellulose acetate compositions, as well as for a variety of noncellulosic membranes, both ionic and neutral in nature. The cellulose acetate data are listed in Table 11.2 where the salt rejection and flux values are obtained at 800 psi pressure and 25°C. These conditions are somewhat of a standard used in projects sponsored by the Office of Saline Water.

The compilation of experimental results on noncellulosic membranes are listed in Table 11.3 and 11.4. There, salt rejection and flux values are obtained at a variety of pressure drop values. If one were to divided the flux values by the pressure drop, the resulting transfer coefficients would still not show any degree of uniformity. Thus, the performance of the respective membranes and the effect of experimental conditions, specifically the occurrence of concentration polarization, are all lumped together. Obviously, the material that is contained in this source is insufficient for attempts at correlations. For details of the listed items, the publications of Walch should be consulted; there, appropriate references are quoted.

Table 11.2 Comparative Performance Data for Cellulosic Membranes[a]

Membrane Material	Degree of Substitution[b]	Desalting Action % Salt Rejection	Water Flux $\times$ 10^5 [g/(cm^2)(sec)]
Cellulose acetate	2.1	90	200
Cellulose acetate-	2.1	93	250
Methacroyl chloride	0.3		
Cellulose acetate	2.5	95	100
Cellulose acetate	2.8	99	50
Cellulose acetate	3.0	99	25
Cellulose acetate-	2.1	96	0.4
Octanoate	0.9		
Cellulose acetate-	2.1	83	0.1
Palmitate	0.9		

[a] References 39, 40. Conversion factor: 1 gal/(ft^2)(day) = 2.12 $\times$ 10^{-4} g/(cm^2) (sec). [b] Hydroxyl groups per β-glucopyranose unit.

The effects of the pressure drop on the solvent permeability and solute permeability are shown in Fig. 11.15 and 11.16. As indicated in Fig. 11.16, the pressure dependence of the solvent permeability is pronounced in the membrane annealed at a lower temperature.

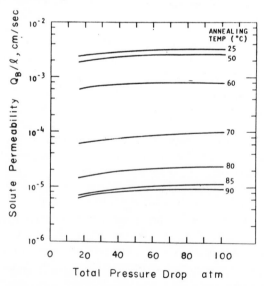

Fig. 11.15 Pressure dependence of the solute permeability (membrane: Loeb-Sourirajan membranes; solution: aq NaCl at 25°C) [38].

Table 11.3 Desalination Properties of Noncellulose Ionic Membranes in Reverse Osmosis[a]

No.	Membrane Material	Desalination Properties		Operating Conditions	
		(SR%)	N_B [g/(cm^2)(sec)]	NaCl (%)	ΔP (atm)
1	Graphite oxide	65	29×10^{-5}	0.5	41
2	Vycor glass	≈55	$≈28 \times 10^{-5}$	0.2	100
3	Zirconium(IV)hydroxide, polyacrylic acid (on 0.05-μ millipore)	82	370×10^{-5}	0.3	68
4	Polyacrylic acid (on cellulose acetate-cellulose nitrate)	80	200×10^{-5}	0.3	102
5	Block copolymer styrene-2-vinyl-pyridine (quaternized)	70	12×10^{-5}	3.5	102
6	2,6-Dimethylpolyphenylene oxide sulfonate	90	$>125 \times 10^{-5}$	≈1.6	75
7	Polyvinyl alcohol, cross-linked with dichlorotriazine (with acid pigment group)	75	17.5×10^{-5}	0.06	41
8	Polyethylene grafted with acrylic acid	68	11×10^{-5}	0.25	27

[a] SR = salt rejection, N_B = water flux (for No. 5 theoretically expected value), NaCl = salt concentration in raw water, ΔP = applied pressure [39, 40].

Table 11.4 Desalination properties of Noncellulose Neutral Membranes in Reverse Osmosis[a]

No.	Membrane Material	Water Sorption (%)	Film Thickness (μ)	Desalination Properties		Operating Conditions	
				SR (%)	N_B [g/(cm²)(sec)]	NaCl (%)	ΔP (atm)
1	Polyacrylonitrile	46.9	53	21	14×10^{-5}	3.5	102
2	Polyvinylene carbonate	17	94	94	0.28×10^{-5}	3.5	102
3	Polyvinylene glycol	—	94	50	6.5×10^{-5}	3.5	102
4	Polethyl acrylate cross-linked with trimethylol-propane trimethacrylate (2/1)	22	175	98	0.19×10^{-5}	1	102
5	Poly(galactose methacry-latemethylmethacrylate, 1/9)	10	20	97	0.28×10^{-5}	1	102
6	Polyvinylpyrrolidone cross-linked with methylene-bis-(4-phenylene isocyanate)	15	75	>99	0.04×10^{-5}	0.8	34
7	Nylon 6 cross-linked with toluene-2,4-diisocyante	—	—	84.3	0.15×10^{-5}	—	—

8	Zein cross-linked with formaldehyde	28	31	92	0.07×10^{-5}	5	100
9	Polyimidazopyrrolone (promellitic dianhydride, 3,3'-diaminobenzidine)	9	2.5	>99	1×10^{-5}	0.5	102
10	Polyamide I-OP (Montecatini Edison)						
	(a)	—	70	99.2	0.23×10^{-5}	0.5	80
	(b)	—	9	98.4	2.65×10^{-5}	0.5	80
11	Polyamide XPA-1112 (Monsanto)	—	—	>99	25×10^{-5}	3.5	102
12	Polyamide hydrazide DP-1 (Du Pont)	—	—	≈99	24×10^{-5}	2.9	68

[a]SR = salt rejection, N_B = water flux (for Nos. 6, 8, 12 theoretically expected values), NaCl = salt concentration in raw water, ΔP = applied pressure [39, 40].

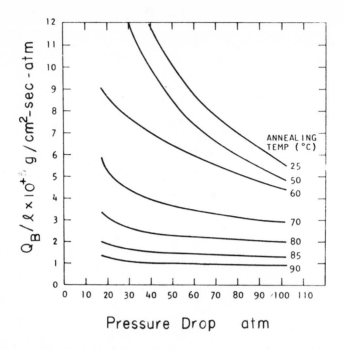

Fig. 11.16 Pressure dependence of the solvent permeability (membrane: Loeb-Sourirajan membranes: solution aq. NaCl at $25°C$) [38].

Ultrafiltration

The molecular weight of the solute to be separated in this process varies from 500 to 100,000. Consequently, a large number of ultrafiltration membranes of different pore sizes are available to separate solutes of different molecular weights over this range. The ultrafiltration membrane is, therefore, often characterized by its average pore size or by the minimum molecular weight of the solute that can be effectively retained by the membrane.

The main feature in transfer coefficients of ultrafiltration membranes is essentially the same as that of reverse-osmosis membranes. For example, the permeation flux of the solvent water and the solute rejection increase generally with the net pressure drop. However, they decrease often beyond the certain values of the pressure drop, when the average pore size is comparable to the solute size. This occurs because the solute plugs the pore in this case. In addition, the plugged solute molecules are squeezed out with the solvent water through the elastic distortion of either the solute molecules or the porous membrane when the pressure drop is high.

Dialysis

The membranes that have been mentioned in previous sections, reverse osmosis and ultrafiltration, can also be used in the process of dialysis. Therefore, a separate discussion of dialysis membranes is not necessary. However, the ion-exchange membrane employed in Donnan dialysis deserves further treatment.

Electromembrane Processes

The transfer coefficients and other parameters to characterize an electro-membrane are obviously the diffusivities of the counterion and co-ion, $\overline{D}_A$ and $\overline{D}_Y$ (or $\overline{D}_+$ and $\overline{D}_-$), the ion-exchange capacity $\overline{C}_R$, the activity coefficient ratio, the solvent diffusivity, and other geometric parameters such as thickness l and porosity ϵ'. When the diffusivities of the ionic solutes and solvent, and also their activity coefficients, are constant and uniform, the solute flux and the electric current density can be evaluated from Eqs. 9.34 and 9.35 using these transfer coefficients. Then, the solvent flux is estimated from Eq. 9.53. However, these equations are too complicated to be practical. Furthermore, the validity of the assumption of constant diffusivities and activity coefficients is often questionable, even at moderate concentration of the contiguous solutions. Therefore, the solute fluxes and the current density are commonly evaluated from the simple equations, Eqs. 9.19 and 9.21, for a neutral membrane and Eqs. 9.43 and 9.47 for an ion-exchange membrane. The solvent flux is then obtained from Eq. 9.54. In this case, the electromembrane is characterized by the transference numbers of ionic species, $\overline{t}_A$ and $\overline{t}_Y$ (or $\overline{t}_+$ and $\overline{t}_-$); the membrane area resistance R_m; the dialysis coefficient U; and the electro-osmotic water-transport number $\overline{t}_v$. Since these transfer coefficients can be regarded as effectively averaged values over some ranges of operating parameters, the characterization is meaningful only in such ranges. The correlation between this set of effective transfer coefficients and the former set of more fundamental characterizing parameters is possible in principle, but has not been well established. Therefore, the discussion will be confined to the latter set of effective transfer coefficients. Also, the characterization by a set of more phenomenological transfer coefficients that are derived from irreversible thermodynamic considerations is not treated here. For a detailed treatment, the reader could refer to the report of Lacey [41].

Transference Number

The effective transference number of an ionic species is usually measured by the emf and Hittorf electrophoretic methods [42]. In the emf method, the concentration potential is measured in the absence of the external electric potential across an electromembrane that separates two solutions of different ionic concentrations. The transference number is then evaluated from the measured potential using Eq. 9.43 with $I = 0$; or the more exact evaluation can be made by substituting the activities for the concentrations in the equation.

In the Hittorf method, on the other hand, the amount of an ionic species transferred by a given current density is measrued directly between the two solutions of the *same* concentration. Then Eq. 9.47 can be used to evaluate the transference number, neglecting the term of the diffusional leakage, that is, the first term on the right-hand side of the equation.

The transference numbers measured by both methods are not true values. Since an appreciable amount of osmotic and electro-osmotic flux cannot be avoided in such experiments, the true transference number with respect to the moving solution is always different from the observed value that is taken with resepct to the stationary membrane (see Chapter IX, Section 3). A correction should, therefore, be made with an additional measurement of the osmotic and electro-osmotic flux. In the Hittorf method, the solution concentrations are polarized near the membrane-solution interface as the ionic species is electrically transferred across the membrane. Thus, the apparent transference number is reduced by the subsequent water splitting at a low concentration and a high current density. But the number is also lowered by the backdiffusional leakage at a high concentration and a low current density. Special care should therefore be taken to minimize the effect of the concentration polarization. A typical example of this effect is shown in Fig. 11.17. It should be noted that even the true intrinsic transference number also depends on the concentration and current density as indicated in Eqs. 9.34 and 9.35. Therefore, the figure actually shows both the intrinsic and extrinsic effects, even if the latter effect apparently dominates over the former effect.

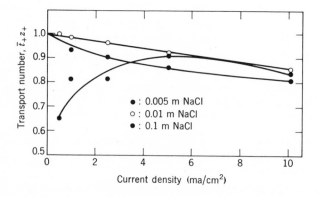

Fig. 11.17 Effect of current density on the transference number of the counterion [44].

Thus, Rosenberg and his co-workers [42] found that the two apparent transference numbers determined by the emf and Hittorf methods were not in good agreement, as shown in Fig. 11.18. Here, the numbers obtained by the Hittorf method are corrected only for the electro-osmotic flux. In most cases, the transference number determined by the emf method can be expected to be closer to the actual value, since the osmotic effect is less significant and the polarization effect is essentially absent. However, the Hittorf method may be useful, since the experimental conditions can be maintained more closely to the operating condition at which the actual separation process is to proceed.

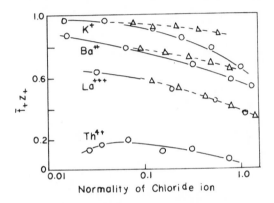

Fig. 11.18 Comparison of Hittorf and emf transport numbers [42] (membrane: Ionic CR-61 cation-exchange membrane; emf $--\triangle--$; Hittorf $-\bigcirc-$). Reprinted by permission of the Electrochemical Society.

As indicated in Fig. 11.18, the transference number of a univalent counterion is higher than the number of a multivalent ion. It may be attributed to the fact that the mobility of a multivalent ion is reduced by the corresponding increase of electrostatic attraction and also specific association with the fixed ion in the membrane. The existence of a significant effect of the association may be confirmed by the abnormally small electro-osmotic flux, or even by negative osmotic flux, in a membrane of such a multivalent ionic form. This subject matter will be treated in detail later in conjunction with the electro-osmotic transport number.

Membrane Area Resistance

The electric resistance or specific conductivity in a membrane is usually measured with an alternating current of high frequency. Since this current does not permit the electric transference of ionic species, the troublesome effect of concentration polarization can be effectively avoided.

As indicated in Eq. 9.6, the specific conductivity increases with the diffusi-vities and concentration of ionic species present in the membrane phase. Since the ionic concentration in the membrane is enhanced by the solution concen-tration, the conductivity increases with the solution concentration as shown in Fig. 11.19. The small conductivities in multivalent ionic forms of the membrane,

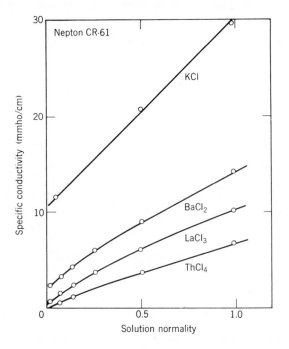

Fig. 11.19 Specific conductance in salt solution [42] (membrane: ionic CR-61 cation-exchange membrane). Reprinted by permission of the Electrochemical Society.

as indicated in the figure, are obviously resulting from the small diffusivities of the multivalent ions. The effective diffusivity increases as the free water content of a membrane increases and also as the effective size of the counterion decreases. This effect on the specific conductivity was reported by Breslau [36] and co-workers [43]. The overall hindrance effect of the membrane matrix on the ionic transport also was investigated by this team. The reduction of the equivalent conductivity $\overline{\Lambda}$ is the membrane compared to that Λ in free solution is 1/5 to 1/10 for univalent inorganic counterions but 1/50 to 1/100 for divalent inorganic ions. The equivalent conductivity was defined by Eq. 9.71 in the solution phase. Similarly, the equivalent conductivity in the membrane phase can be defined as the specific conductivity divided by the equivalent concentration in the membrane. It can also be expected that, as temperature increases, the conductivity increases with an activation energy that is almost the same as the energy of the diffusivity of the counterion in a high-capacity ion-exchange membrane.

Dialysis Coefficients

The solute flux across a neutral membrane is given by Eq. 8.14. However, the ionic flux cannot be expressed as a linear function of the concentration difference across a membrane if there exists interionic interaction. This case has already been treated in the section on Donnan dialysis in Chapter VIII, Section 1. For example, the flux is better expressed by Eq. 8.77 in Donnan dialysis with a neutral membrane, but by Eqs. 8.84 or 8.87 in Donnan dialysis with a high-capacity ion-exchange membrane. In an electromembrane process, on the other hand, the ionic flux can be reasonably expressed with Eq. 9.47. Thus, it is more convenient to define the dialysis coefficient by Eq. 9.83. However, it is customary to define the coefficient as the solute permeability Q'_A divided by the thickness of the membrane l (see Eq. 8.14). In this case, the concentration difference across the membrane should be specified if there is some interionic interaction. In any case, the dialysis coefficient is closely related to the effective diffusivity of the solute species in the membrane, which has already been discussed in Section 3, the subsection on solute diffusion coefficients, p. 279.

Electro-osmotic Water-Transport Number

The osmotic flux in the presence of the external electric potential is given by Eq. 9.54, where the electro-osmotic transport number $\bar{t}_v$ is defined. Since most inorganic ionic species are transported in their hydrated form, the experimentally observed number is usually comprised of the true electro-osmotic number and the contribution of such hydration, as shown in Eq. 9.56. Also, the effect of concentration polarization cannot be excluded in measuring the number in practice.

Some of the multivalent cations are apparently associated with the fixed ion of the cation-exchange membrane. Therefore, the effective ion-exchange capacity is reduced, or the membrane even exhibits the anionic character as shown in Fig. 11.20. Since the Donnan exclusion is less effective at a higher concentration, the apparent transport number decreases as the solution concentration increases. The decrease may be ascribed to the increasing tendency of association for multivalent ions. The apparent transport number is commonly reported to decrease with the current density. This phenomenon is generally attributed to concentration polarization.

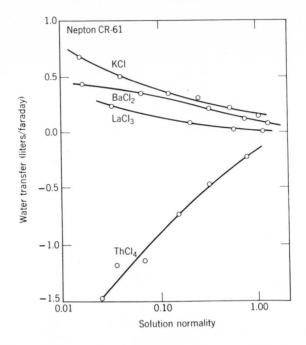

Fig. 11.20 Water transfer in membrane (membrane: Ionic CR-61 cation-exchange membrane) [42]. Reprinted by permission of the Electrochemical Society.

Nomenclature

Symbol	Meaning
A, B	Constants defined in Eq. 11.10
$\overline{C}_A$	Concentration of counterion in the membrane
$\overline{C}_R$	Ion-exchange capacity
D_A	Diffusivity of solute in the solution phase
$\overline{D}$	Diffusivity in membrane phase
$\overline{D}_0$	Pre-exponent coefficient defined in Eq. 11.6
$\overline{D}_A$	Diffusivity of solute or counterion in membrane phase
$\overline{D}_v$	Diffusivity of solvent in membrane
$\overline{D}_Y$	Diffusivity of co-ion in membrane
$\overline{D}_+, \overline{D}_-$	Diffusivities of cation and anion
E_0	Activation energy at $0°K$
E_D	Activation energy, observed
E_Q	Activation energy, permeation

F	Reduction factor of diffusivity in the membrane $(\overline{D}_A/D_A)$
ΔH	Heat of solution
$\Delta H^{\ddagger}$	Enthalpy of activation
I	Electric current density
L_A, L_v	Transfer coefficients defined in Eqs. 8.19 and 8.20
M	Molecular weight
N_A, N_B	Fluxes of solute and solvent across membrane
$\Delta \mathcal{P}$	Total Pressure drop (see Eq. 8.12)
Q	Permeability
Q^0	Permeability at zero pressure
Q_0	Pre-exponent coefficient defined in Eq. 11.8
Q'_A	Solute permeability defined by Eq. 8.14
Q_B	Solvent permeability defined by Eq. 8.11
R	Gas constant
R_m	Membrane area resistance
S	Membrane area
S_m	Solubility in a membrane
S_0	Pre-exponent coefficient in Eq. 11.7
$\Delta S^{\ddagger}$	Entropy of activation
SR	Solute rejection defined by Eq. 8.15
T	Absolute temperature
T_c	Absolute critical temperature
U	Dialysis coefficient
V_c	Critical volume
a, b, c	Constants defined by Eq. 11.12
d	Molecular or ionic diameter
f	Partition function of normal state
$f^{\ddagger}$	Partition function of activated state
h	Planck constant
k	Boltzmann constant
l	Membrane thickness
n	Space factor (exponent)
r	Pore radius of a membrane
$\overline{t}_A, \overline{t}_Y$	Transference numbers of counterion and co-ion in membrane phase
$\overline{t}_v$	Electro-osmotic water-transport number
$\overline{t}_v'$	Apparent electro-osmotic water-transport number (see Eq. 9.56)
$\overline{t}_+, \overline{t}_-$	Transference numbers of cation and anion in membrane phase
z_+	Electrochemical valence of cation

Symbol	Meaning
Δ	Constant defined in Eq. 11.10
Δ'	Driving force across a membrane
Π	Gas Permachor
Λ	Equivalent electric conductance in solution phase, defined by Eq. 9.71
$\overline{\Lambda}$	Equivalent electric conductance in membrane phase
a	Volume fraction of amorphous polymer
ϵ	Lennard-Jones parameter
ϵ^*	Interaction energy, gas-solid
$\epsilon^{\ddagger}$	Activation energy, surface diffusion
ϵ'	Porosity
λ	Distance between activated sites
λ'	Scale factor characterizing heterogeneity of a membrane
$\overline{\kappa}$	Specific electric conductivity
σ	Reflection coefficient defined in Eqs. 8.19 and 8.20

References

1. H. Eyring, *J. Chem. Phys.*, **4**, 283 (1936); J. O. Hirschfelder, D. P. Stevenson, and H. Eyring, *ibid.*, **5**, 896 (1937).
2. T. Nakagawa, H. B. Hopfenberg, and V. Stannett, *J. Appl. Polymer Sci.*, **15**, 231 (1971).
3. A. S. Michaels, and H. J. Bixler, in *Progress in Separation and Purification*, Vol. 1, E. S. Perry, Ed., Wiley-Interscience, New York, 1968 p. 149.
4. J. E. Jolly and J. H. Hildebrand, *J. Amer. Chem. Soc.*, **80**, 1050 (1958).
5. S.-T. Hwang and K. Kammermeyer, *Can. J. Chem. Eng.*, **44**, 82 (1966).
6. N. N. Li and E. J. Henley, *A.I.Ch.E. J.*, **10**, 666 (1964).
7. A. S. Michaels, and H. J. Bixler, *J. Polymer Sci.*, **50**, 393, 413 (1961).
8. G. Gee, *The British Rubber Producers Research Assn.*, No. 86 (1947).
9. C. H. Klute and P. J. Franklin, *J. Polymer Sci.*, **32**, 161 (1953).
10. F. J. Norton, *J. Appl. Polymer Sci.*, **7**, 1649 (1963).
11. S. A. Stern, J. T. Mullhaupt, and P. J. Gareis, *A.I.Ch.E. J.*, **15**, 64 (1969).
12. D. F. Othmer and G. J. Frohlich, *Ind. Eng. Chem.*, **47**, 1034 (1955).
13. K. Kammermeyer, "Gas and Vapor Separations by Means of Membranes," in *Progress in Separation and Purification*, E. S. Perry, Ed., Vol. 1, Wiley-Interscience, New York, 1968, p. 335.
14. S.-T. Hwang and K. Kammermeyer, *Separation Sci.*, **1**, 629 (1966).
15. S.-T. Hwang, *A.I.Ch.E. J.*, **14**, 809 (1968).
16. R. M. Barrer, *Inst. Chem. Eng. (London), No. 1*, 1 (1965).

17. H. E. Huckins, and K. Kammermeyer, *Chem. Eng. Progr.*, **49**, 180, 294, 517 (1953).
18. E. R. Gilliland, R. F. Baddour, and J. L. Russell, *A.I.Ch.E. J.*, **4**, 90 (1958).
19. K. Higashi, H. Ito, and J. Oishi, *J. Nucl. Sci. Tech. (Japan)*, **1**, 298 (1964).
20. R. A. Malachowski, Ph.D. thesis, University of Iowa, 1971.
21. M. Salame and J. Pinsky, *Mod. Packaging*, **36**, 153 (Nov. 1962).
22. M. Salame, *Polymer Preprints* (ACS), **8**, 137 (1967).
23. M. Salame, A.C.S. Meeting, New York, August-September 1972.
24. M. Salame, 67th National Meeting, Am. Inst. Chem. Engrs, Atlanta, Ga., February 1970.
25. M. Salame and J. Pinsky, *Mod. Packaging*, **37**, 131 (1964).
26. M. Salame and J. Pinsky, *SPE Trans.*, **1**, 153 (1961).
27. H. J. Bixler, A. S. Michaels, and M. Salame, *J. Polymer Sci. A*, **1**, 895 (1963).
28. K. Higashi, H. Ito, and J. Oishi, *J. Atom. Energy Soc., Japan*, **5**, 846 (1963).
29. V. Stannett, M. Szwarc, R. L. Bhargava, J. A. Meyer, A. W. Meyers, and E. E. Rogers, "Permeability of Plastic Films and Coated Paper to Gases and Vapors", Technical Association of the Pulp and Paper Industry, 1962.
30. R. D. Marshall and J. A. Storrow, *Ind. Eng. Chem.*, **43**, 2934 (1951).
31. S. T. Hwang, T. E. S. Tang, and K. Kammermeyer, *J. Macromol. Sci.–Phys. B*, **5**, 1 (1971).
32. R. L. Riley, G. R. Hightower, and C. R. Lyons, "Thin film composite membrane for single-stage sea-water desalination by reverse osmosis," paper presented at the joint NASA Ames Research Center/University of Califormia at San Diego Conference, Nov. 29-Dec. 1, 1972, Moffet Field, Calif.
33. A. Wheeler, *Advan. Catal.*, **3**, 249 (1951).
34. H. Faxen, *Ann. Phys.*, **68**, 89 (1922).
35. J. Happel and H. Brenner, *Low Reynolds Number Hydrodynamics*, Prentice-Hall, Englewood Cliffs, N. J., 1965.
36. B. R. Breslau and I. F. Miller, *I.E.C. Fund.*, **10**, 554 (1971).
37. F. Helfferich, *Ion Exchange*, McGraw-Hill, New York, 1962.
38. H. K. Lonsdale, "Theory and Practice of Reverse Osmosis and Ultrafiltration," in *Industrial Processing with Membranes*, R. E. Lacey and S. Loeb, Eds., Wiley-Interscience, New York, 1972.
39. A. Walch, *CZ Chemie-Technik*, **2**, 7 (1973).
40. A. Walch, *Battelle Inform.*, **14**, 20 (1972).
41. R. E. Lacey, *U.S. Off. of Saline Water Res. Dev. Prog. Rept.*, **343** (1968).
42. N. W. Rosenberg, J. H. B. George, and W. D. Potter, *J. Electrochem. Soc.*, **104**, 111 (1957).
43. B. R. Breslau, I. F. Miller, C. Gryte, and H. P. Gregor, Preprint at MESD Biennial Conference, A.I.Ch.E., 1970, p. 363.
44. N. Lakshminarayanaiah, *Chem. Rev.*, **65**, 491 (1965).
45. S. T. Hwang, C. K. Choi, and K. Kammermeyer, *Sepn. Sci.*, **9**, 461 (1974).

Chapter XII

METHODS AND APPARATUS FOR THE DETERMINATION OF TRANSFER COEFFICIENTS

Transfer coefficients, in one way or another, express a capacity parameter for a given set of circumstances. Thus, some such coefficients represent basic values that are adequately dimensioned, and they can be used with the fundamental rate equations. Other numerical coefficients have been obtained under rather specific experimental or pilot-plant conditions, and are therefore of more limited utility.

1 GASEOUS DIFFUSION

Gaseous-diffusion processes are characterized for the most part by the use of permeability coefficients. These are obtained by volumetric and gravimetric methods.

Although many of the gaseous-membrane separations can be computed on the basis of a constant permeability, this is usually not the case when operating conditions are such that the gaseous phase becomes vaporlike in behavior. Thus, low temperatures and elevated pressures will affect the permeability coefficients so that they become concentration dependent.

Volumetric Methods

Brief discussions of the basic methods used to make permeability measurements were included in the treatment of gas flow and water-vapor transfer through polymers in Chapter V.

In essence, there are three types of procedures. In the gravimetric methods, the amount of vapor permeating through a film is measured by a change in weight. The volumetric methods are represented by two versions, where measurements depend on a change in gas pressure (variable-pressure method) or make use of a gas-flow measuring device (variable-volume method). A third type is essentially an isostatic method where a specific detector is used to measure the amounts which permeate.

Variable-Pressure Method

This method has become well known through the publications of Barrer and co-workers. It is frequently called the "time-lag" method. Originated by Daynes [1], it was brought into prominence by Barrer [2], to an extent that it often is labeled the Barrer method. Figure 12.1 typifies the type of data that are obtained with this method [3].

The pressure-time plot contains readings of the pressure change on the downstream side of the membrane due to permeating gas. Usually, the operating conditions are selected in such a way that the change in that pressure is held below about 1 mm of mercury, and thus the pressure on the high-pressure side remains essentially constant. The mathematical interpretation for a flat film results in the expression:

$$D = \frac{l^2}{6t_0} , \qquad (12.1)$$

where D is the diffusion constant; l is the film thickness in centimeters; and t_0 is the time lag in seconds obtained from extrapolation of the straight portion

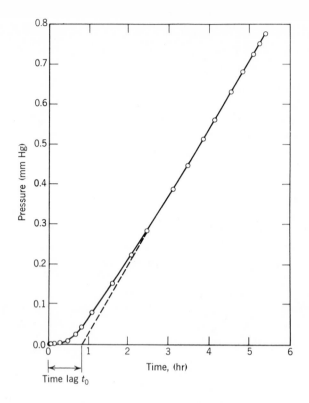

Fig. 12.1 "Time-lag" plot, pressure versus time for CO_2 through Mylar A-100 at $30^\circ C$ and 714 mm Hg pressure [3].

of the function.

Permeability Q is obtained from the slope by transforming the pressure change into a volume reading, and the solubility S_m can be calculated from the well-known relationship

$$Q = DS_m . \tag{12.2}$$

Although any desired pressure level can be used on the low-pressure side of the membrane, most often a vacuum is employed. The degree of sophistication or, on the other hand, of simplicity that may be introduced into the structuring of the equipment is subject to considerable variation. A well-designed assembly representing an apparatus useful for relatively low pressures is pictured in Fig. 12.2 [4].

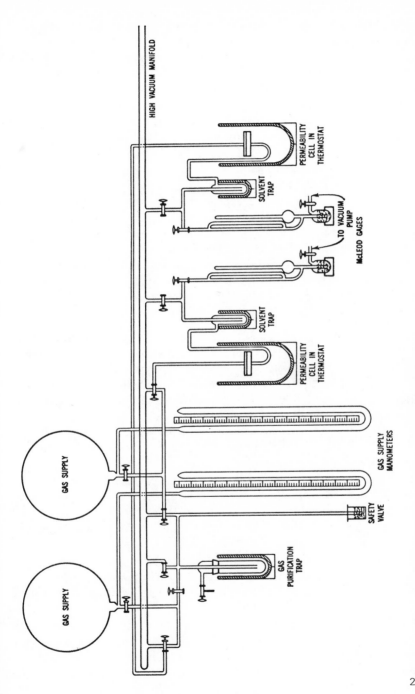

Fig. 12.2 Gas permeability apparatus [4] : variable-pressure method.

299

Equipment that permits considerable latitude of operation is that recommended by ASTM in their standard for gas-permeation measurements [5], ASTM method D-1434-66 (reapproved 1972). A choice of methods is provided, that is, a variable-pressure method, designated as "manometric" method, and a variable-volume method, called a "volumetric" method.

The apparatus recommended for the *variable-pressure* method can be operated by visual observation or by use of a recording device. This apparatus was developed by Brown and Sauber [6] and is often referred to as the "Dow cell." It is commercially available from Custom Scientific Instruments. In order to assure validity of measurements, it is recommended that the directions listed by ASTM method D-1434-66 (reapproved 1972) be consulted.

A versatile design of a highly adaptable piece of equipment was described by Major [7]. It concerns a cell assembly where the film to be tested can be clamped into position, and rapid determinations can be carried out with a rather surprising degree of accuracy. The permeation flux through the membrane is measured by the rise of a mercury column (or other confining fluid) in a vertical glass capillary. The original publication contains adequate details so that a reasonably competent laboratory shop facility can construct the apparatus.

Variable-Volume Method

Originally proposed by Todd [8], this method has found extensive use in a number of modifications. In principle it measures the flow of a gas or vapor (under noncondensible conditions) through a polymer film, by means of a sufficiently sensitive flow meter. Because the method is dependent on the creation of a measurable volume change, it has the disadvantage that film materials of very low permeability are difficult to test.

A well-known and quite versatile modification was developed by Brubaker [9]; the diagram in Fig. 12.3 brings out the rather simple type of equipment that is needed.

Experience has shown that the capillary tubing containing the mercury drop should be vibrated lightly, in order to prevent sticking of the moving mercury. With all equipment where a flat sheet of film is fastened between flanges, care must be exercised to prevent cutting of the film by the edges of the gaskets. This is easily accomplished by a slight rounding off of what are likely to be sharp edges created in the cutting of the gaskets. When the equipment is to be submerged in a constant temperature bath, the capillary can be fashioned with a bend to bring its end above the level of the bath liquid in a snorkel-type arrangement. If a very fast film is to be tested, that is, a film with high permeability, it is possible to use multiple thicknesses of film and larger capillary tubing to slow the gas flow down to a level commensurate with the capacity of the flowmeter. It has been found that even a moderate pressure drop across a film sandwich will eliminate potential gas pockets between film layers and yield the straight-line inverse effect of film thickness.

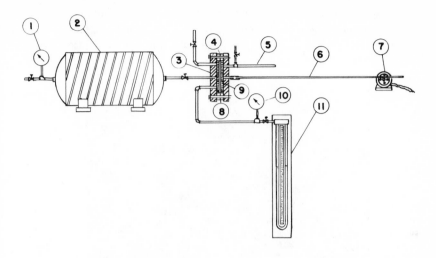

Fig. 12.3 Diagram of variable volume-type permeability apparatus. 1: 600-lb pressure gage; 2: gas storage tank; 3: 11-in. steel flange; 4: rubber gasket; 5: thermometer; 6: glass capillary tube; 7: motor-driven vibrator; 8: plastic film; 9: filter paper film support; 10: 100-lb pressure gage, in 1-lb graduations; 11: 50-in. manometer.

The ASTM apparatus described for the variable-volume method was originally proposed by Stern [10] and is frequently called the "Linde cell." It also is marketed by Custom Scientific Instruments. Here too, the ASTM directions are essential to good operation [5].

A novel and very handy apparatus was built about 10 years ago by Major [11] as a *portable* model. It is illuminating to quote the requirements that were to be met by this apparatus:

1. Mercury should not be used as the containing fluid (to eliminate the nuisance of special mercury handling and cleaning problems).

2. The film holder should be small, light, easy to assemble and disassemble, and should not require the use of special tools.

3. No fragile glass components should be used.

4. No vacuum source and no utilities should be required.

5. The unit should be completely portable.

These aims were actually achieved, so that the film holder and measuring system weighed less than 1 lb and measured less than 6 in. in diameter and 3 in. in height. With the use of a small cylinder as the source of permeating gas (lecture-bottle type), the entire unit becomes completely portable, with no electrical connections for vacuum or other utilities being required. No tools are needed for assembling or disassembling the cell to change samples. Thus, the unit can be used for on-the-spot tests in the plant, in the warehouse, or even in an office.

Fig. 12.4 View of portable gas-permeability apparatus; variable volume type [11]. (Courtesy of *Modern Packaging Magazine.*)

A photograph of the unit is pictured in Fig. 12.4 as it is connected to a cylinder containing the gas whose permeability is to be determined. Actually, with the help of the original publication, this unit can readily be built in in-house facilities. Using plastic tubing, instead of glass, for the flow meter, it is possible to employ water as the slug material. A hypodermic syringe will serve to inject slugs of water into the flow meter tube, which for the sake of compactness is coiled around the cell.

Gravimetric Methods

The gravimetric procedures were developed primarily to measure *water-vapor transmission* (WVT). They can, however, be used equally well for vapor, that is, solvent transfer.

A brief reference to the basic gravimetric method of measuring WVT was given in Chapter V. To repeat, the plastic film is fastened over a cup in which there is liquid water (or any solvent) or a desiccant. The weight loss of water, or weight gain of desiccant, is followed as a function of time, and the WVT coefficient can be calculated directly from the data. This procedure as well as alternate methods were treated in a comprehensive review paper by Newns [12].

Today's version is that recommended by ASTM method E96-66 [5]. The method in itself is basically simple, but there are a fair number of variations and six procedures are described to take care of specific conditions that are most frequently encountered. It would be redundant to list these details, and as with the gas permeation method it is recommended that the ASTM directions be consulted directly. The type of cups and the sealing procedures used can be obtained from the ASTM publication.

Probably the simplest apparatus for obtaining WVT as well as *vapor-transfer measurements* in general, is the procedure called a "dynamic sorption method" by Osburn et al. [13]. It is actually a convenient modificaiton of a method described by Hennessy et al. [14] and attributed to Jeffs and E. P. Noble, who reported on it in unpublished papers. A plastic packet, sometimes referred to as a pouch, is suspended from a balance beam. This arrangement measures transfer from the outside vapor atmosphere to an adsorbent contained inside of the packet, resulting in a weight increase. If desired, the liquid, water or solvent, can be placed inside the packet and the weight decrease be measured.

With a material such as polyethylene the change in weight represents permeated water vapor, as a negligible amount of water will dissolve in the polymer. Data in the original paper [13] show that the rate of weight gain is the same whether a desiccant is inside the packet and saturated water vapor on the outside, or liquid water is inside the packet and dry air on the outside. When sorption of water vapor, or of any vapor for that matter, is significant, two sets of measurements must be made. The upper curve is obtained with a desiccant, that is, the total weight gain, and the middle curve is in essence a blank determination without desiccant, so that the bottom curve is the net gain and its slope measures the WVT. The use of this method with organic vapors was reported by Laine and Osburn [15].

Instrumental Detector Methods

It is difficult to ascertain just when such methods came into use. One of the earlier reports is by Fricke [16] who described the use of a gas chromatograph to detect permeated gas. To do this, a sweep or carrier gas has to be employed to move the permeated stream to the GC unit.

An apparatus specifically intended for oxygen permeation is the OX-Tran unit developed by Modern Controls [17]. An oxygen-specific coulometer detector is used, and the equipment utilizes a carrier gas-flow scheme. Also, provision is made to introduce water vapor for testing moist gas transmission.

The same company markets a gas-permeability meter manufactured in Sweden. It is called an *Isostatic Gas Permeability Meter* GPM-200 [18]. Figure 12.5 is a schematic of the cell construction. Two samples of the same film are mounted in the cell, so that one is a test sample, and the other one serves as a reference. The apparatus is started up by flushing both chambers (and films) with a reference gas as, for instance, hydrogen or helium, until steady-state conditions are reached. Then the gas on the outer side of the film (away from thermistor chamber) is quickly switched over to test gas, and the previously balanced voltage bridge indicates the change in gas composition.

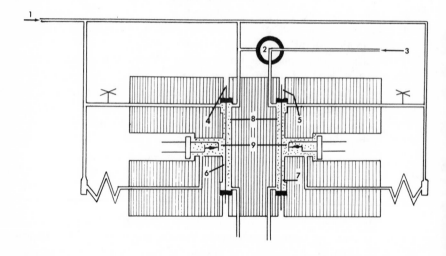

Fig. 12.5 Isostatic gas permeability meter [18]. 1: reference gas; 2: main valve; 3: test gas; 4: reference foil; 5: test foil; 6: reference chamber; 7: test chamber; 8: outer chambers; 9: thermistors.

The *Infrared Diffusiometer* described by Wood [19] is primarily intended for water-vapor-transmission measurements. The sensing element is an infrared hygrometer, that is, a narrow wavelength beam of light that is susceptible to attenuation by water vapor. Normally, the instrument is used to determine the relative merit of a film with respect to a reference standard, as, for instance, Goodyear Pliofilm. Although the cell can be operated at 100% relative humidity (RH) by placing water-soaked sponges in the humid compartments, it is often used with salt solutions to give lower RH conditions. Thus, saturated solutions of ammonium sulfate and zinc sulfate would provide 81 and 90% RH, respectively. The lower RH conditions may be needed with some films that are prone to excessive swelling

Apparatus for Separation Experiments

The primary objective in the construction of membrane separation is to crowd as much area as possible into a minimum of volume. As many investigators will be interested in laboratory or semiscale size equipment, some such devices will be described.

Osburn [20] developed the concept of using a coiled continuous sheet of tubular film with inner and outer porous liners, f.i. paper, which serve as a flow path for gases. This is a rather simple piece of equipment and can readily be constructed on a home-made basis.

Major [21] discovered that a film of silicone rubber could be deposited on a porous paper substrate, such as paper toweling, without complete penetration of the paper matrix. Thus, the uncoated portion of the paper serves as a flow path for the gases. The coated paper sections can be assembled in a multistack unit. A "space" version developed for carbon dioxide removal in space cabins consists of a multistack packet with silicone rubber as the membrane. The pack is potted into a convenient size, and the thickness of the polymer encapsulation is great enough to eliminate gas leakage through it. Also to be mentioned should be configurations suggested by Michaels et al. [22], which represent variations of the Osburn and Major designs.

The well-known spiral-wound artificial kidney for hemodialysis as discussed by Kolobow et al. [23] was developed for liquid systems. Compact, ingenious engineering designs of commercial equipment are the reverse-osmosis stacks [24]. Also, many examples of flat membrane units are represented by the variety of dialysis cells which are in common use. Even though these latter devices were primarily intended for liquid separations, their construction lends itself for application to gas separation.

Diffusion Cell — Fine-Bore Tubing

The steps involved in making a diffusion cell from silicone-rubber tubing are outlined below. If any other tubing or hollow fibers are used, the given method will work when the silicone-rubber caulk adheres adequately to that material or when an appropriate potting compound is used, as, for instance, methylmethacrylate polymer, epoxy compositions, and the like.

The procedure was developed and used successfully by Blaisdell [25]. A diagram of his cell assembly is shown in Fig. 5.20.

Procedure. Steps *a* and *e* are illustrated in Fig. 12.6*a* to *e*. Briefly the steps are as follows:

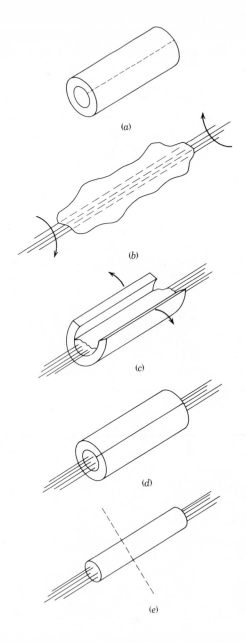

Fig. 12.6 (*a*) to (*e*) Illustration of steps to be followed in preparing a tubular bundle for gas-diffusion cell; bundle is inserted in a cylindrical sleeve for mounting.

1. A piece of Tygon is cut to serve as mold for plug.
2. Tubes are laid out so as not to touch. Silicone-rubber caulk is applied to tubes by spreading motion with a spatula. Caulked tubes are then wound up to eliminate gap.
3. Tygon mold is coated on inside liberally with silicone-rubber caulk, and caulked bundle is depressed into this coating.
4. Tygon mold is closed slowly and tightly. Some caulk should squeeze out of crack in mold. Spatula is run down mold crack to ensure that no tubes are caught in the crease.
5. After 12 to 16 hr plug is dry and mold is removed. Plug is then cut to desired fit.

Hollow Fibers

Hollow-fiber testing calls for fairly sophisticated equipment as the trend is to operation at high pressures. A view of a Du Pont test stand is given in Fig. 12.7. Accessories such as pumping equipment, though not shown, are, of course, needed. For short-time testing a pressurized tank can be used for the supply of feed gas.

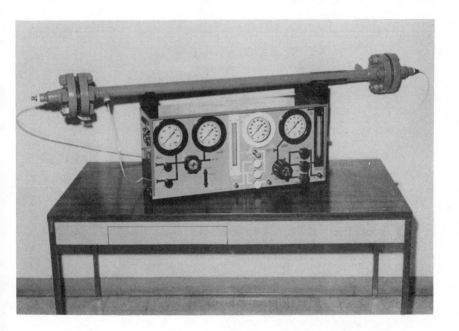

Fig. 12.7 Test module for hollow-fiber evaluation. (Courtesy Du Pont Company.)

2 PERVAPORATION

Measurements in pervaporation are the same as in the usual permeability determinations. Even in experiments on separations the flow in the liquid phase is not critical, and there is little or no change in composition of the liquid phase. Consequently, the volumetric methods of permeability determination and, in particular, the gravimetric methods are utilized as described for gaseous-diffusion coefficients.

Equipment for testing the performance of membranes is relatively simple and can easily be constructed on a home-made basis [26]. What is required is a film holder, usually circular flanges, and a container where the assembled film can be immersed in the liquid to be tested. The simplest way to carry out experiments is to have one side of the membrane exposed to the liquid, while the other side is connected to a vacuum pump. Ahead of the pump, there will be a cold trap to freeze out the permeated vapors.

This scheme is illustrated in Fig. 12.8, where the liquid side of the holder consists of an open ring only. If it is desired to operate with elevated pressures on the liquid side of the membrane, the membrane holder is closed on both sides with solid flanges, and the liquid is circulated over the membrane by means of inlet and outlet ports in the flange.

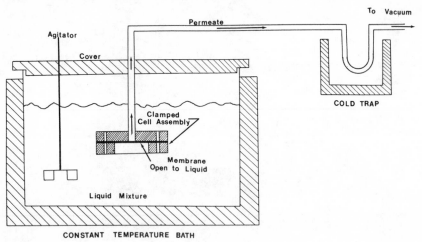

Fig. 12.8 Schematic of laboratory setup for pervaporation experiments; single-stage operation.

3 LIQUID-PHASE PROCESSES

Conventionally, the determination of transfer coefficients is carried out in batch or continuous test cells. Usually, the data are then reported as functions of operating parameters.

The information that has been and continues to be published, especially under the aegis of the Office of Saline Water, is simply overwhelming. However, a considerable amount of the technical aspects cover operating data that are quite specific to the prevailing local conditions, and structural and design details. Although this information is certainly useful and can be applied to similar situations, it is not suited for developing basic correlations of general utility.

Some data can be evaluated in terms of dialysance and clearance that have the character of a basic coefficient. However, the inevitable effect of boundary-layer conditions that exert a very great effect on membrane performance in liquid flow makes such coefficients quite sensitive to equipment design. Readings are taken in terms of flow quantities, that is, volumes or weights, and respective changes in solute compositions of feed and product streams. It is somewhat difficult to categorize equipment for liquid-phase processes in a completely satisfactory manner. This is so because some items can readily be used for more than one process. For instance, laboratory equipment for dialysis is likely also to be serviceable in ultrafiltration experiments. Therefore, the arrangement decided upon is somewhat arbitrary, but it does present the desirable information in an adequate manner.

Methodwise, the procedures in liquid-phase processes, of course, consist simply in contacting the liquid mixture to be separated with an appropriate membrane under operating conditions that will establish the proper drivign force. Separation should then occur under the influence of a concentration gradient and usually the concomitant application of pressure in varying degrees. As in any other testing method one can operate in steady state, that is, carry out a continuous flow test, or in unsteady state. The latter would be essentially a batch experiment where conditions in the equipment will vary with time and tend to approach an equilibrium state.

Ultrafiltration and Dialysis

The equipment for these two processes, particularly the laboratory-scale devices, is very much alike, and the discussion is thus facilitated when both of them are treated together. There is a more definitive difference when it comes to reverse osmosis; a separate section will cover it.

Membranes

Although the subsequent Chapter XIV on membrane preparation treats this subject quite thoroughly, there are some specific aspects, especially in regard to laboratory equipment, that need to be emphasized.

A wide variety of membranes for dialysis and ultrafiltration is available from the Amicon Corporation [27]. The so-called DIAFLO ultrafilters are described as anisotropic membranes with a very thin (0.1-1.5 μ), dense "skin" of extremely fine, controlled pore texture upon a much thicker (50-250 μ), spongy layer of the same polymer wtih increasingly larger openings. This type of structure is well known to give high selectivity and through-put, at modest pressure with minimal clogging because retained substances are rejected at the surface. This same type of membrane is available in hollow-fiber form, called DIAFIBER. This configuration consists of fine cylindrical tubes, each with a passage of uniform diameter of about 0.2 mm. The polymer composition imparts inertness and nonionic character. The fibers have the same anisotropic (skinned) structure as the DIAFLO membranes. The retentive side of the membrane is on the fiber interior through which the process stream flows, and permeating species may pass the fiber wall in either direction.

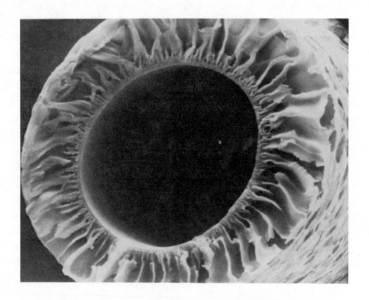

Fig. 12.9   Hollow-fiber micrograph. (Courtesy Amicon Corporation.)

A photomicrograph of the fiber cross section is pictured in Fig. 12.9. For use in dialysis or ultrafiltration equipment, many fibers (typically 1000 or more) are sealed in a disposable cartridge, providing very large area per unit volume. Information regarding the compositions of the various membrane types is available only in general terms. The DIAFLO and DIAFIBER configurations are of noncellulosic nature, and are reported to be polyelectrolyte types (see Chapter XIV). Another Amicon membrane is the DIAPOR microporous filter, which consists of a cellulosic polymer on heavy filter paper with uniform 70 to 80% porosity. A considerable variety of all three types of membranes is available and details for their proper use and selection for molecular weight cutoff points are clearly presented in several Amicon bulletins [27, 28].

Membranes manufactured by Dorr-Oliver are stated to be made of a substituted aromatic polymer (noncellulosic). Evidently, their configuration is anisotropic, that is, a thin skin 0.1 to 10 μ in thickness and a porous substrate in the 5-to10-mils range. The membranes are flat sheets that are used in a parallel-plate arrangement and intended primarily in ultrafiltration as so-called IOPOR membranes [29, 30].

Cellulose acetate hollow fiber membranes are marketed by the Dow Chemical Company. A summary article by Mahon and Lipps [31] contains a fair amount of details on fiber compositions and spinning techniques. The membranes are tailored for various molecular-weight cutoff points. So, the ultrafiltration membrane works in the 20,000 to 40,000 mol wt range, and the dialysis modification will pass materials below 5000 mol wt. An osmosis membrane is also available and its molecular-weight cutoff range is approximately 100 to 300. Wall thickness of the hollow fibers is 20 to 30 μ, and inside diameters are quoted to vary from 180 to 230 μ. The membrane used in dialysis is a cellulose composition, and the osmosis and ultrafiltration membranes are made of cellulose acetate.

The well-known microporous membrane filters manufactured by the Gelman Instrument Company and by Schleicher & Schuell are primarily intended for a host of filtration operations. Available types are numerous and suitable for use with gases and liquids.

Gelman's *Metricel* membranes with pore ratings of 5 μ down to 75 Å are made in a variety of compositions, such as cellulose triacetate, mixed cellulose esters, cellulose acetate, regenerated cellulose, nitrocellulose, and cellulose acetate blend, polyvinyl chloride, copolymer of acrylonitrile and polyvinyl chloride, all of unsupported type [32]. *Acropor* filters consist of a polyvinyl chloride-acrylonitrile copolymer membrane coated on a nylon web. They are also made as ion-exchange resin membranes. A Metricel filter for protein enrichment is considered to be an ultrafiltration membrane.

The Schleicher & Schuell *Selectron* membrane filters likewise serve for a great variety of applications, from air sampling to biological and medical laboratory uses [33]. Compositions of the membranes are cellulose based, such as nitrocellulose or cellulose acetate. Porosities vary from 8 μ down to 50 Å and the use of a 0.2-μ filter results in cold pasturization of a fluid, as does the corresponding Gelman filter. There is a special collodion bag filter, which is made of nitrocellulose with pores of apporximately 5 nm. This membrane is suitable for ultrafiltration in the enrichment of proteins having molecular weights of 25,000 and larger [34].

The *Millipore* filters are in the same category as Gelman's and S & S products. An electron micrograph of their surface structure is presented in Chapter XIV. Some of their membranes are specifically designated for ultrafiltration uses [35].

Laboratory Equipment

The simplest laboratory dialysis procedure is to place the solution to be dialyzed into a sausage type of membrane as, for instance, a seamless cellulose tube, tie off both ends, and submerge the encased sample into a beaker. Usually, a stirring device is used to provide a rotating type of agitation. The dialysis will proceed to an equilibrium state. Normally, a timing schedule is worked out for routine testing. Rather convenient devices are the multiple dialyzer models of Pope Scientific [36], where the dialyzer bags can be fastened to hooks on a rotating shaft for agitation. It is recommended that the reader procure company catalogs, which contain excellent illustrations.

Somewhat more elaborate equipment would be represented by the well-known rectangular dialysis cells, often built of transparent plastic, which have been available from numerous apparatus dealers for many years. Diagrams and pictures of these devices can be found in any number of catalogs usually available in laboratories.

Testing equipment is offered by *Amicon* in the following types: stirred cells, thin-channel systems, hollow-fiber systems, and nonagitated systems. For each type several different models are available. Concentration polarization effects, of course, depend on the type, so that the thin-channel and hollow-fiber systems give a high degree of control, the stirred system a moderate degree, with no control in the nonagitated system [27, 28].

Description of the equipment is worthwhile. A rather unique item is the MINICON Series designed for microscale operation. This apparatus has the dimensions of 4.5 in. length, 2 in. height, and 1.5 in. width at base. The item is considered a disposable unit. It finds extensive use in concentrating and purifying solute mixtures; in serum and plasma studies; urine, cerebrospinal fluid, and other biological media. Samples as small as 1 ml can be processed in each cell, of which there are 8 or 10 in each unit depending on the respective model configuration. The cells are backed by a membrane, ultrafiltration or selective permeability type, and the membrane is in contact with an absorbent pad. Water and permeating solutes pass through the membrane into the adsorbent, and the sample volume drops until the level reaches the impermeable area of the membrane, or with viscous materials the viscosity will reduce the flow, due to the falling hydraulic head, to a negligible level.

The *stirred cell* assembly contains a built-in magnetic stirrer. Thin-channel and hollow-fiber modules are available. The latter units can be operated at pressures up to 15 to 25 psi, and the DIAFIBER membrane cartridges may be reused many times, or for very long, continuous operation. They are easily cleaned between uses.

A scaled-up system is also available. Here, three Diafiber cartridges are arranged for parallel operation with 10 ft^2 of effective area in each cell. A diaphragm pump forces the feed stream through a prefilter to the cells, and

manifolding permits solution concentrating, ultrafiltration, or dialysis operation with pressure ranging up to 25 psig. Capacity of the unit is such that
multiliter quantities can easily be processed at a rate of about 1 liter/min with
dilute solutions. Specifications given by Amicon are as follows [28]:

Size	33 X 10 X 35 in.
Weight	125 lb
Hollow-fiber Cartridges	
Length	26¾ in. overall; 24 in. fiber length
Cutoff (nominal)	10,000 mol wt
Area (3 cartridges)	30 ft^2
Internal diameter	0.020 in.
No. of fibers per cartridge	1050
Prefilter	100 μ mesh
Pump	Air-operated, dual-diaphragm type
Compressed air requirement	1 CFM at 50 psi
Maximum inlet pressure	25 psig
Minimum process volume	10 liters
Minimum hold-up volume	Approximately 200 ml

It is evident from the catalog views that all of the above items of equipment
are used at very low overpressure conditions. Thus, for operation at pressures
up to 100 psi a recirculating type of separator is available. Built of stainless steel
with a filter diameter of 6 in. the unit provides an effective membrane area of
0.15 ft^2.

Ultrafiltration at very low pressures, that is, less than about 15 psi can be
carried out in a hollow-fiber device available from the Dow Chemical Company
[37]. This is a small laboratory-type apparatus, and essentially the same
configuration is usable for dialysis, ultrafiltration, and osmosis. The Dow Hollow
Fiber Devices, as they are known, are covered by U.S. Patents [38], and these
should be consulted if greater details are wanted.

In principle, each unit is made up of a hollow-fiber bundle and is quite like
an artificial kidney assembly. Several arrangements are available, and the most
common laboratory unit consists of a U-shaped fiber bundle. A straight through-
flow model is also available, constructed on the shell and tube principle.
Dimensions are the same for all uses, but the fiber type varies with the
application [31, 37]. What appear to be identical models are also available
from BIO-RAD Laboratories [39].

Rather recent items of equipment are the *Dowex* version of hollow-fiber
design [37]. Figure 12.10 shows that they are suitable for very small scale
experimentation. These miniature units are made as hollow-fiber devices, with
fiber mounted in nontoxic polyurethane tube sheets and contained in Pyrex
or Polycarbonate jackets. The fibers are of nominal 200 μ i.d. size, and a variety

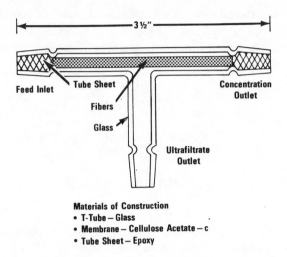

Materials of Construction
- T-Tube — Glass
- Membrane — Cellulose Acetate — c
- Tube Sheet — Epoxy

Fig. 12.10 Dimensional diagram of Dowex small-scale modules [37].

of compositions are available to fit different applications. Thus, fibers of silicone rubber, silicone-rubber copolymer, and several types of cellulose acetate can be used for gas permeation (gas to liquid), dialysis, ultrafiltration, concentration and purification of biologicals, and cell-culture studies. Somewhat larger-scale units are also available to bridge the gap to beaker scale operation.

An *Iopor* Laboratory unit is supplied by Dorr-Oliver for ultrafiltration experiments [29]. The unit comes as a compact test stand with three series-mounted test cells built of Lucite material. All necessary accessories are incorporated in the assembly. Continuous processing of a 3-gal sample can be conducted with 0.15 ft^2 of total membrane area giving an ultrafiltration rate of 2 to 4 cc/min. Thus, three different membranes can be evaluated simultaneously under the same processing conditions. Maximum allowable pressure is 100 psi.

Millipore Corporation markets ultrafiltration cells with membrane areas of 0.7, 3, 3.9, and 11.3 cm^2. The capacity of the largest unit is 90 ml, with internal piston, and construction is rugged to permit pressures up to 275 psig [35]. Mountings for the *Gelman* filters are all of conventional filter-holder design [32]. The *Schleicher & Schuell* collodion bags for enrichment of protein solutions are conveniently mounted in glass unit designed like a vacuum flask [34].

Bench-Scale Equipment

Bench-scale equipment for the more usual operating pressures in ultrafiltration, that is, about 50 psi, is available from *Abcor* as tubular units [40]. For laboratory use the module is 54 in. long, and for pilot scale and commercial installations the tubes are 110 in. in length. The 54-in. tube provides a transfer area of 1.1 ft^2 and handles approximately 0.15 gal/ft^2. Operating temperature should be maintained between 4 and 60°C. Complete pilot units also are available. One such unit in tubular construction contains 22 ft^2 of membrane area and can process 50 to 250 lb of feed per hour, depending on feed characteristics, concentration desired, and fraction of membrane area in active use. A larger unit containing 480 ft^2 is also marketed. It can be used as a pilot plant with a capacity of 3000 lb/hr, and it is also utilized as the basic building block for full-scale plants. The tubular membranes are described as seamless and integrally bonded to the inside of porous polyolefin tubes. Typical pressure drops are 20 to 35 psi.

For bench scale testing, as well as industrial installations, the *Amicon* membranes have been incorporated into appropriate equipment modules. These are marketed by Romicon [41]. A convenient unit for a flow capacity of 5 gal/min and an operating pressure of 100 psi consists of a cartridge mounted on a test stand. The cartridge contains 7 membrane-tube assemblies fixed between two headers. Each tube consists of a splined core wrapped with a membrane strip that is sealed along the length of the tube. The length of the cartridge is 43 in., and the active membrane area amounts to 1.6 ft^2. Any of the membrane types of Amicon's production are available, and Table 12.1 lists some characteristics, including molecular weight cutoffs. If larger capacities are needed, modules at 13 and 45 gal/min are available. For details on materials of construction and recommended operating parameters, the various bulletins should be consulted.

Table 12.1 Romicon Ultrafiltration Membranes[a]

DIAFLO	Nominal Molecular Weight Cutoff	Apparent Pore Dia. A	Water Flux (GSFD) (at 55 psi)
UM 05	500	21	5
UM 2	1,000	24	9
UM 10	10,000	30	40
PM 10	10,000	38	360
PM 30	30,000	47	540
XM 50	50,000	66	180
XM 100A	100,000	110	675
XM 300	300,000	480	1350
HM	—	—	180

[a] Reference 41.

Dorr-Oliver's *Iopor* system is made in a 2 ft² module size. The unit can be obtained as a complete test stand, or only as a membrane cartridge mounted in a stainless-steel housing [29]. A view of the cartridge, using flat sheets of membrane is shown in Fig. 12.11. Ultrafiltration flux rate is 0.5 to 1.5 gal/hr, and maximum system pressure is 50 psi.

Fig. 12.11   Iopor module, flat-sheet cartridge. (Courtesy Dorr-Oliver, Inc.)

Reverse Osmosis

Testing equipment for reverse osmosis is quite a bit more substantial than the usual permeability apparatus. Pressures are likely to go as high as 1500 psi, and the equipment thus falls in the category of high-pressure apparatus Consequently, the required sturdiness and the pumping accessories call for careful design and appropriate safety features.

In their pioneer studies Reid and Breton [42] developed a rather simple apparatus as shown in Fig. 12.12. It consisted of a 2-ft length of 3-in. schedule

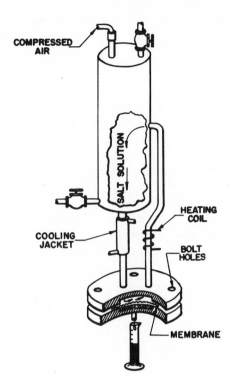

Fig. 12.12 Reverse-osmosis apparatus of Reid and Breton [42].

80 Monel pipe with a capacity of 2.75 liters. The approximate dimensions of the membrane holder and connections can be inferred from the photograph. They are detailed in Reid's paper. The title of the original figure was "Apparatus for determining the semipermeability of *imperfect* osmotic membranes," and the terminology indicates the degree of originality of the investigation. Certainly, this model is still good enough for investigational studies and has the advantages that it can be constructed in most laboratory facilities.

Bench scale and pilot plant units are manufactured by *Abcor* covering a capacity range of 200 to 600 gal/day. The bench scale unit is described as follows [40]:

". . .reverse osmosis membranes are seamless, one-half inch in inside diameter and 60 inches long. They are inserted into porous fiberglass/FDA-approved epoxy-impregnated support tubes, fourteen of which are arranged in a circular pattern and connected by stainless steel headers in each module. A clear plastic shroud encases each module for collection of permeate. The effective membrane area in each module is 9.5 sq ft. Volume is approximately 0.75 gallon."

The pilot-unit module contains 14 ft^2 of membrane area, is 60 in. long, with 20 tubes per module. Operating pressures in both modules are in the 300-to- 600-psi range, with a maximum of 1500 psi.

Equipment that is suitable for reverse osmosis, ultrafiltration, and hyper- filtration is offered by *De Danske Sukkerfabrikker*. Their Module 20, having an active area of 0.36 m^2, is designed for an operating pressure of 80 atm (1200 psi). A smaller unit of one liter capacity is also available. It is of the usual batch type and built for a working pressure range up to 150 atm (2200 psi). The arrangement of the Module-20 unit is shown schematically in Figure 12.13. The column section thus consists of layers of plates, frames, and membranes stacked on the center bolt and compressed by end flanges. For the laboratory and production units, eleven membrane types are available. Details of their composition are not disclosed, except for the fact that they are cellulose acetate [45].

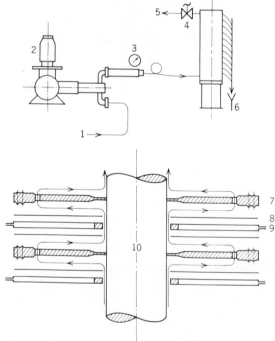

Fig. 12.13 Diagram of internal arrangement of De Danske module for laboratory- and large- scale work. 1: Inlet; 2: pump; 3: manometer; 4: pressure relief valve; 5: concentrate outlet; 6: permeate outlet; 7: membrane spacer; 8: membrane; 9: membrane support plate; 10: center bolt.

Liquid Films and Reactive Membranes

The investigations in this area have been discussed in Chapter VIII. So far, the work has been mainly on a laboratory scale and perhaps some scaleup to pilot-plant size with liquid film.

There is no particular equipment available as yet from suppliers. Actually, for laboratory work with liquid films it is essentially a matter of setting up beakers, stirrers, and separating funnels to create the liquid-film-encapsulated phase, contacting it with a bulk phase, separation, and finally breaking the liquid-film phase.

With reactive membranes the procedures so far are strictly of a research nature. Therefore, the investigator will have to resort to respective references in order to develop means of creating and testing such membranes.

4 ELECTROMEMBRANE PROCESSES

These processes comprise electrodialysis, electro-osmosis, and variations or combinations thereof. Activities, however, seem to be limited to electrodialysis, and thus only equipment for this process will be covered.

The carrying out of electrodialysis experiments simply consists of charging the respective liquid into a cell provided with proper membrane and electrodes, and conducting the operation under a variety of parameters. Thus, one can use equipment ranging from the most elementary cell combination to sophisticated instrumentation.

It seems that laboratory-scale apparatus is not available to a satisfactory degree. Some rather nice items of equipment that were available several years ago have gone off the market. The only readily available item that we could ascertain is the Bradfield Electrodialyzer of Arthur H. Thomas Company [43]. It is described as consisting of a three-compartment cell of borosilicate glass. Any desired type of membrane can be utilized. Platinum anode and nickel cathode are welded into glass tubes and mounted adjustably in the ends of the apparatus. The whole assembly is supported in a frame with provision for electrical insulation.

While laboratory experiments in electrodialysis can be conducted with fairly simple apparatus, testing on a bench scale calls for more substantial equipment. A very good example of the type of equipment that is available is the Stackpack module of Ionics, Inc. Their model, a bench-scale assembly, has a hydraulic capacity up to 0.8 gal/min.

The multicell unit is a membrane stack with dimensions of 9 by 10 in., and normally contains alternate cation and anion membranes. For testing purposes the pumping and piping equipment is kept as simple as possible, even to open solution tanks. Each unit incorporates an electron-arc rectifier to convert line ac power to controllable variable voltage dc power.

An Ionics bulletin [44] states that electrical requirements can be computed from voltage, current density, and pressure-drop readings. Composition of the electrodialyzed solution, of course, is measured in a conventional manner. Membrane area requirements are computed from changes in solution composition, current density, and flow rate. As with any bench-scale equipment appropriate variations in process parameters will yield information on process optimization.

An exploded view of the cell schematic as shown in Fig. 12.14 is informative. This view illustrates the manifold holes punched through the membranes and the spacers. In this specific arrangement there are alternate cation and anion membranes. In other types of operation the cell assembly may have only anion or cation membranes. An example of a cell using anion membranes only is presented in Chapter XV for citrus juice processing.

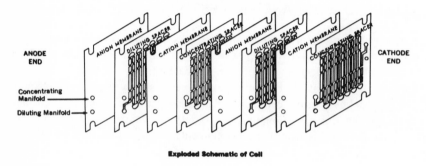

Fig. 12.14   Open view of electrodialysis cell assembly [44].

References

1. H. A. Daynes, *Proc. Royal Soc. A,* **97**, 286 (1920).
2. R. M. Barrer, *Trans. Faraday Soc.,* **35**, 628 (1939).
3. V. Stannett and H. Yasuda, "Permeability," in *Crystalline Olefin Polymers,* Part II, R. A. F. Raff and K. W. Doak, Eds, Wiley-Interscience, New York, 1964, Chap. 4.
4. C. Rogers, J. A. Meyer, V. Stannett, and M. Szwarc, *Trappi,* **39**, 737 (1956); also Ref. 3.
5. ASTM Standard Methods, E96-66, D434-66 (reapproved 1972) "Part 27" 1972 Annual Book of Standards; American Society for Testing and Materials.
6. E. Brown and W. J. Sauber, *Modern Plastics,* **36**, 107 (1959).
7. C. J. Major and K. Kammermeyer, *Modern Plastics,* **39**, 135 (1962).

8. H. R. Todd, *Mc ern Packaging*, **18**, 124 (1944).

9. D. W. Brubaker and K. Kammermeyer, *Ind. Eng. Chem.*, **44**, 1465 (1952).

10. S. A. Stern, P. J. Gareis, T. F. Sinclair, and P. H. Mohr, *J. Appl. Polymer Sci.*, **7**, 2035 (1963).

11. C. J. Major, *Modern Packaging*, **36**, 119 (1963).

12. A. C. Newns, *J. Textile Inst. Trans.*, **41**, T269 (1950).

13. J. O. Osburn, K. Kammermeyer, and R. Laine, *J. Appl. Polymer Sci.*, **15**, 739 (1971).

14. B. J. Hennessy, J. A. Mead, and T. C. Stening, *The Permeability of Plastic Films*, The Plastics Institute, London, 1966.

15. R. Laine and J. O. Osburn, *J. Appl. Polymer Sci.*, **15**, 327 (1971).

16. H. L. Fricke, *Package Eng.*, **7**, 51 (1962).

17. The OX-TRAN, Modern Controls bulletin.

18. Permeability Meter GPM-200, bulletin: Incentive Research & Development AB, Box 11074, Bromma 11, Sweden U.S. Agent: Modern Controls.

19. R. Wood, *Modern Converter*, Sept., Oct. (1970).

20. J. O. Osburn and K. Kammermeyer, *Ind. Eng. Chem.*, **46**, 739 (1954).

21. C. J. Major and R. W. Tock, *Atmosphere in Space Cabins and Closed Environment*, K. Kammermeyer, Ed., Appleton-Century-Crofts, New York, 1966, Chap. 6, p. 120.

22. A. S. Michaels, H. J. Bixler, and P. N. Rigopulos, World Petr. Congr. Proceedings, *New Concepts on Techniques for Hydrocarbon Separation*, 21 (1967).

23. T. Kolobow, W. Zapol, and J. Marcus, *Organ Perfusion and Preservation*, J. C. Norman, Ed., Appleton-Century-Crofts, New York, 1968, Chap. 15.

24. S. Sourirajan, *Reverse Osmosis*, Academic, New York, 1970, Chap. 8.

25. C. T. Blaisdell and K. Kammermeyer, *A.I.Ch.E. J.*, **18**, 1015 (1972).

26. S. N. Kim and K. Kammermeyer, *Separation Sci.*, **5**, 679 (1970); S. N. Kim, Ph.D. thesis, University of Iowa, 1970.

27. Amicon Corp. Publication Nos. 426 and 427.

28. Amicon Corp. Publications 431-433; 435.

29. Dorr-Oliver, Inc., Bulletin No. 10-3.

30. I. Bermberis, P. J. Hubbard, and F. B. Leonard, Paper presented at Winter Mtg. *Am. Soc. Argric. Engrs*, Chicago, Ill. (Dec. 10, (1971).

31. H. I. Mahon and B. J. Lipps, "Hollow-Fiber Membranes," in *Encyclopedia of polymer Science and Technology*, Vol. 15, Section H, Wiley, New York, 1971, p. 258.

32. Gelman Instr. Co., Company bulletins.

33. Schleicher & Schuell, Inc., Catalog No. 100.

34. Schleicher & Schuell, Inc., Bulletin No. 145.

35. Millipore Corp., Catalog MC/1.

36. Pope Scientific, Inc., Company bulletin.

37. The Dow Chemical Co. bulletins; general info., b/HFD-1 dialyzer, b/HFU-1 ultrafilter, b/HFD-1 osmolizer, Membrane Systems.

38. U.S. Pattents 2,972, 349 and 3,228,876, The Dow Chemical Co.

39. BIO-RAD Laboratories, Richmond, Calif.

40. Abcor, Inc., Cambridge, Mass., Company bulletins.
41. Romicon, Inc. (Rohm and Haas & Amicon Corp.), Company bulletins.
42. C. E. Reid and E. J. Breton, *J. Appl. Polymer Sci.*, 1, 133 (1959).
43. Arthur H. Thomas Co., Catalog.
44. Stackpack Electrolysis Plant, Bulletin L-2 (2nd Edition), Ionics, Inc.
45. A/S De Danske Sukkerfabrikker, Catalogue "Reverse Osmosis, Ultrafiltration, Hyperfiltration."

Chapter **XIII**

ENGINEERING ASPECTS OF MEMBRANE SEPARATION

A considerable variety of factors can be lumped under the designation of engineering aspects. To keep the presentation to a manageable extent, only a few major considerations will be dealt with in some detail. Thus, the subjects of available operating schemes, cascade operation, and the effect of boundary-layer phenomena in liquid-phase separation will receive major attention.

1 OPERATIONAL SCHEMES

It is evident that methods of membrane separation will in the main be analogous to those established for conventional separation processes. A few specific arrangements have, however, been proposed.

Thus, possible operating schemes are as follows:

Single-stage operation
  1. Countercurrent flow
  2. Cocurrent flow
  3. Cross flow
  4. Steady-cycle pulsing

Multistage operation
  1. Nonrecycling cascades
  2. Cascades with recycling
  3. Two-dimensional cross-flow cascade

The single stage methods are pictured diagrammatically in Fig. 13.1*a* to *d*. Diagrams 13.1*a*, 13.1*b*, and 13.1*c* are self-explanatory. To obtain an understanding of Diagram 13.1*d*, the original publication by Paul [1] should be consulted. As this scheme represents at present only a suggested method of operation, a detailed treatment is not warranted. The diagrams for multistage operations are treated in detail in the section on cascade operation. They appear as Figs. 13.5 to 13.10.

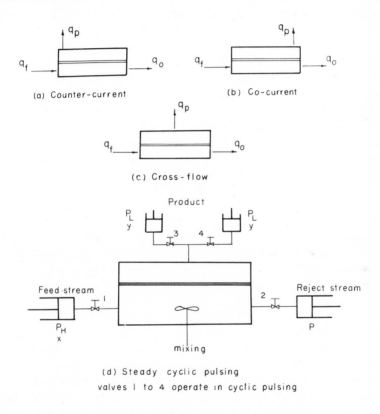

(a) Counter-current

(b) Co-current

(c) Cross-flow

(d) Steady cyclic pulsing
valves I to 4 operate in cyclic pulsing

Fig. 13.1 (*a*) to (*d*) Diagrammatic presentation of single-stage flow schemes.

2 CASCADE OPERATION

Separation processes by means of membranes result, in general, in a *partial* enrichment or depletion of the permeating stream. There are a few known exceptions where complete or almost complete separation of some specific component can be attained. Two of them are the separation of helium from natural gas by means of a silica barrier, and hydrogen purification by means of palladium foils. These processes will be discussed in a later chapter.

Usually, the change in composition from a given feed stream to that of the desired product is greater than that which can be accomplished in a single stage. Consequently, it becomes necessary to go to multistage operation, which is commonly referred to as *cascade* operation. Of course, the cascade scheme can be, and is being applied in any multistage process, as, for instance, distillation, extraction, absorption, and related mass-transfer operations. The basic design concepts consider each stage as a unit in which "equilibrium" conditions are attained and deviations from this idealized condition are handled by introducing a stage efficiency factor. Many good textbooks cover stagewise operation in an authoritative manner and in considerable detail, as, for instance, the works of Smith [2] and of Henley and Staffin [3].

Most of the available textbooks, however, deal with the rather conventional processing, where staging is quite simple and straightforward. Cascading, however, requires some special considerations that are normally not treated. The book by Benedict and Pigford on nuclear chemical engineering [4] contains several chapters on isotope separation with reference to cascading. The most authoritative presentations, however, are those of Cohen [5] and Rozen [6], both dealing primarily with isotope separation.

The cascade relationships deserve treatment in considerable detail as they apply to any and all multistage operations, irrespective of the mechanisms that control the individual process.

In principle, there are two possible methods of cascade operation. One method calls for straight-through passage of product streams without recycling of any intermediate streams. The other method utilizes recycling of intermediate reject streams.

Simple Cascades

The simplest procedure is that pictured in Fig. 13.2. Obviously, as the permeated stream V will always be only a fraction of the feed to each cell, the final product stream will rapidly become very small in amount. Of course, one can use cells of different sizes. A more convenient way to overcome this difficulty is to use a tapered-cascade arrangement where each stage is of the same size. By a proper choice of multiple parallel stages it is then possible to obtain a reasonable amount of product. This scheme is shown in Fig. 13.3. The quantity of final enriched product V_1, of course, depends on the cut $\theta = V/L$, which is used in the various stages. For instance, if $\theta = 0.5$, then an 80-moles feed to Stage 1 whould yield only 5 moles of product V_1.

It is evident that the nonrecycle schemes involve the discharge and nonutilization of considerable material in the partially enriched reject streams. This is a serious disadvantage and thus one would use such schemes only when relatively few stages are involved. Again, it must be noted that each V stream is to be recompressed before going as feed to the next stage.

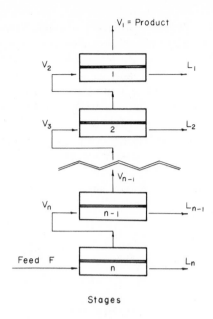

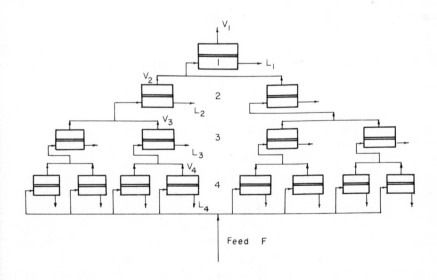

Fig. 13.2 Diagram of simple once-through cascade. F = feed; V = permeated stream; L = Reject stream. Each V stream must be recompressed before entering the next stage.

Fig. 13.3 Diagram of tapered cascade; no recycling.

Because of the potentially high inefficiency of simple cascade operation, the recycling of partially depleted streams is employed. This becomes an absolute necessity when the feed material is a fairly valuable product. If the starting material is of little value as, for instance air would be in the production of an oxygen-enriched stream, one could afford to discard all reject streams. Almost any other material, however, would not permit this and recycling becomes necessary.

Wankat [7] proposed as a new cascade design, to arrange separation cells in a square arrangement with flow of separating agents in a cross-flow pattern. The scheme is diagrammed in Fig. 13.4. The concept was developed primarily to be used in liquid-liquid extraction, and in essence it represents an extension of the

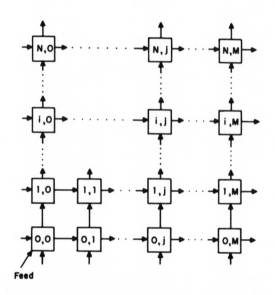

Fig. 13.4 Two-dimensional cross-flow cascade. Reprinted from Ref. 7, by courtesy of Marcel Dekker, Inc.

Craig countercurrent extraction scheme to two dimensions. The basic requirement is that distribution coefficients for components to be separated will differ when exposed to the cross flow of two separating agents. Wankat emphasizes that this scheme could be particularly useful in separating multicomponent mixtures. Although as stated, the scheme was developed for extraction, it should be suitable in principle with arrangements where the individual units are composed of membrane-separation cells. A possible application might thus be in sweep diffusion, where the use of two different sweeping agents could be considered.

The subject of *cascade* operation is of rather fundamental importance, and, therefore, it is treated in depth in the following section.

Recycling Cascades

The conventional McCabe-Thiele diagram for a distillation column represents a well-known example of a recycle cascade. There, the reflux ratio remains under external control, but the internal overflow ratio is in principle not subject to manipulation by the operator.

In gaseous diffusion, however, the overflow ratio, which has previously been termed the cut, remains under the control of the operator, and individual cells can be operated with varying cuts. The subject of the operating lines in cascade operations was treated in considerable detail by Hwang [8], and the essential features are presented here.

Equilibrium Curve

The "equilibrium curve" as applied to cascade operation expresses the composition relationship between the enriched stream (diffused or product stream) and the depleted stream (undiffused or bottom stream) of any one single stage of the cascade. Obviously, the meaning of equilibrium is not the same as that used in thermodynamics. The expression "equilibrium curve" is retained in this book because it is commonly accepted. It should be noted that a number of individuals disapprove of this usage.

For the gaseous-diffusion cascade, the stage equilibrium curve is established by Weller and Steiner [9] and by Naylor and Backer [10] assuming various models of cell operation. In any case, the equation of the equilibrium curve is a function of concentrations, pressures, and the interrelated mass flow quantities, that is, the "cut."

In the case of a series of columns each column in the series corresponds to a single stage in the cascade.. Consequently, when the operating conditions for each column are specified, such as reflux ratio or number of stages, an equilibrium equation can be established for the respective concentrations of the outlet streams.

Operating Lines

The operating line or lines depend on the operating scheme used in a cascade. It is possible to operate cascades with either constant cut for all stages, or a variable cut. Also, as in any multistage process, one can use a complete cascade consisting of enriching and stripping sections, or use only one or the other of these sections. Figures 13.5 to 13.7 picture conditions of variable cut, or overflow, for enriching and stripping, respectively, and the case of constnat overflow.

Referring to Fig. 13.5, the total material balance around top and nth stage gives

$$V_{n+1} = L_n + D \ ; \tag{13.1}$$

component balance:

$$y_{n+1}V_{n+1} = x_nL_n + y_DD \ . \tag{13.2}$$

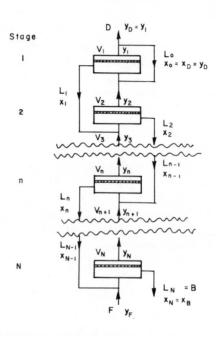

Fig. 13.5 Cascade of enriching section.

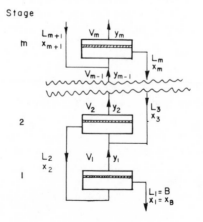

Fig. 13.6 Cascade of stripping section.

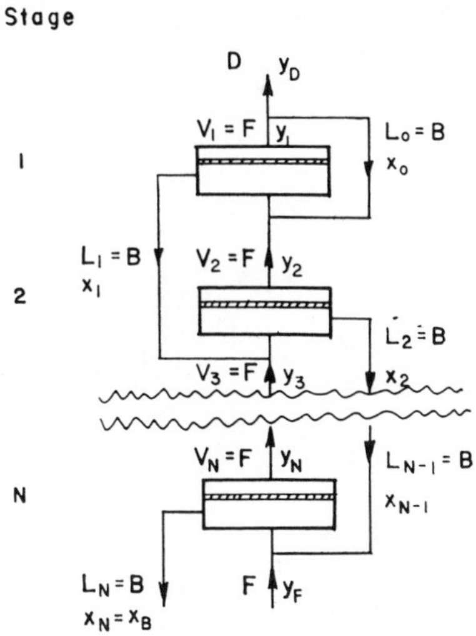

Stage

Fig. 13.7 Cascade for a constant overflow.

Then, the equation of the operating line becomes

$$y_{n+1} = \left(\frac{L_n}{V_{n+1}}\right) x_n + \left(\frac{D}{V_{n+1}}\right) y_D . \qquad (13.3)$$

Define "cut" at nth stage as

$$\theta_n = \frac{V_n}{V_{n+1} + L_{n-1}} . \qquad (13.4)$$

In order to express Eq. 13.3 in terms of the cut θ_n, it is necessary to proceed as follows: From Eq. 13.1,

$$L_{n-1} = V_n - D .$$

Substituting into Eq. 13.4 and solving for $V_{(n+1)}$ gives

$$V_{n+1} = \left(\frac{1 - \theta_n}{\theta}\right) V_n + D. \qquad (13.5)$$

This equation permits the calculation of successive streams starting from the top and proceeding to the bottom stage. Introducing a ratio γ defined as follows:

$$\gamma_n = \frac{1 - \theta_n}{\theta_n} , \qquad (13.6)$$

Eq. 13.5 becomes

$$V_2 = \gamma_1 V_1 + D \;,$$

$$V_2 = \gamma_2 V_2 + D = \gamma_2(\gamma_1 V_1 + D) + D \;,$$

$$\cdots$$

$$V_{n+1} = \gamma_n(\gamma_{n-1} \cdots (\gamma_1 V_1 + D) \cdots + D) + D$$

$$= V_1 \prod_{i=1}^{n} \gamma_i + D \left[\sum_{i=2}^{n} \left(\prod_{j=i}^{n} \gamma_i \right) + 1 \right] . \tag{13.7}$$

Likewise, L_n can be established from Eq. 13.1 and 13.7 so that

$$L_n = V_{n+1} - D$$

$$= V_1 \prod_{i=1}^{n} \gamma_i + D \sum_{i=2}^{n} \left(\prod_{j=i}^{n} \gamma_i \right) .$$

Finally, Eq. 13.3 becomes

$$y_{n+1} = \left[\frac{V_1 \prod_{i=1}^{n}\gamma_i + D \sum_{i=2}^{n}\left(\prod_{j=i}^{n}\gamma_i \right)}{V_1 \prod_{i=1}^{n}\gamma_i + D \left\{ \sum_{i=2}^{n}\left(\prod_{j=i}^{n}\gamma_i \right) + 1 \right\}} \right] x_n$$

$$+ \left[\frac{D}{V_1 \prod_{i=1}^{n}\gamma_i + D \left\{ \sum_{i=2}^{n}\left(\prod_{j=i}^{n}\gamma_i \right) + 1 \right\}} \right] y_D . \tag{13.8}$$

Equation 13.8 represents the generalized equation for the operating line in the enriching section of a cascade. V_1 can be any number greater than or equal to D. The cut θ_n may be changed from stage to stage or it can be held constant over the enriching section. In any case, Eq. 13.8 gives a straight line or a set of straight lines, all of which intersect at the product point (x_D, y_D). This fact is also stated in Rozen's book [6].

The equation for the stripping section is developed in an entirely analogous manner. Reference to Fig. 13.6 shows that the following equations will hold:

Total material balance:

$$V_m = L_{m+1} - B . \tag{13.9}$$

Component balance:

$$V_m y_m = L_{m+1} x_{m+1} - B x_B \qquad (13.10)$$

Cut:

$$\theta_m = \frac{V_m}{V_{m-1} + L_{m+1}} \qquad (13.11)$$

With proper combination of these equations and using the ratio

$$\gamma_m = \frac{1 - \theta_m}{\theta_m}, \qquad (13.12)$$

the final generalized equation for the operating line in the stripping section becomes

$$y_m = \left[\frac{\sum\limits_{i=1}^{m} \left(\prod\limits_{j=i}^{m} \frac{1}{\gamma_i} \right) + 1}{\sum\limits_{i=1}^{m} \left(\prod\limits_{j=i}^{m} \frac{1}{\gamma_j} \right)} \right] x_{m+1} - \left[\frac{1}{\sum\limits_{i=1}^{m} \left(\prod\limits_{j=i}^{m} \frac{1}{\gamma_j} \right)} \right] x_B . \qquad (13.13)$$

Just as in the case of the enriching lines, all operating lines in the stripping section will intersect at the bottom product point (x_B, y_B).

When a cascade consists of both enriching and stripping sections, the intersection point of the operating lines must be matched to correspond to the feed composition and stream. This is accomplished by varying θ_n, or θ_m, or both, until the desired intersection is obtained.

From the difinition of the cut it is evident that the permissible value of θ, for both the enriching and stripping sections, is confined between zero and unity:

$$0 \leqslant \theta \leqslant 1 ,$$

which simply means that θ cannot be greater than 1 (see definition) and that it must be positive.

Specific Methods of Operation

The treatment of several specific cases that are frequently encountered is of definite interest. Consequently, three cases that cover the most pertinent situations are discussed. The cases selected are as follows:

Case 1: Constant cut θ and constant overflow
Case 2: Constant cut θ and varying overflow
Case 3: Varying cut θ and varying overflow

In all cases only the enriching section will be discussed as the procedures for the stripping section will be analogous, in that Eq. 13.3 is used.

CASE 1:   CONSTANT CUT θ AND CONSTANT OVERFLOW

When both the cut and the overflow are held constant throughout any section of the cascade, the derivation of the equations and the graphical procedure become rather simple.

The conditions for the enriching section that is shown diagrammatically in Fig. 13.5 are

$$V_1 = V_2 = \cdots = V_{n+1} = F$$

and

$$\gamma_{n+1} = \gamma_n = \cdots = \gamma.$$

Then the overhead product becomes

$$D = (1 - \gamma)F = \frac{2\theta - 1}{\theta} F, \tag{13.14}$$

and Eq. 13.2 reduces to

$$y_{n+1} = \frac{1 - \theta}{\theta} x_n + \frac{2\theta - 1}{\theta} y_D. \tag{13.15}$$

Equation 13.15 gives a unique straight operating line. It is important to keep in mind that this is valid if and only if V_n remains constant as shown in Fig. 13.5. The constancy of the cut θ alone, in general, does not determine the L/V ratio.

The permissible range of θ can be obtained by inspection of Eq. 13.14 as follows:

$$0 \leqslant \frac{2\theta - 1}{\theta} \leqslant 1$$

or

$$\frac{1}{2} \leqslant \theta \leqslant 1.$$

The step-by-step graphical procedure for obtaining the number of theoretical stages in an enriching cascade is shown in Fig. 13.8. The amount of top product D is calculated from Eq. 13.14. This particular case has thus the advantage of requiring only simple calculations and also, that the number of stages that are needed for any given separation will be less than the number called for in the following Case 2.

The advantages are off-set at least in part by the fact that Case 1 requires a larger amount of membrane area (or a larger number of columns) because each stream remains constant. The arrangement does not result in a tapered cascade as is usually encountered in isotope separation. Additionally, the amount of overhead product obtainable in Case 1 (see Eq. 13.14) is less than that produced in Case 2 (see Eq. 13.17).

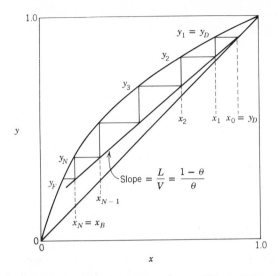

Fig. 13.8 Cascade operation; graphical procedure for conditions of Case 1. This figure is reprinted from Ref. 8, through courtesy of the *Canadian Journal of Chemical Engineering*.

CASE 2: CONSTANT CUT AND VARYING OVERFLOW

When θ is constant, γ becomes constant and the term

$$\sum_{i=2}^{n} \left(\prod_{j=i}^{n} \gamma_j \right)$$

in Eq. 13.8, simplifies to a geometric series. By using the series summation:

$$\sum_{i=2}^{n-2} \gamma_i = \frac{1 - \gamma^{n-1}}{1 - \gamma} \, ,$$

Eq. 13.8 reduces to

$$y_{n+1} = \left[\frac{V_1 \gamma^n + D \left(\dfrac{\gamma - \gamma^n}{1 - \gamma} \right)}{V_1 \gamma^n + D \left(\dfrac{1 - \gamma^n}{1 - \gamma} \right)} \right] x_n + \left[\frac{D}{V_1 \gamma^n + D \left(\dfrac{1 - \gamma^n}{1 - \gamma} \right)} \right] y_D \tag{13.16}$$

This equation gives a set of straight operating lines as shown in Fig. 13.9. The slope of any one line is determined by fixing the values of θ_n, V_1, and n. For example, when $V_1 = D$ ($L_0 = 0$, no reflux at top stage), then the respective slope of the operating line at the nth stage is determined as

$$\frac{\gamma - \gamma^{n+1}}{1 - \gamma^{n+1}} \, .$$

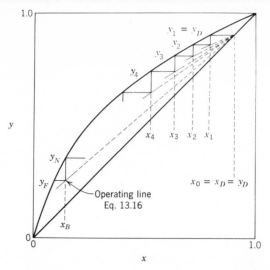

Fig. 13.9 Cascade operation; same as Fig. 13.8, but for condition of case 2. This figure is reprinted from Ref. 8, through courtesy of the Canadian Journal of Chemical Engineering.

In order to calculate the amount of product, one considers that $V_{n+1} = F$, and that γ remains constant; then

$$D = (F - V_1 \ \gamma^N) \ \frac{1 - \gamma}{1 - \gamma^N} \ . \tag{13.17}$$

This equation also permits the determination of the required number of ideal stages for a certain amount of top product D. If $V_1 = D$, then Eq. 13.17 becomes

$$D = \frac{1 - \gamma}{1 - \gamma^{N+1}} \ F \ , \tag{13.18}$$

and the number of stages will be

$$N + 1 = \frac{\ln \ [1 - (1-\gamma) \ F/D]}{\ln \ \gamma} \ .$$

In this case γ could be any positive number, that is,

$$0 \leqslant \theta \leqslant 1 \ .$$

The restriction for the value of θ is less than for the previous Case 1. The efficiency of separation is better than in the case of constant overflow, Case 1, but it is not ideal due to the constancy of ϑ. Improvement of the efficiency can be attained by varying θ from stage to stage, and this situation is treated as Case 3.

CASE 3: VARYING CUT θ AND VARYING OVERFLOW

The fraction diffused, that is, the cut θ can be changed from one stage to another for the purpose of a better separation, or for the adjustment of compositions and stream quantities near the feed stage when a complete cascade is used. The procedure is not as simple as those discussed above, because the change of θ causes a corresponding change in the equilibrium curve. However, if a digital computer is available, such a calculation can be carried out without much difficulty. Therefore, formalization of equations involved might be helpful.

Ideal Cascade

The operations stated in Case 1 and Case 2 do not guarantee that the compositions of two streams, which are combined before feeding to every stage, are exactly the same. In any kind of separation process, if a combination of streams results in a mixing of two different concentrations, the separation efficiency decreases due to the extra entropy production of mixing. Therefore, if it is at all possible, such mixing has to be avoided. The definition of an ideal cascade is that of a cascade in which no mixing takes place. Referring to Fig. 13.15 the conditions for any stage must be

$$x_{n-1} = y_{n+1} \ . \tag{13.19}$$

A graphical illustration is shown in Fig. 13.10 with the assumption that the equilibrium relationship does not change too much by varying the value of ℓ. This simplification is made for the purpose of illustration. The more exact

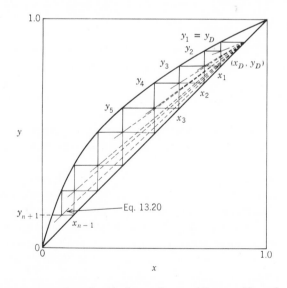

Fig. 13.10 Graphical procedure for ideal cascade, requiring matching of recycle stream compositions. This figure is reprinted from Ref. 8, through courtesy of the *Canadian Journal of Chemical Engineering*.

method can readily be handled with a digital computer program. In this case the reflux at the top stage should be zero, otherwise mixing takes place, unless $x_0 = y_2$, but this is impossible because $x_0 = y_1 \neq y_2$. Therefore, with $V_1 = D$, Eq. 13.8 becomes

$$y_{n+1} = \left[\frac{\sum\limits_{i=1}^{n}\left(\prod\limits_{j=i}^{n}\gamma_j\right)}{\sum\limits_{i=1}^{n}\left(\prod\limits_{j=i}^{n}\gamma_j\right) + 1}\right] x_n + \left[\frac{1}{\sum\limits_{i=1}^{n}\left(\prod\limits_{j=i}^{n}\gamma_j\right) + 1}\right] y_D . \qquad (13.20)$$

The equilibrium relationship at the nth stage can be expressed as

$$y_n = f(x_n, \theta_n), \qquad (13.21)$$

or solving for x_n gives

$$x_n = g(y_n, \theta_n). \qquad (13.22)$$

These equations represent the general case. As shown in the section on the equilibrium curve (Section 2, p. 329), the pertinent model will determine if θ enters as a factor or not.

Now, one can start from the top stage successively downward. At the very first stage θ_1 is arbitrary, but this should be determined by economic optimization. When θ_1 is fixed and y_1 is known, x_1 can be calculated from Eq. 13.22. By substituting these values into Eq. 13.20 the value of y_2 is obtained. Then θ_2 is not an arbitrary number because it should yield $x_2 = y_3$. After a value θ_2 is assumed, x_2 is obtained from Eq. 13.22. Then from Eq. 13.10, y_3 can be calculated, and compared with x_1. If the calculared y_3 is different from x_1, another θ_2 must be assumed and the calculations repeated until y_3 equals x_1. By repeating this procedure to the feed stage, one can determine each θ_n, V_n, L_n, and the total number of theoretical stages required. This kind of computation can be carried out easily with a computer. The graphical procedure becomes very involved because of the change of equilibrium equation every time that θ is changed.

The curved operating line shown by Rozen [6] and Benedict and Pigford [4] is actually a trajectory of operating points. For the case of isotope separation this particular curve can be expressed by a relatively simple equation. When separation factors are large, however, this in general will not be feasible. Also, although the equilibrium curve for systems of small separation factors will change very little with changes in cut, this is not so when the separation factor is large.

3 EXAMPLES OF CASCADE OPERATION

Gaseous Diffusion

It is, of course, well known that the most remarkable achievement of cascade operation is that of the UF_6 isotope separation of the U.S. Atomic Energy Commission's gaseous diffusion plants. National security reasons still call for a classified status of that technology.

A fair number of numerical examples of cascade operation can be found in the literature. On closer examination, however, they turn out to be strictly computer simulations. Of course, as such they are valuable and instructive, but they do not reflect results from experimentation or operation. Chapter XVI goes into more details.

Gaseous Diffusion — Computed Examples

Weller and Steiner [9] discuss computed multistage operation for oxygen enrichment from air, helium separation from natural gas, hydrogen separation from hydrogenation tail gas, and hydrogen enrichment in refinery gas. Naylor and Backer [10] present detailed calculations for hydrogen enrichment from a H_2-N_2 mixture and show a cascade diagram for this operation. Stern and co-workers [11] published a rather complete treatise, including economic considerations, for obtaining an enriched helium stream from a feed consisting of natural gas.

A somewhat unusual application of membrane separation involving a computed cascade operation was reported by Blumkin [12]. It concerns the removal of krypton and xenon from the atmosphere of a reactor containment vessel after an accidental release of fission product gases. A General Electric Company dimethyl silicone membrane was considered for the separation process. An anomalous situation is reported in that permeability values for a specific test package obtained by General Electric and by the Oak Ridge National Laboratory were in considerable disagreement, as shown in Table 13.1.

Table 13.1 Permeability Data for Silicone Rubber

	General Electric Co.		Oak Ridge National Lab.	
Gas	*Permeability* $\frac{(cc)(cm)}{(sec)(cm^2)(cm\ Hg)}$	*Max. Point Sepn. Factor (Refd. To N_2)*	*Permeability* $\frac{(cc)(cm)}{(sec)(cm^2)(cm\ Hg)}$	*Max. Point Sepn. Factor (Refd. to N_2)*
Kr	98×10^{-9}	3.4	45×10^{-9}	4.5
O_2	60×10^{-9}	2.1	16×10^{-9}	1.6
N_2	29×10^{-9}	1.0	10×10^{-9}	1.0

There is no explanation for the disagreement. However, numerous other reports have substantiated the values of oxygen and nitrogen as given by G. E. The increased selectivity of krypton resulted in an 8-stage design compared to 17 stages based on the original G. E. data.

A rather complete engineering analysis of the proposed application was carried out by Blumkin et al. [13]. In their report, published in 1971, comparison is also made with other possible methods of decontamination. Some operating data of estimated performance are listed in a recent book by Lacey and Loeb [14].

Gaseous Diffusion — Experimental

Experimentally, there seems to be only one publication that describes a laboratory-scale cascade. Higashi and co-workers [15] constructed a 10-stage tapered cascade utilizing sintered nickel porous disks of 50 mm diameter in the diffusion cells. A view of the assembly is shown in Fig. 13.11. The investigation was carried out in Professor J. Oishi's Department of Nuclear Engineering at Kyoto University and checked out by separating nitrogen isotopes. The publication gives only limited information about detailed operating conditions. but it is possible to estimate the N^{15} concentration in the unpermeated streams coming from each cell. The data are shown in Fig. 13.12.

Fig. 13.11 View of 10-stage cascade for isotope separation [15]. (Courtesy Professor J. Oishi.)

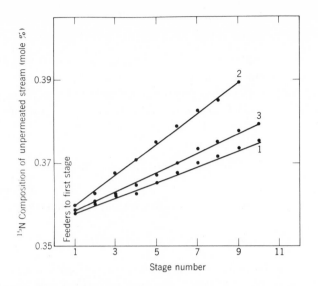

Fig. 13.12 Separation of nitrogen isotopes in apparatus of Fig. 13.11. Composition of unpermeated streams under steady-state conditions [15]. 1: Ordinary operation, θ 0.5; 2: total reflux operation; 3: operation with limited product removal.

A very up-to-date description of a two-stage cascade operation for helium extraction from natural gas has been presented by Litz and Smith of Union Carbide [16]. Details of the operation are reported in Chapter XV, as the scale of operation is of industrial magnitude.

4 QUALITY RATING FOR PERMEATION UNIT

In the design of any permeation equipment one of the criteria is to optimize the amount of product with minimum cost. In order to achieve this, we have to examine the permeation equation for a species A:

$$N_{A} = S_{T}Q_{A}\ \frac{\Delta P'}{l}\ .\tag{13.23}$$

This equation holds for only pressure-driven processes, specifically gaseous diffusion or osmotic processes. It is obvious that the maximum amount of product N_{A} can be obtained if the area S_{T} is maximized while the thickness l is minimized for a given membrane and pressure drop $\Delta P'$. This could be done easily by adding more membranes, but then the total cost will rise. Therefore, a more convenient form would be the flow equation for a unit volume of equipment:

$$\frac{N_A}{V'} = \frac{S_T}{V' l} Q_A \ \Delta P' \ ,$$
(13.24)

where V' is the volume of the equipment. Thus, rather than examining the ratio of the area over the thickness of a given membrane, the ratio of the area over thickness per unit volume of equipment should be optimized. This concept was introduced first by McCormack in a U.S. Patent on capillary membranes [18]. He defined the "efficiency number" as

$$\text{Efficiency number} = \frac{S_T}{V' l \pi} \ ,$$
(13.25)

which can be multiplied by the pressure drop to yield the productivity.

5 STAGING IN LIQUID-PHASE PROCESSING

Experimental investigations of multistaging in liquid phase apparently have not been reported. The probable reason for this situation is that desired information can be obtained from testing in single units. Also, the recycling of reject streams has not been considered as an attractive procedure, particularly in reverse osmosis for desalination where the reject streams are not valuable enough for reuse. However, multistaging may be necessary even in liquid-phase processes in cases (a) where the reject stream is valuable, (b) where a desired degree of separation cannot be achieved with a single stage, and (c) where the pressure loss across a single stage is too large for effective separation in a pressure-driven process. It has repeatedly been stressed that the troublesome effect of concentration polarization can be significantly reduced by a high recirculation rate of the reject stream.

Actually, there has been some straightforward multistaging in large-scale installation of pressure-driven processes, as described in Chapter XV. Two of the examples in that chapter (whey processing and sewage benefication) use a series application of ultrafiltration followed by reverse osmosis as an operating choice and this, in a manner of speaking, does represent a staging scheme. Sourirajan [17] discusses multistaging and cascading for reverse osmosis and presents a multiplicity of equations for design purpose. He also gives a numerical example, which, however, is entirely computational in nature.

In electromembrane processes, a multimembrane stack with a single pair of electrodes usually constitutes a membrane module. In order to increase the capacity of the module, each repeating unit of the membrane stack can be operated in parallel with or without recirculation. Also, a higher degree of separation or ion exchange can be achieved by employing an internal or external multistaging scheme as shown in Fig. 13.13 and 13.14. For a detailed discussion of these schemes, the article by Mintz [19] should be consulted.

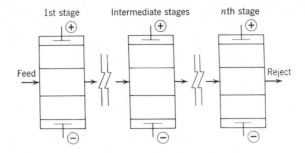

Fig. 13.13 External multistaging of electromembrane modules.

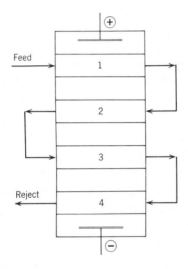

Fig. 13.14 Internal multistaging of electromembrane modules.

6 BOUNDARY-LAYER PHENOMENON

As one molecular or ionic species is separated from another species in a membrane process, accumulation or depletion of the species always takes place near the membrane surface due to permselectivity of the membrane. This phenomenon of concentration polarization is negligible in gaseous systems. However, it is usually of considerable importance in liquid-phase processes, including pervaporation.

In general, the polarization phenomenon has an adverse effect on separation. It reduces the flux of the more permeable molecular or ionic species and enhances the leakage of the less permeable species. The accumulation of a certain species at the membrane surface often deteriorates the polymeric membrane. In electromembrane processes, depletion of electrolyte solutes results in a high ohmic electric resistance across the boundary layer where the concentration is polarized. Further depletion of the electrolyte solutes induces the undesirable electrolysis of the solvent water, changing the pH value in the solution. Therefore, control of concentration polarization is required to obtain a high efficiency in a membrane module.

An obvious way of control in a batch system is vigorous agitation of the solution. In a continuous system, the polarization can be reduced by using a *high* longitudinal velocity in very *thin* channels for the solution streams. In order to promote local turbulent mixing, one may also effectively employ tortuous paths and certain turbulence promoters within the channel; apply ultrasonic vibration; or introduce solid particulate materials in the solution. However, such control measures always sacrifice simplicity of the membrane module and also require high pumping energy for the solution stream. The hollow-fiber type of membrane module is a typical example. Therefore, an *adequate* control of the boundary layer is essential to balance the two opposing effects.

MASS-TRANSFER COEFFICIENT—FORCED CONVECTION

When one slowly introduces into a closed conduit a solution stream whose velocity profile is originally uniform, the velocity near the wall of the conduit is instantaneously distributed at the entrance because of the frictional drag at the wall. In the laminar flow regime, the velocity boundary layer, where the velocity is polarized, gradually grows in thickness as the stream flows downstream. Then the thickness usually approaches a certain constant asymptotic value at the far-downstream end. In the turbulent flow regime, the thickness assumes the constant value much faster than in the laminar flow regime. If the closed conduit is comprised of a permselective membrane, the concentration profile also develops in a way similar to that of the longitudinal velocity as shown in Fig. 13.15. Since the momentum is generally transferred much faster than the mass by the molecular diffusion in most liquid-phase processes, the concentration boundary layer is much thinner than the velocity boundary layer, but the concentration gradient across the concentration boundary layer is steeper than the velocity gradient. Thus, it can be expected that, in the bulk solution outside the (concentration) boundary layer, the concentration profile is almost uniform in the direction normal to the membrane surface. Transport of a molecular or ionic species occurs in the longitudinal direction parallel to the membrane largely because of the longitudinal convective motion. Within the boundary layer, on the other hand, the transport is mainly due to molecular or eddy diffusion and the convective transport

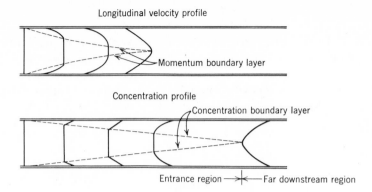

Fig. 13.15 Profiles of longitudinal velocity and concentration in a developing boundary layer.

in the direction normal to the membrane. In the presence of an electric current across the membrane, the electric transference is also important. The longitudinal transport by either the molecular or convective diffusion is, however, negligible within the boundary layer. The longitudinal motion of the bulk stream influences the mass transport in the normal direction only through keeping the boundary layer thin. This Nernst film concept is a good approximation, particularly when the boundary layer is very thin and uniform as in the turbulent-flow regime. Even in the laminar-flow regime, the concept can be effectively used with suitably averaged quantities.

In the case where the Nernst film concept is valid, it also implies that, at steady state, the normal fluxes of solute and solvent are constant in the boundary layer, that is, the same as the permeating fluxes, N_A and N_B, through the membrane. Then, the mass conservation over an infinitesimal volume within the boundary layer of thickness δ^I in the absence of an electric current yields (see Fig. 13.16)

$$N_A = \frac{C_A}{C} (N_A + N_B) + D_A \frac{dC_A}{dz} \qquad (13.26)$$

with boundary conditions

$$C_A = C_{Aw}^I \quad \text{at } z = 0$$
$$= C_{Ab}^I \quad \text{at } z = \delta^I , \qquad (13.27)$$

where

$$C = C_A + C_B. \tag{13.28}$$

Here D_A is the diffusion coefficient of solute in the solution phase, and C_A and C_B are molar concentrations of solute A and solvent B, respectively. Also, subscripts W and b refer to the membrane surface and the bulk solution. It is assumed in the mass conservation, Eq. 13.26, that there exists a laminar subboundary layer even in the turbulent-flow regime of the bulk solution stream.

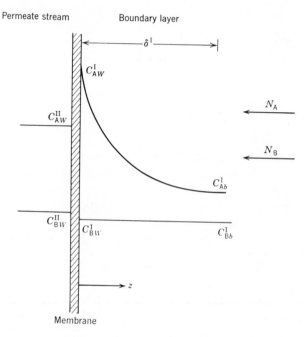

Fig. 13.16   Concentration polarization in an osmotic membrane module.

The solutions to be processed are usually so dilute that the total molar concentration C is almost constant and close to the molar concentration of the solvent. In this case, Eq. 13.26 can be integrated to obtain

$$N_A + N_B = Ck^I \ln \left[\frac{C_{AW}^I - \dfrac{CN_A}{(N_A + N_B)}}{C_{Ab}^I - \dfrac{CN_A}{(N_A + N_B)}} \right]. \tag{13.29}$$

Here, the mass-transfer coefficient k^I in the solution I is introduced as

$$k^{\mathrm{I}} = \frac{D_{\mathrm{A}}}{\delta^{\mathrm{I}}} \ . \tag{13.30}$$

When the membrane is permselective with respect to the solvent, that is, $N_{\mathrm{A}} \ll N_{\mathrm{B}}$, the molar solute concentration $C_{\mathrm{A}W}^{\mathrm{II}}$ at the membrane surface in the solution II can be approximated as

$$C_{\mathrm{A}W}^{\mathrm{II}} = \frac{CN_{\mathrm{A}}}{N_{\mathrm{A}} + N_{\mathrm{B}}} \ . \tag{13.31}$$

Then, Eq. 13.29 becomes

$$N_{\mathrm{B}} = Ck^{\mathrm{I}} \ \ln \frac{C_{\mathrm{A}W}^{\mathrm{I}} - C_{\mathrm{A}W}^{\mathrm{II}}}{C_{\mathrm{A}b}^{\mathrm{I}} - C_{\mathrm{A}W}^{\mathrm{II}}} \ , \tag{13.32}$$

or the polarization $C_{\mathrm{A}W}^{\mathrm{I}}/C_{\mathrm{A}b}^{\mathrm{I}}$ can be expressed in terms of the solute rejection SR defined in Eq. 8.15 as

$$\frac{C_{\mathrm{A}W}^{\mathrm{I}}}{C_{\mathrm{A}b}^{\mathrm{I}}} = \frac{\exp\left(\dfrac{N_{\mathrm{B}}}{k^{\mathrm{I}}C}\right)}{\mathrm{SR} + (1-\mathrm{SR}) \ \exp\left(\dfrac{N_{\mathrm{B}}}{k^{\mathrm{I}}C}\right)} . \tag{13.33}$$

In particular, if the membrane is ideally osmotic (SR=1), Eq. 13.32 is simplified to

$$N_{\mathrm{B}} = Ck^{\mathrm{I}} \ \ln \frac{C_{\mathrm{A}W}^{\mathrm{I}}}{C_{\mathrm{A}b}^{\mathrm{I}}} \ . \tag{13.34}$$

In a dialytic process, where the separation is achieved by the solute transport across the membrane, the following condition is usually satisfied:

$$\frac{CN_{\mathrm{A}}}{N_{\mathrm{A}} + N_{\mathrm{B}}} \gg C_{\mathrm{A}} \ . \tag{13.35}$$

In this case, integration of Eq. 13.26 simply yields

$$N_{\mathrm{A}} = k^{\mathrm{I}} \ (C_{\mathrm{A}b}^{\mathrm{I}} - C_{\mathrm{A}W}^{\mathrm{I}}) \ . \tag{13.36}$$

In the presence of an electric current I in the direction normal to the membrane surface, the ionic solute flux N_{A} is supplemented by an additional contribution of electric transference (see Chapter IX), so that

$$N_A = k^I (C_{Ab}^I - C_{AW}^I) + t_A \frac{I}{\mathcal{F}} . \qquad (13.37)$$

Here, the transference number t_A is assumed to be constant. It should be noted that the diffusion coefficient D_A of ionic species A is not a "self-diffusion coefficient," but an effectively averaged quantity including the interionic effect as discussed in Chapter IX.

In the approximation of the Nernst boundary film, the mass-transfer coefficient k can thus be determined from Eq. 13.30. With the boundary-layer thickness δ, then, the effect of concentration polarization may be obtained from Eq. 13.32 in an osmotic process, and from Eqs. 13.36 or 13.17 in a dialytic process. In reality, the longitudinal convective transport, however, cannot be safely neglected even within the boundary layer. In the turbulent-flow regime, the eddy diffusion also contributes to the mass transfer. Therefore, instead of the conceptual quantity of the boundary-layer thickness, an effective mass-transfer coefficient k is usually defined directly through Eq. 13.32 for an osmotic process, or through Eqs. 13.36 or 13.37 for a dialytic process. It is then common practice to introduce the dimensionless Sherwood number Sh to express the coefficient such that

$$\text{Sh} = \frac{4k \ r_h}{D_A} , \qquad (13.38)$$

where r_h is the equivalent hydraulic radius, that is,

$$r_h = \frac{b}{2} \quad \text{for the sheet membranes of width } b$$

$$= \frac{r}{2} \quad \text{for the tubular membrane of radius } r.$$

In the Nernst-film theory, the Sherwood number can be interpreted as the ratio of the characteristic dimension of the channel to the boundary-layer thickness.

Obviously, the Sherwood number depends in a complicated manner on the longitudinal velocity of the bulk stream u; the equivalent hydraulic radius of the channel r_h; kinematic viscosity v; diffusivity of the solute D_A; and selectivity of the membrane. In a developing boundary layer, the Sherwood number is a function of the distance x from the entrance of the channel. The most common type of correlation of the Sherwood number is the Nusselt type:

$$\text{Sh} \sim \text{Re}^{n_1} \ \text{Sc}^{n_2} \left(\frac{4r_h}{x} \right)^{n_3} , \qquad (13.40)$$

where the dimensionless Reynolds number Re and Schmidt number Sc are introduced as

$$Re = \frac{4r_h u}{\nu} ,$$
(13.41)

$$Sc = \frac{\nu}{D_A} ,$$
(13.42)

and n_1, n_2, and n_3 are constants. The constant exponent of the Schmidt number n_2 is in the range of 0.25 to 0.33, and n_3 is about 0.33 in the developing boundary layer. The exponent of the Reynolds number n_1 is in the range of 0.3 to 0.4 in the laminar-flow region and 0.5 to 0.8 in the turbulent-flow region. In this type of correlation, an ideal permselective membrane is implicitly assumed, and its validity has been relatively well established. The detailed discussion and the more rigorous treatment will be made subsequently in conjunction with the individual membrane processes.

Natural Convection and Concentration Polarization

The previous discussion treated the case where the mass transfer in the bulk solution stream occurs only through longitudinal forced convection. When the concentration of solute is polarized to such an extent that a significant density distribution occurs near the membrane surface the buoyancy effect becomes also important however, as in the heat-transfer case. The natural or free convection induced by the buoyancy greatly influences the mass transfer, particularly in inorganic aqueous solutions whose density is very sensitive to the solute concentration. In a horizontal sheet-membrane configuration, where two membranes are separated by a distance b, the natural convection would thus dominate over the forced convection, if the following condition is satisfied:

$$\frac{g |\rho_W - \rho_b| \delta^2}{\rho_b \nu u} \gg 1 ,$$
(13.43)

where ρ_b and ρ_W are the solution densities in the bulk solution and at either membrane surface, respectively, and g is the gravitational constant. In this case, the upward solute transfer against the gravity is retarded, whereas the downward transfer is accelerated. It can, therefore, be expected that the polarized boundary layer at the lower membrane surface is stabilized, and the concentration polarization is augmented. On the other hand, the boundary layer at the upper membrane surface is not stable and is depolarized by natural convection. Near the upper membrane, an earlier transition of the stream will consequently occur from the laminar-flow regime to the turbulent regime, resulting in a further higher solvent flux. The effect of the natural convection is thus more important at the upper part of the sheet membranes than at the lower part. This phenomenon is illustrated in Fig. 13.17, which was prepared by Sloan and Harshman [20]. As shown in the figure, the discrepancy between the rejected saline fraction at the upper membrane and the fraction rejected by the lower membrane gradually increases as the boundary layer is

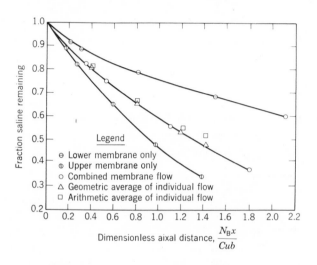

Fig. 13.17 Effect of natual convection on separation in a horizontal osmotic sheet membrane at 455 mm Hg transmembrane pressure drop [20];

developing. This is implicitly indicated by the condition of Eq. 13.43. The overall rejected fraction with the two combined membranes thus assumes an average of the two values. The situation is more complicated and difficult than in heat transfer, and thus the theoretical treatments have not been developed rigorously. Only a few investigators have dealt with this subject [20, 21]. However, similarly to the case of heat transfer, the Sherwood number may be approximately expressed in terms of the Grashof number Gr, as

$$Gr = \frac{g \ |\rho_W - \rho_b| \ b^3}{\rho_b \ v^2} , \tag{13.44}$$

and the Schmidt number Sc as

$$Sh \sim Gr^{n_4} \ Sc^{n_5} \left(\frac{b}{x} \right)^{n_6} \tag{13.45}$$

or

$$\text{Sh} \sim \text{Gr}^{n_4} \text{Sc}^{n_5} \left\{ \frac{N_B(x)}{N_B(0)} \right\}^{n_7} . \qquad (13.46)$$

Here n_4, n_5, n_6, and n_7 are constants, and $N_B(0)$ and $N_B(x)$ are the solvent fluxes at the entrance of the channel and at a distance x from the entrance, respectively. Indeed, Sloan and Harshman could correlate their experimental data with Eq. 13.46 resonably well. The exponents n_4, n_5, and n_7 all were found to be alike, with a value of essentially $1/3$ when Gr Sc $> 10^4$. This result is actually very similar to that encountered in the ionic transport near the Electrode [22].

In a vertical-sheet membrane, the situation is more complicated. However, the semiempirical expression for the Sherwood number in a horizontal-sheet membrane may also be approximately valid. In the ionic transport near a vertical electrode, which is similar to this case, the exponents n_3 and n_4 are reported to be in the range of 0.25 to 0.28 [22].

If forced convection is superimposed on natural convection, the problem is naturally still more complicated. But one may expect that the transition region between predominance of the natural convection and that of the forced convection is narrow as in heat transfer or in the electrode reaction.

Reverse Osmosis

When solvent permeates selectively through a membrane, the solute accumulates near the ingoing interface between the membrane and the reject stream. The concentrated solute depletes then by diffusion to the bulk phase of the reject stream. Since the accumulation rate is greater than the depletion rate in most osmosis units, a net concentration buildup of the solute is commonly observed. As the concentration is thus polarized, the solute flux through the membrane is enhanced, and thus the solute rejection is reduced. Also the osmotic pressure of the solution increases with the polarization, and this reduces the solvent permeation flux through the membrane. The reduction of the permeation flux is crucial, particularly in the operation where the pressure gradient across the membrane is comparable to the gradient of the osmotic pressure. Furthermore, the solute is often locally concentrated to the extent of the saturation point and a layer of its precipitate or gel is formed on the membrane surface, thus further reducing the solvent permeation flux. The concentration polarization can also be expected near the membrane surface on the side of the product stream. However, the salt rejection of the reverse-osmosis membrane is usually close to 1 (i.e., the ideal semipermeable membrane), and thus the concentration of the solute in the product stream is very small. Therefore, the polarization on the side of the product stream is not so significant as that on the side of the reject stream.

As mentioned earlier, the concentration polarization increases with a decrease of the stream velocity and diffusivity, but with an increase of the channel width or radius. In the developing boundary layer, the polarization is progressively enhnaced as the fluid flows downstream. Therefore, the polarization can be considerable in the configuration of the osmosis unit where the dimension r_h is large, such as in a plate-and-frame or a spiral-wound module, On the other hand, it is negligible in a unit of a bundle of hollow fibers with small r_h. One may also minimize the concentration polarization by sweeping away the polarized layer employing a high flow rate of the feed stream. However, the high flow rate induces a considerable pressure loss in the osmosis unit and so requires a high pumping power. Therefore, one must balance the two opposing effects.

If the flow is laminar, one may solve a system of nonlinear partial-differential equations consisting of equations of consrevation of mass and momentum in order to obtain the complete analysis of the concentration polarization. This procedure is, however, very complicated, and the exact analytical solution is not available. Several approximate solutions have been obtained and especially the elegant Dresner solution [23] is very simple and accurate enough for design purposes. For turbulent flow, the theory has not been well established, but the analysis is ironically simpler and easier, if one accepts the validity of the Nernst-film theory or the eddy-diffusion model. Gill [24] presented a lucid summary of the development of the various analyses.

The separation can also be achieved in a batch operation. This operation is, however, usually not practical for large-scale or industrial operation, but is mainly of theoretical interest in studying the performance of a given membrane. Therefore, the batch procedure will not be discussed, and only the continuous operation will be covered.

Laminar Flow

In most membrane processes the solvent permeation flux is very small compared to the axial flow rate of the feed stream. Thus, the velocity field is almost independent of the concentration distribution. Generally, with the approximate analyses, an appropriate velocity field has, therefore, been assumed a priori, and then the equations of mass conservation have been solved by suitable mathematical techniques. The velocity fields commonly used in those analyses are essentially based on the perturbation solution for a pure-component system obtained by Berman [25] in porous parallel plates, and by Yuan and Finkelstein [26] in porous circular tubes, through which the permeation flux is kept constant.

Dresner [23] noted that in desalination of brackish water the pressure gradient is usually much larger than the gradient of the osmotic pressure, and the solvent permeation flux is nearly constant. Since, the salt rejections of most osmotic membranes are close to unity, the concentration of the product stream is almost zero. He further recognized that since the transport of the solute by the molecular diffusion is very small compared to that by the viscous bulk flow, the diffusional boundary layer remains very thin even at a considerable distance from the inlet of the feed stream, and thus the stream can be considered as a semi-infinite medium (see Fig. 13.15). In the entrance region, the axial velocity profile can also be expected to be linear in the transverse direction. By means of some mathematical approximation with this notion, Dresner obtained a simple solution in the entrance region of the sheet membranes as

$$\frac{C_{Aw}}{C_{Ab}} = 1 + 1.536 \; \xi^{1/3} \qquad\qquad \text{if } \xi \leqslant 0.02 \qquad (13.47)$$

$$= \xi + 6 - 5 \; \exp\left[- \left(\frac{\xi}{3}\right)^{1/2} \right] \quad \text{if } \xi > 0.02, \qquad (13.48)$$

where

$$\xi = \frac{2N_B Pe_W^2 x}{3uCb} , \qquad\qquad (13.49)$$

and the so-called wall Peclet number Pe_W is defined as

$$Pe_W = \frac{N_B b}{2CD_A} . \qquad\qquad (13.50)$$

Here u and N_B are the longitudinal velocity of the feed stream and the solvent flux, respectively, at the entrance.

For the far-downstream region away from the inlet, on the other hand, Dresner further postulated that even the (concentration) boundary layer would fill the entire channel. Then, the following asymptotic solution in a sheet membrane was obtained using the Berman velocity field:

$$\frac{C_{Aw}}{C_{Ab}} = 1/3 \; Pe_W^2 + 1 . \qquad\qquad (13.51)$$

Sherwood and his co-workers [27] independently carried out the Graetz-type analysis for the same problem and obtained a very accurate but complicated solution. It was found that the simple Dresner solutions, Eqs. 13.47, 13.48, and 13.51, were in excellent agreement with their solution, as shown in Fig. 13.18. Later, Fisher et al. [28] modified the Dresner solution for a tubular

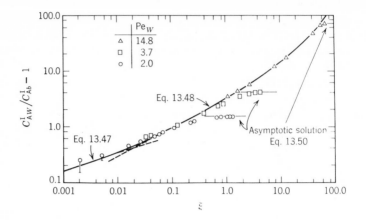

Fig. 13.18 Graetz-type analysis for laminar flow in a reverse-osmosis sheet membrane. Reprinted by permission from Ref. 27. Copyright 1965 by the American Chemical Society.

membrane of radius r. Even in this configuration, Eqs. 13.47 and 13.48 are applicable if parameters

$$\xi = \frac{N_B Pe_W^2 x}{uCr} \tag{13.52}$$

and

$$Pe_W = \frac{N_B r}{2CD_A} \tag{13.53}$$

are substituted for those defined by Eqs. 13.49 and 13.50, and in the far downstream region the asymptotic equation becomes instead of Eq. 13.51,

$$\frac{C_{AW}}{C_{Ab}} = Pe_W^2 + 1 \ . \tag{13.54}$$

Then, they concluded that the modified Dresner solution is also in good agreement with the result of the accurate Graetz-type analysis.

Gill et al. [29] and Mastromonico [30] analyzed the problem for an annulus configuration. Brian [31] later solved the equations of mass conservation using the finite difference method in Berman's velocity field by allowing for the variation of the solvent permeation flux in the axial direction in the case where the salt rejection is not complete (i.e., SR < 1). He concluded from the analysis that the Dresner equation could still be applied in this case with reasonable accuracy if one uses the length-averaged solvent permeation rate in the Dresner equation. Recently, Gill and co-workers [32] have studied the more general case of simultaneous velocity and concentration profile with variable solvent-permeation flux, but complete salt rejection.

For a perfect osmotic membrane, the Dresner solution can more conveniently be presented in terms of the Sherwood number by substituting Eqs. 13.47 to 13.54 into Eq. 13.34. The Sherwood number in the entrance region is then given as

$$Sh = \frac{4Pe_W}{\ln (1 + 1.536\xi^{1/3})} \qquad \text{if } \xi \leqslant 0.02 \qquad (13.55)$$

$$= \frac{4Pe_W}{\ln \left\{ \xi + 6\text{-}5 \ \exp \left[- \left(\frac{\xi}{3} \right)^{1/2} \right] \right\}} \qquad \text{if } \xi > 0.02 , \qquad (13.56)$$

and in the far downstream region as

$$Sh = \frac{4Pe_W}{\ln [1 + 1/3 \ Pe_W^2]} \qquad \text{for a sheet membrane} \qquad (13.57)$$

$$= \frac{4Pe_W}{\ln (1 + Pe_W^2)} \qquad \text{for a tubular membrane} \qquad (13.58)$$

The local Sherwood number can thus be estimated at an arbitrary distance from the inlet of the channel using this Dresner solution in the following procedure. For a given membrane module and solution to be separated, the wall Peclet number Pe_W is evaluated from Eqs. 13.50 or 13.53. With a given longitudinal velocity u of the feed stream at the entrance, the parameter is also calculated at an arbitrary distance x using Eqs. 13.49 or 13.52. Then, the corresponding local Sherwood number can be estimated from Eqs. 13.54 or 13.55. If the estimated Sherwood number is smaller than the asymptotic value given by Eqs. 13.57 or 13.58, the subsequent local Sherwood number may be approximated as the asymptotic value.

It is well known that the convective mass transfer is very analogous to the convective heat transfer in a closed conduit. Therefore, one may use the following simpler but reasonably accurate Leveque solution in the entrance region instead of Eqs. 13.55 and 13.56:

$$Sh = 1.23 \left[Re \ Sc \left(\frac{2b}{x} \right) \right]^{1/3} \qquad \text{for sheet membrane} \qquad (13.59)$$

$$= 1.077 \left[Re \ Sc \left(\frac{2r}{x} \right) \right]^{1/3} \quad \text{for tubular membrane} \quad (13.60)$$

$$\text{if } 100 < Re \ Sc \left(\frac{4r_h}{x} \right) < 5000.$$

This solution is essentially the same as in convective heat transfer [33].

When the channel of the membrane module is not long and the recovery of the product is relatively small, the following length-averaged mass-transfer coefficient $\bar{k}$ is usually introduced in order to make a short-cut estimation of the overall transfer rate:

$$\bar{k} = \frac{1}{L} \int_0^L k(x) \ dx \ , \quad (13.61)$$

where L is the channel length. Then the corresponding average Sherwood number

$$\overline{Sh} = \frac{4\bar{k}r_h}{D_A} \quad (13.62)$$

can be obtained from Eqs. 13.59 or 13.60 as

$$\overline{Sh} = 1.86 \left[Re \ Sc \left(\frac{2b}{L} \right) \right]^{1/3} \quad \text{for sheet membrane} \quad (13.63)$$

$$= 1.62 \left[Re \ Sc \left(\frac{2r}{L} \right) \right]^{1/3} \quad \text{for tubular membrane} . \quad (13.64)$$

Sourirajan [17] compared the Leveque solution with the more accurate numerical solution for the desalination with the anisotropic cellulose acetate membrane. Except for the constant numerical coefficients, the two solutions were found to be in good agreement. Table 13.2 contains a convenient summary of all present equations.

If the membrane is not perfect (SR < 1), Gill [24] recommended to modify the expressions for the mass-transfer coefficient by using Eq. 13.33 in order to evaluate the concentration polarization C_{Aw}/C_{Ab}.

Turbulent Flow

As mentioned earlier, the concentration polarization can be reduced by sweeping away the polarized layer with a high flow rate of the feed stream. When the flow rate is sufficiently high (Re > 2000), the stream becomes

Table 13.2 Sherwood Number for Laminar Flow

Region	Solution		
Entrance region (short or moderately long conduit)	Dresner solution	$\text{Sh} = \dfrac{4\text{Pe}_W}{\ln(1 + 1.53\xi^{1/3})}$ $\xi \leqslant 0.02$	(13.55)
		$= \dfrac{4\text{Pe}_W}{\ln[\xi + 6 - 5\exp(\sqrt{\xi}/3)]}$ $\xi > 0.02$	(13.56)

where

$$\text{Pe}_W = \frac{N_B b}{2CD_A}; \quad \xi = \frac{2N_B \text{Pe}_W^2 x}{3ubC} \qquad \text{for sheet membrane}$$

$$\text{Pe}_W = \frac{N_B r}{2CD_A}; \quad \xi = \frac{2N_B \text{Pe}_W^2 x}{urC} \qquad \text{for tubular membrane}$$

	Lebeque solution	$\text{Sh} = a_2 \left[\text{Re Sc } \dfrac{4r_h}{x}\right]^{1/3}, \quad 100 < \text{Re Sc } \dfrac{4r_h}{x} < 5000$	(13.59) or (13.60)
		$\overline{\text{Sh}} = a_3 \left[\text{Re Sc } \dfrac{4r_h}{x}\right]^{1/3}, \quad 100 < \text{Re Sc } \dfrac{4r_h}{x} < 5000$	(13.63) or (13.64)

where

$$\text{Re} = \frac{4ur_h}{\nu}; \quad \text{Sc} = \frac{\nu}{D_A}$$

$r_h = b/2; \quad a_2 = 1.23 \; ; \quad a_3 = 1.85 \quad$ for sheet membrane
$r_h = r/2; \quad a_2 = 1.077; \quad a_3 = 1.62 \quad$ for tubular membrane

	Modified Lebeque solution by Sourirajan	$a_2 = 1.49; \quad a_3 = 2.24 \qquad$ for sheet membrane	
		$a_2 = 1.30; \quad a_3 = 1.95 \qquad$ for tubular membrane	

Table 13.2 Sherwood Number for Laminar Flow (continued)

Region	Solution	
Far-downstream region (very long conduit)	Dresner solution	$\mathrm{Sh} = \dfrac{4\,\mathrm{Pe}_W}{\ln[1 + a_1 \mathrm{Pe}_W^2]}$ (13.57) or (13.58)
		where
		$a_1 = \dfrac{1}{3}$ $\mathrm{Pe}_W = \dfrac{N_{\mathrm{B}} b}{2C D_{\mathrm{AB}}}$ for sheet membrane
		$a_1 = 1$ $\mathrm{Pe}_W = \dfrac{N_{\mathrm{B}} b}{2C D_{\mathrm{AB}}}$ for tubular membrane

Table 13.3 Sherwood Number for Turbulent Flow

Region	Solution		
Short conduit	Linton and Sherwood [34]	$\overline{Sh} = 0.276\ Re^{0.58}\ \left(Sc\ \dfrac{2r}{L}\right)^{1/3}$ for tubular membrane	(13.65)
	Sourirajan [17]	$\overline{Sh} = 0.44\ Re^{7/12}\ \left(Sc\ \dfrac{b}{2L}\right)^{1/3}$ for sheet membrane	(13.66)
Long or moderately long conduit	Chilton and Colburn [24]	$\overline{Sh} = 0.04\ Re^{3/4}\ Sc^{1/3}$	(13.67)
	Gill and Sher [35]	$\overline{Sh} = 0.18n\ Re^{7/8}\ Sc^{0.25}$	(13.68)
		$n = 0.127 \left(1 - \dfrac{60}{Re^{0.875}}\right)^{0.5}$	
	Dittus and Boelter [36]	$\overline{Sh} = 0.023\ Re^{0.8}\ Sc^{1/3}$ if $\dfrac{L}{2r} > 60$	(13.69)

turbulent. In turbulent flow, a number of correlation equations are available for the convective heat transfer in a closed conduit. One may directly use one of the equations considering the above-mentioned analogy between the concentration polarization and convective heat transfer. The numberical results obtained from the equations may be expected to agree within 20% of each other. The partial list of the correlation equations is presented in Table 13.3.

Ultrafiltration

The detailed treatment of the occurrence of concentration polarization in the preceding section applies essentially even to ultrafiltration. However, the diffusivity of a solute of large molecular weight (say, greater than 500) in a solution to be separated in ultrafiltration is extremely low. Consequently, the adverse effect of the concentration polarization in the process is more severe than that observed in reverse osmosis. The local concentration is frequently polarized to such an extent that a gel slime is formed on the ingoing membrane surface, which considerably reduces the ultrafiltrate rate as indicated in Fig. 13.19. The gel point or the saturated concentration for gelation depends only

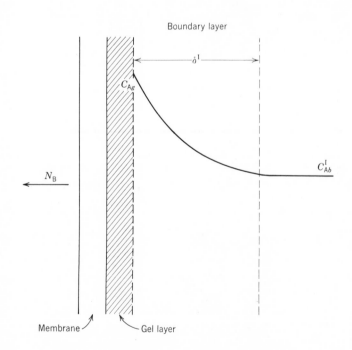

Fig. 13.19 Concentration polarization in an ultrafiltration membrane.

on the chemical and physical properties of the solution. It is also commonly observed that the transport resistance through the polarized boundary layer is much larger than that through the gel slime and the membrane itself. Indeed, experimental data in a large stirred ultrafiltration unit show in Fig. 13.20 that the rate increases greatly with the stirring speed at a given pressure gradient. At any fixed stirring speed, the rate also increases with the gradient as predicted by Eq. 8.11. However, the rate usually approaches a constant value at a sufficiently high pressure gradient for a given bulk concentration.

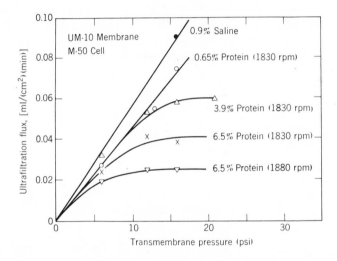

Fig. 13.20 Effect of transmembrane pressure on ultrafiltration flux for bovine serum albumin solutions in a stirred batch cell [38]. By permission from *Chem. Tech.*, the multidisciplinary monthly of the American Chemical Society.

The reason for this behavior can be deduced from the following argument [37]. At a sufficiently high stirring speed, the bulk concentration becomes quite homogeneous. Also the boundary-layer thickness and the concentration profile in the layer become almost uniform over the entire surface of the membrane at steady state. If the solvent permeation flux is enhanced by a high pressure gradient such that the gel slime is formed, the concentration at the membrane surface assumes a constant value, that is, the gel point, irrespective of the pressure gradient and the bulk concentration. Also, the transport resistance through the polarized boundary layer becomes primarily controlling. The high pressure gradient is almost dissipated over the gel slime, and, therefore, the solute rejection is practically independent of the pressure gradient. It can be seen from Eq. 13.32 that in this case the ultrafiltration rate is constant at a given bulk concentration regardless of the pressure gradient. If the solute rejection is unity, (meaning no solute passes), one may rearragne the equation to yield

$$N_{\mathrm{B}} = k \ln \frac{C_{\mathrm{A}g}}{C_{\mathrm{A}b}} , \tag{13.70}$$

where $C_{\mathrm{A}g}$ is the gel point of the solution, and the mass transfer coefficient k depends only on the stirring condition. The argument can also be expected to be valid for the configuration of the thin channel as long as an appropriately averaged value of the mass-transfer coefficient is used and the channel is not too long. The validity of the logarithmic dependence of the ultrafiltration rate on the solute concentration is indeed illustrated in Fig. 13.21.

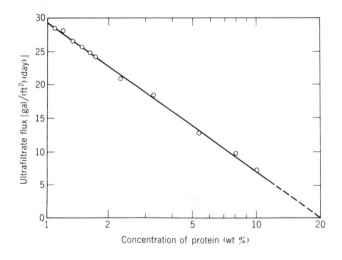

Fig. 13.21 Effect of bulk concentration on ultrafiltration flux for bovine serum solutions with PM-30 (10-mil channels, recirculation rate of 3000 ml/min) [39]. By permission from *Chem. Tech.*, the multidisciplinary monthly of the A.C.S.

Recently, Porter [40] discussed the concentration polarization in ultra-filtration using the results of extensive experimental studies conducted at Amicon Corporation. The studies show that the ultrafiltration rate is proportional to the power of 0.33 to 0.5 of the circulation rate in laminar flow, instead of the 0.33 power as predicted by the Leveque solution, Eqs. 13.59 or 13.60. However, the numerical values estimated by the equation agree within 15 to 30% of experimental data. In turbulent flow, the slope in the log-log plot of the ultrafiltration rate versus the circulation rate is in the range of 0.75 to 1.0, rather than the fixed value of 0.8 predicted by the Dittus and Boelter equation, Eq. 13.69 (see Table 13.3), but the agreement of numerical values is much better than in the case of laminar flow. The proportionality of the ultrafiltration flux to the 0.67 power of the circulation rate in both equations could also be confirmed in experiments with aqueous solution of human albumin and whole bovine serum.

In colloidal suspensions, however, it was observed that the experimental ultrafiltration flux is much higher by the factor 20 to 30 in laminar flow than predicted by the Leveque equation. The factor in turbulent flow was also in the range of 8 to 10 (see Fig. 13.22). The gross discrepancies could not be explained by adjustment of molecular properties such as kinematic viscosity, diffusivity, or the gel point. Porter finally attempted to interpret the anomalous observation

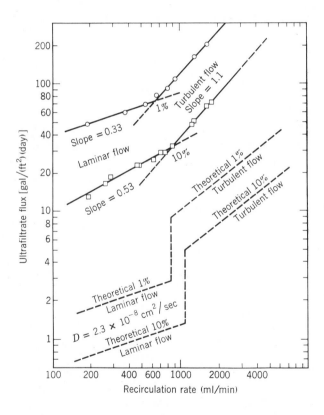

Fig. 13.22 Ultrafiltration of styrene-butadiene latex (30-mil channel, PM-30 membrane, 40-psig average transmembrane pressure drop). Reprinted by permission from Ref. 40. Copyright 1972 by the American Chemical Society.

in terms of the so-called tubular pinch effect, that is, the eccentric concentration of suspended fine particles along the line parallel to the axis located at a radial position between the wall and the axis of a small channel. The radial migration of suspended particles away from both the wall and the axis of the channel is pronounced at a high axial velocity of the stream and a high ratio of the particle size to the dimension of the channel. The exact cause of this phenomenon is unknown. Rubinow and Keller [41] postulated the slip-spin Magnus effect, while Staffman [42] considered the slip-shear inertia force, In any event, the migration and rotation of the particles would be expected to create a secondary flow near the wall in the ultrafiltration of colloidal suspensions. Then, the flow would depolarize the concentrated boundary layer and consequently enhance the ultrafiltration rate. Indeed, the increase in the ultrafiltration flux is commonly observed with whole blood containing red cells, skimmed milk, casein, and clay suspensions. Interestingly, Bixler and Rappe [43] patented an ingenious technique to promote the ultrafiltration flux by introducing fine solid particles into the feed stream.

Dialysis and Electrodialysis

In dialysis, where separation is achieved by utilizing the selective transport of a solute across a dialytic membrane, the solute is generally depleted at the ingoing surface of the membrane, but accumulated at its outgoing surface. This concentration polarization occurs because the solute transport across either boundary layer is hampered by the slow process of the molecular diffusion. Figure 13.23 illustrates the phenomenon in ordinary dialysis. If the

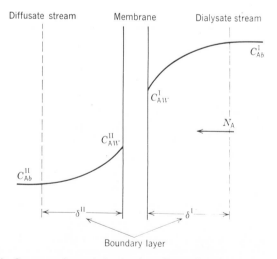

Fig. 13.23 Concentration polarization in ordinary dialysis.

solute is a nonelectrolyte, the solute flux is expressed by Eq. 13.36 in terms of the mass-transfer coefficient. Transport of an ionic species is, however, greatly influenced by the interionic interaction through the concentration (electric) potential generated by the concentration distribution of other ionic species as discussed in Chapter VIII. When the solution contains a single electrolyte comprised of a cation and an anion, the electrolyte solute as a whole transports thus with an average diffusivity of self-diffusion coefficients of the two ions (see Eq. 8.57). In this case, Eq. 13.36 can also be applied with this effective diffusivity.

When the solution contains more than two ionic species, the situation becomes complicated, even in the absence of an external electric potential. In principle, the effect of the interionic interaction can be expressed by integrating a system of the Nernst-Planck equations for all ionic species subject to the electroneutrality condition and the charge conservation. The integrated solution is, however, hardly available in a closed form even for a ternary ionic solution, which corresponds to the simplest case of Donnan dialyses. Therefore, the multi-ionic case is not treated here. If there exists only a trace of an ionic species in the multi-ionic solution, the concentration potential is quenched as in a supporting solution, and its transport is determined largely by its own concentration gradient with its self-diffusion coefficient. In this case, the solute flux is simply given by Eq. 13.36.

Consider the process of electrodialysis of a solution that contains a single electrolyte solute. When an external electric potential is applied across an ion-selective membrane, the electric current transfers the counterion across the membrane. As the electric transference of the counterion proceeds, the co-ion is momentarily accumulated at the ingoing surface of the membrane. However, the accumulated co-ion immediately migrates away from the surface in order to satisfy the electroneutrality condition. Thus, the electrolyte solute as a whole is depleted at the surface as discussed in Chapter IX (see Figure 13.24). Similarly the electrolyte solute is accumulated at the outgoing surface.

An electrodialysis module commonly operates so that the concentration in the transfer stream is higher than in the dialysate stream. In this case, the concentration polarization enhances the backpermeation due to the concentration gradient from the transfer stream into the dialystate stream. Therefore, the polarization reduces the efficiency of the module. In general, the concentration depletion has a worse effect than the accumulation. Since the electric conductivity of an electrolyte solution is proportional to its electrolyte concentration, the concentration depletion increases the electric resistance across the boundary layer, so that a considerable portion of the electric energy is wasted into heat dissipation. Further increase in the external electric potential beyond a certain value induces electrolysis of the solvent water or "water splitting." Consequently, the pH value of the solution is changed, and

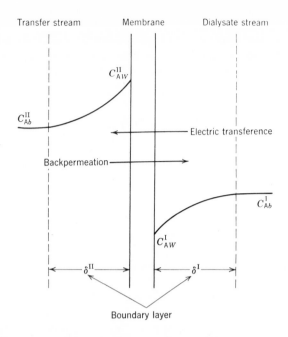

Fig. 13.24 Concentration polarization in electrodialysis.

the electrolyzed hydrogen or hydroxyl ion competes with the original counterion in electric transference. The resulting membrane deterioration by the pH change has already been described in Chapter IX, where the concentration polarization was discussed using the Nernst-film concept and characterized by the parameter of the limiting current density I_l. The parameter is actually defined by Eq. 9.87 in terms of the mass-transfer coefficient k and the bulk concentration. In the absence of any obstruction or turbulence promoter in the channel, the mass-transfer coefficient, or the Sherwood number, can be expected to be valid. Thus, the limiting current density for a given bulk concentration is approximately proportional to the 0.33 power of the longitudinal-stream velocity in the laminar-flow regime and to its 0.8 power for the tubulent-flow regime. The more rigorous theoretical analysis was made by Sonin and Probstein [44].

Turbulence Promoters

The adverse effect of concentration polarization may be minimized by employing suitable turbulence promoters in the flow channels. In the configuration of the *tubular* membrane, common types of promoters are the plain spiral; detached spiral (the wire spiral-wound around runners attached to a central support rod); and spheres spaced along a central support rod. In the *sheet* membranes, various types of screens are extensively used such as cross straps, expanded, and perforated and corrugated types. For the tubular type of reverse-osmosis membrane, Savage [45] summarized the improvement of the mass-transfer rate with the detached spiral promoter in Fig. 13.25. Indeed, the

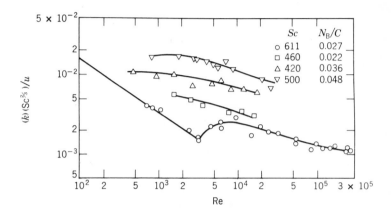

Fig. 13.25 Effect of spiral promoter on the mass-transfer coefficient in tubular type of reverse-osmosis membrane. ○: no turbulence promoters; △□: with turbulence promoters [45].

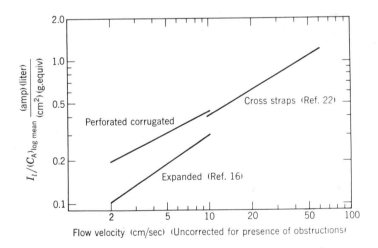

Fig. 13.26 Effect of longitudinal velocity on the limiting current density for commercial screen types of turbulence promoters. Reprinted by permission from Ref. 46. Copyright 1965 by the American Chemical Society.

mass-transfer coefficient with the promoter is 4 to 10 times that of the coefficient in the smooth tube. This effectiveness prevails even in the nominal laminar flow regime as shown in the figure. It is interesting that the mass-transfer coefficient is not quite proportional to the $\frac{1}{3}$ power of the Schmidt number as predicted by the Leveque solution, Eq. 13.60, and the Dittus and Boelter equation, Eq. 13.69.

In the sheet membranes of electrodialysis, Mandersloot and Hicks [46] summarized the dependence of the limiting current density on the longitudinal stream velocity for some commercial screen types of promoters as shown in Fig. 13.26. The current density is found to be proportional to the 0.5 to 0.7 power of the stream velocity. Recently, Belfort and Guter [47] made an extensive hydrodynamic study on various commercial types of promoters.

The turbulence promoters are also expected to retard the membrane fouling. However, the promoters always increase the longitudinal pressure drop, as illustrated in Fig. 13.27. Therefore, one must adequately balance the two opposing effects.

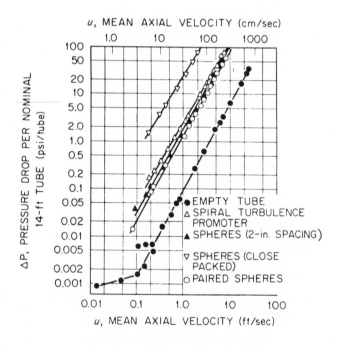

Fig. 13.27 Longitudinal pressure loss with turbulence promoters in tubular type of reverse-osmosis module [45].

7 DESIGN OF GASEOUS-DIFFUSION MEMBRANE MODULES

A membrane module of gaseous diffusion can be designed and operated such that the permeate (low-pressure) and reject (high-pressure) streams flow either parallel or normal to the membrane surface. In this section, the following two flow modes shall be considered: cross flow and parallel flow. In cross flow, the normal velocity component dominates over the longitudinal velocity component in the permeate stream, the reverse situation holding in the reject stream (see Fig. 13.28a). The cross flow usually takes place when the permeate stream is pulled into a high vacuum, or the permeation rate across the membrane is large. In parallel flow, on the other hand, the longitudinal velocity components are larger than the normal velocity components in both the permeate and reject streams. Also, the cocurrent and countercurrent flow must be considered separately in the parallel-flow situation depending on the direction of the longitudinal velocities, as shown in Fig. 13.28b.

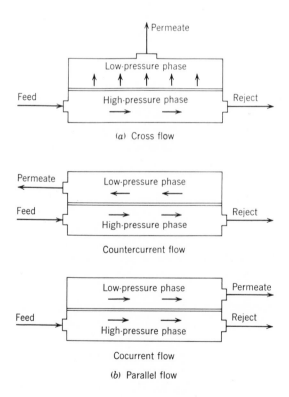

Fig. 13.28 Cross and parallel flows in a membrane module.

In a system where a molecular species is transported selectively as in a membrane module, the concentration of the species is generally distributed in every contiguous phase of the system. However, its concentration gradient in a gas phase is usually not so steep as in a liquid phase when the momentum of the species is transferred at the same rate. It is because the diffusivity of the species in the gas phase is much greater than in the liquid phase. The ratio of the gas to liquid diffusivities is often in the order of 10^4, and the Schmidt number in the gas phase is usually higher by thousands of times that in the liquid phase. Consequently, the concentration polarization in the direction normal to the membrane surface is practically absent in the gaseous stream, even though the longitudinal velocity may be polarized by the viscous drag at the membrane surface. Thus, the concentration profile direction normal to the membrane is always almost uniform even when the gaseous stream flows parallel to the membrane surface. However, a concentration distribution may still be established in the parallel direction, if the longitudinal velocity u and the length L of the membrane channel are so large that the convective transport in this direction dominates over the molecular diffusion. Or, if the Peclet number defined as

$$\text{Pe}_L = \frac{uL}{D_A}$$

is much larger than unity, the gas stream behaves like a "plug flow." It should be emphasized that the term "plug flow" is referred to only with respect to the concentration distribution. The distribution is particularly pronounced when the permeation rate and the channel width are small, but the selectivity of the membrane and the fraction of recovery are large.

Thus, several cases can be postulated depending on conditions of design and operation of a membrane module: cross flow and parallel flow; complete mixing and plug flow. Here, the term "complete mixing" is used in the sense that the concentration is uniform throughout the phase under consideration. When both the permeate and reject streams are in plug flow, the countercurrent and cocurrent cases must also be considered separately. A membrane module that is designed and operated in plug flow gives a superior degree of separation over one operated under complete mixing conditions. Of all cases, the countercurrent situation is the most efficient one. As mentioned earlier, the plug-flow situation can be realized in a stream when a high longitudinal velocity is used in a thin channel. In this case, a considerable pressure loss occurs usually along the channel due to the hydrodynamic drag and therefore a high pumping power is required. Also, the pressure loss reduces the driving force, frequently to a significant extent, for permeation across the membrane. The efficiency of the module is thus decreased. Therefore, adequate design and operation of the module are necessary to balance the two opposing effects.

Complete Mixing Case — Weller-Steiner Case I [9]

First consider the simplest case where the concentration and the total pressure are almost uniform in both the permeate and reject streams, as shown in Fig. 13.29. The mass conservation over the overall membrane surface area S_T for a binary mixture A and B leads to (see Chapter IV)

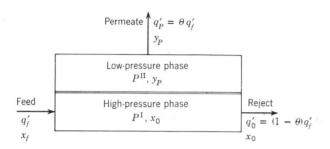

Fig. 13.29   Complete mixing (Weller-Steiner Case I).

$$q'_p y_p = \frac{Q_A P'^I S_T}{l} (x_0 - \text{Pr } y_p), \tag{13.71}$$

$$q'_p (1-y_p) = \frac{Q_B P'^I S_T}{l} [1 - x_0 - \text{Pr } (1-y_p)], \tag{13.72}$$

where Q_A and Q_B are permeabilities of species A and b, respectively, across the membrane of thickness l, and q' is the total molar flow rate. P'^I and P'^{II} are the total pressures of the reject (high-pressure) and permeate (low-pressure) streams, respectively, and Pr is the ratio of the low to high pressure, defined as

$$\text{Pr} = \frac{P'^{II}}{P'^I}. \tag{13.73}$$

Also, x_A and y_A are mole fractions of species A in the reject and permeate streams, respectively, and the subscripts f, p, and 0 refer to the feed, permeate and reject streams respectively. Combining Eqs. 13.71 and 13.72, one obtains

$$\frac{y_p}{1 - y_p} = a^* \frac{x_0 - \text{Pr } y_p}{(1 - x_0) - \text{Pr } (1 - y_p)}, \tag{13.74}$$

and the total membrane area becomes

$$S_T = \frac{l \; q'_f}{Q_A P'^I} \; \theta \; \frac{y_p}{x_0 - \text{Pr} \; y_p} \tag{13.75}$$

Here, θ is the cut defined as

$$\theta = \frac{q'_p}{q'_f} \; , \tag{13.76}$$

and the ideal separation factor a^* is introduced as

$$a^* = \frac{Q_A}{Q_B} \; . \tag{13.77}$$

The ideal separation factor a^* should not be confused with the actual separation factor a, which is defined as an overall quantity by

$$a = \frac{y_p}{1 - y_p} \; \frac{1 - x_0}{x_0} \tag{13.78}$$

Considering Eq. 13.74, the two separation factors are not the same unless the total pressure of the permeate stream is zero, that is, $\text{Pr} = 0$. In what follows, a convention will be used such that the ideal separation factor a^* is always greater than unity, that is, $Q_A > Q_B$. Consequently, the mole fractions x and y always refer to the higher permeable species A.

Membrane Area

The total membrane area S_T can then be evaluated from Eqs. 13.74 and 13.75 with the overall material balance

$$x_f = \theta y_p + (1 - \theta) x_0 \; . \tag{13.79}$$

It is very convenient to introduce the dimensionless membrane area s_T as

$$s_T = \frac{Q_A P'^I}{q'_f} \; S_T \; . \tag{13.80}$$

Then the dimensionless area s_T becomes, from Eq. 13.75,

$$s_T = \theta \; \frac{y_p}{x_0 - \text{Pr} \; y_p} \; . \tag{13.81}$$

It is interesting that the quantity s_T is very similar to the "number of transfer units," which is widely used in the design of the continuous fractionating column, when the transfer unit is defined as

$$s^* = \frac{l \, q_f'}{Q_A P'^{\mathrm{I}}},$$ (13.82)

so that the actual total membrane area S_T becomes

$$S_T = s_T \; s^*$$ (13.83)

This transfer unit can also be considered as a capacity factor.

Thus, a membrane module in the complete mixing case is characterized by seven variables: x_f, x_0, y_p, θ, a^*, Pr and s_T, four of which are independent variables. Then, the actual size of the module is determined simply with Eq. 13.83 through the capacity factor s*, which depends on the total feed-flow rate and the total pressure P'^{I} as indicated by Eq. 13.82. Consequently,

$$\frac{7!}{3!4!} = 35$$

choices are possible. Among the choices, the following two situations will be treated here:

1. To find s_T, x_0, and y_p for given x_f, θ, a^*, and Pr
2. To find s_T, θ, and y_p for given x_f, x_0, a^* and Pr

The procedural details and the corresponding programs for the two design problems are presented in Appendix B. The programs that are written in FORTRAN IV actually solve the system of the algebraic equations, Eqs. 13.74, 13.79, and 13.81.

Minimum Reject Concentration

Obviously, the inequality

$$x_f \leqslant y_p$$ (13.84)

is always valid. The equal sign corresponds to the case where all the feed stream is permeated, that is, $\theta = 1$. This is the limiting case where the membrane area S_T becomes infinitely large. Substituting this inequality into Eq. 13.74, one may obtain the minimum reject concentration for a given x_f value as

$$x_{0M} = \frac{x_f \, [1 + (a^* - 1)\mathrm{Pr}(1 - x_f)]}{a^* \, (1-x_f) + x_f}.$$ (13.85)

Thus, a feed stream of concentration x_f cannot be stripped beyond the value x_{0M} in a single unit, even with an infinitely large membrane area. In order to strip the feed stream beyond the limit, one must resort to a cascade system if both streams are completely mixed. However, the single module could be designed such that the mixing is not complete but plug flow is controlling.

Cross Plug Flow — Weller-Steiner Case II and Naylor-Backer Case [9, 10]

In this case, the longitudinal velocity and the channel length of the high-pressure (reject) stream are so large that the Peclet number Pe_L is great enough for the stream to be in plug flow. Then, the material balance over a differential membrane area dS leads to (see Fig. 13.30)

$$-d(q'^{I}x_A) = \frac{Q_A P'^{I}}{l} dS (x_A - Pr\ y_A) , \qquad (13.86)$$

$$-d\ [q'^{I}(1-x_A)] = \frac{Q_B P'^{I}}{l} dS\ [1 - x_A - Pr\ (1-y_A)] . \qquad (13.87)$$

Here, the superscript I refers to the high-pressure stream and mole fractions, x_A and y_A, are values at the point of the cumulative area S.

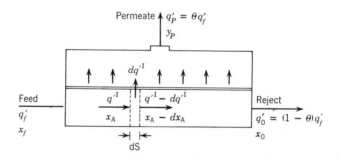

Fig. 13.30 Cross flow (Weller-Steiner Case II).

On the other hand, the low-pressure (permeate) stream is almost pulled into vacuum such that the flow is mainly in the direction normal to the membrane surface. The permeated molecular species are therefore immediately swept away by convection from the outgoing membrane surface, and the transport by molecular diffusion in this direction is negligible. Then, the mole fraction y_A of the permeate stream can be approximated as

$$y_A = \frac{d(x_A q'^I)}{d(x_A q'^I) + d[q'^I(1-x_A)]} = \frac{d(x_A q'^I)}{dq'^I}. \tag{13.88}$$

This is very similar to the Raleigh relationship in continuous distillation. Rearranging Eqs. 13.86 to 13.88 and then expressing in dimensionless forms, a system of ordinary differential equations can be obtained:

$$\frac{dq^I}{ds} = - [x_A - Pr \, y_A + \frac{1}{a^*}\left\{1 - x_A - Pr(1 - y_A)\right\}], \tag{13.89}$$

$$q^I \frac{dx_A}{ds} = - [x_A - Pr \, y_A + x_A \frac{dq^I}{ds}], \tag{13.90}$$

$$\frac{y_A}{1-y_A} = a^* \frac{x_A - Pr \, y_A}{1-x_A-Pr(1-y_A)}, \tag{13.91}$$

where the following dimensionless variables are introduced:

$$q^I = \frac{q'^I}{q'_f} \tag{13.92}$$

and

$$s = \frac{Q_A \, P'^I}{l \, q'_f} S, \tag{13.93}$$

Thus, the membrane module essentially consists of an infinite number of infinitesimally small modules of the complete mixing case. Hence, an arbitrary concentration of the reject stream can be obtained up to zero, by controlling the cut or by enlarging the total membrane area. This is in contrast to the complete mixing case.

The system of differential equations, Eqs. 13.89 to 13.91, can be solved by an appropriate scheme of numerical integration with the boundary conditions $x_A = x_f$, $q^I = 1$ at $s = 0$, which allows us to find $x_A = x_0$ and $q^I = 1-\theta$ at $s = s_T$. In most practical design problems, the total dimensionless area s_T is, however, not specified. The boundary-value problems can be converted into an initial, value problem, which is much easier to solve. Some examples of such problems are illustrated in Appendices B and C. Also, in the Appendices, the computer program for the numerical integration of the corresponding system of differential equations is given.

Weller-Steiner Solution

After some ingenious transformations, Weller and Steiner [9] obtained the analytical solution for Eqs. 13.89 and 13.90 with Eq. 13.91 as follows:

$$\frac{q^l(1-x_A)}{t-x_f} = \left(\frac{t_f - a_1/a_2}{t - a_1/a_2}\right)^{n_1} \left(\frac{t_f - a^* + a_3}{t - a^* + a_3}\right)^{n_2} \left(\frac{t_f - a_3}{t - a_3}\right)^{n_3}, \qquad (13.94)$$

where

$$t = -a_1\zeta + (a_1\zeta^2 + 2a_2\zeta + a_3^2)^{0.5},$$

$$\zeta = \frac{x_A}{1-x_A},$$

$$a_1 = 0.5 \,[(1-a^*)\,\text{Pr} + a^*],$$

$$a_3 = -0.5 \,[(1-a^*)\,\text{Pr} - 1],$$

$$a_2 = -a_1 a_3 + 0.5a^*,$$

$$n_1 = \frac{1}{2a_1 - 1},$$

$$n_2 = \frac{a^*\,(a_1-1)+a_3}{(2a_1-1)(0.5\ a^* - a_3)},$$

$$n_3 = \frac{1}{1-a_1-a_2/a_3}, \qquad (13.95)$$

and t_f is the value at $\zeta = x_f/(1-x_f)$.

Then the mole fraction x_0 of the reject stream can be obtained from Eq. 13.94 with $q^l = 1 - \theta$ and the value t at $\zeta = x_0/(1 - x_0)$. The mole fraction y_p of the permeate stream is then claculated from the overall material balance, Eq. 13.79. The dimensionless total area s_T is also obtainable from:

$$s_t = a^* \int_{\zeta_0}^{\zeta_f} \frac{q^l(1-x_A)d\zeta}{\left\{f(\zeta)-\zeta\right\}\left\{\dfrac{1}{(\zeta+1)} - \dfrac{\text{Pr}}{[1+f(\zeta)]}\right\}}, \qquad (13.96)$$

where

$$f(\zeta) = a_1\zeta - a_3 + (a_1^2\zeta^2 + 2a_2\zeta + a_3^2)^{0.5}. \qquad (13.97)$$

Here, ζ_f and ζ_0 and values at $x_A = x_f$ and $x_A = x_0$, respectively. Even in this solution, numerical integration is thus required to calculate the dimensionless total area s_T. In order to avoid such numerical integration, Weller and Steiner recommended a short-cut approximation to calculate an average value of s_T:

$$s_T = \frac{\theta}{2} \left(\frac{y_f}{x_f - \text{Pr } y_f} + \frac{y_0}{x_0 - \text{Pr } y_0} \right) , \qquad (13.98)$$

where y_f and y_0 and evaluated from Eq. 13.90 at $x_A = x_f$ and $x_A = x_0$, respectively.

Naylor-Backer Solution

In the cross-flow case, the pointwise separation factor is defined as

$$a = \left(\frac{y_A}{1 - y_A} \right) \left(\frac{1 - x_A}{x_A} \right) , \qquad (13.99)$$

From Eq. 12.91, it becomes

$$a = \frac{a^* + 1}{2} - \frac{\text{Pr}(a^* - 1)}{2x_A} - \frac{1}{2x_A} +$$

$$\left[\left(\frac{a^* - 1}{2} \right)^2 + \frac{(a^* - 1) - \text{Pr}(a^{*2} - 1)}{2x_A} + \left\{ \frac{\text{Pr}(a^* - 1) + 1}{2} \right\}^2 \right]^{0.5} .$$

Naylor and Backer [10] noted that the separation factor a is not sensitive to the mole fraction x for a small value of the pressure ratio Pr. With the approximation of a constant a, they integrated Eq. 13.88 to obtain

$$y_p = x_0^{-n_4} \left(\frac{1 - \theta}{\theta} \right) \left[(1 - x_0)^{n_5} \left(\frac{x_f}{1 - x_f} \right)^{n_5} - x_0^{n_5} \right] , \qquad (13.101)$$

where

$$n_4 = \frac{1}{a - 1} ,$$

$$n_5 = \frac{a}{a - 1} . \qquad (13.102)$$

Therefore the dimensionless total area s_T becomes

$$s_T = \int_{x_0}^{x_f} \left(\frac{1}{x_A - \text{Pr } y_A} \right) \left(\frac{x_A}{x_f} \right)^{n_4} \left(\frac{1 - x_f}{1 - x_A} \right)^{n_5} \left(\frac{n_5}{1 - x_A} \right) dx_A . \qquad (13.103)$$

This Naylor-Backer solution is much simpler than that of the Weller-Steiner II equation. However, both methods give results that are in excellent agreement. Although the two solutions were apparently derived from two different models, it was shown by Hwang [48] that the two models became identical for the case where $Pr = 0$. In general then, the Naylor-Backer solution is just an approximate form of the Weller-Steiner solution.

Parallel Plug Flow — Blaisdell Case [49]

When the longitudinal velocities and the ratios of length to width of the channels of both the permeate and reject streams are large, both streams are almost in plug flow. In this case, the cocurrent and countercurrent flows must be considered, depending on the location of the permeate take-off as shown in Fig. 13.31. Oishi et al. [50] numerically solved the system of ordinary differential equations that consist of the material balance in parallel flow. The

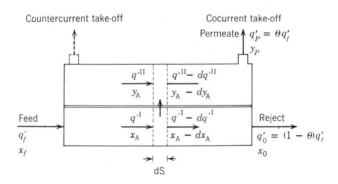

Fig. 13.31   Parallel plug flow (Blaisdell case).

resulting solution shows that the parallel-flow situation is superior even to the cross plug flow in separation. Later, Blaisdell [49] improved the method of numerical integration and demonstrated the superiority with experiments for the first time. Actually, he constructed a module in the shell-and-tube configuration with a bundle of hollow fibers of silicone rubber as shown in Fig. 13.32. This construction permitted experimental investigation of the cocurrent and counter current flows in the module.

Cocurrent Plug Flow

The material balance over a differential area dS in dimensionless forms (see Fig. 13.31) gives

$$-d(q^{\mathrm{I}}x_{\mathrm{A}}) = d(q^{\mathrm{II}}y_{\mathrm{A}}) = ds\,(x_{\mathrm{A}} - \mathrm{Pr}\ y_{\mathrm{A}}), \qquad (13.104)$$

$$-d[q^I(1-x_A)] = d[q^{II}(1-y_A)] = \frac{ds}{a^*} [1-x_A-\text{Pr}(1-y_A)]. \quad (13.105)$$

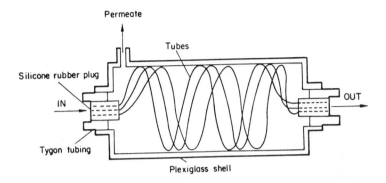

Fig. 13.32 Spiral hollow-fiber membrane module [49]. Courtesy *Chemical Engineering Science.*

Rearranging these equations yields the cross plug flow Eqs. 13.89 and 13.90. But Eq. 13.91 becomes

$$y_A = \frac{x_f - x_A q^I}{1 - q^I}, \quad q^I \neq 1, \quad (13.106)$$

$$y_A = y_f, \quad q^I = 1, \quad (13.107)$$

where y_f is evaluated from Eq. 13.91 with $x_A = x_f$. Here, Eq. 13.107 is obtained by a limiting process of the L'hospital rule as $q^I \to 1$. It implies that over a first differential membrane area the module behaves like the complete mixing case.

Again, the system of differential equations can be integrated by an appropriate scheme of numerical integration with initial conditions of

$$x_A = x_f, \quad q^I = 1, \quad y_A = y_f \quad \text{at } s = 0, \quad (13.108)$$

so that one can find

$$x_A = x_0, q = 1-\theta, \quad y_A = y_p \quad \text{at } s = s_T, \quad (13.109)$$

The detailed transformation of the system of differential equations for design

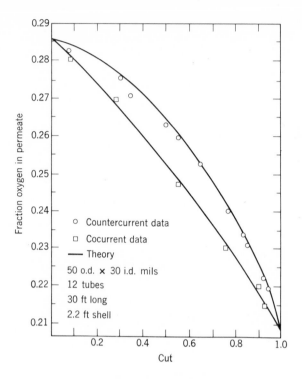

Fig. 13.33 Oxygen enrichment of air in the permeate (silicone-rubber tubing, Medical Eng. Sr. 2000; $a^* = 2.05$, Pr = 0.359, $x_f = 0.209$) [49]. Courtesy *Chemical Engineering Science.*

problems is discussed in Appendices B and C, where the corresponding computer programs are also presented.

It should be noted that there exists a minimum stripping concentration x_{oM} of species A in the reject stream for x_f, a^*, and Pr similar to the complete mixing case. This corresponds to the case where the mole fraction x_A of the high-pressure phase reaches a minimum at the outlet of the module. When such a minimum level of stripping is attained, that is,

$$\frac{dx}{ds} = 0,$$

the follosing relationship can be derived from Eqs. 13.104 and 13.105:

$$\frac{d(q^I x)}{d \ [q^I(1-x)]} = \frac{x_{oM}}{1-x_{oM}}$$

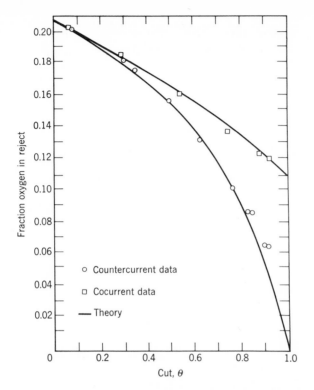

Fig. 13.34 Oxygen stripping of air in the reject (silicone-rubber tubing, Medical Eng. Sr. 2000; $a^* = 2.05$, Pr $= 0.359$, $x_f = 0.209$) [49]. Courtesy *Chemical Engineering Science.*

$$= a^* \frac{x_{oM} - \text{Pr } y_A}{1 - x_{oM} - \text{Pr}(1 - y_A)} . \qquad (13.110)$$

Furthermore, since $y_A \geqslant x_f$, the equation reduces to

$$\frac{x_{oM}}{1 - x_{oM}} = a^* \frac{x_{oM} - \text{Pr } x_f}{1 - x_{oM} - \text{Pr}(1 - x_f)} . \qquad (13.111)$$

This also corresponds to the case where $\theta = 1$ and $s_T \to \infty$. In other words, stripping cannot be acieved within a single module beyond $x_o = x_{oM}$ for given x_f, a^*, and Pr, even if an infinitely large membrane area is used.

Blaisdell compared the numerical solution with the experimental data obtained with his own module and found excellent agreement as shown in Fig. 13.33 and 13.34. Thus, the plug flow can easily be obtained in the hollow-fiber-type membrane modules.

Countercurrent Plug Flow

Similarly to the cocurrent plug flow, the material balance yields Eqs. 13.104 13.105 in which the flow rate q^{II} always possesses a negative value. Therefore, Eqs. 13.89 and 13.90 are still valid, but Eq. 13.105 is replaced by

$$y_A = \frac{x_f - \theta y_p - x q^I}{1 - \theta - q^I}. \tag{13.112}$$

Here, y_A is singular at $q^I = 1 - \theta$, that is, $s = s_T$. Also, the mole fraction y_p of the permeate stream and the cut θ are not known a priori if the feed mole fraction x_f is specified. Consequently, it is convenient to integrate Eqs. 13.89 and 13.90 backward from $s = s_T$ to $s = 0$ with known x_0 instead of x_f. In this case, the permeate mole fraction y_A can be evaluated, instead of using Eq. 13.112, from its equivalent form

$$y_A = \frac{x_A q^I - x_0(1-\theta)}{q^I - (1-\theta)}, \qquad q^I \neq 1-\theta,$$

$$y_A = y_0 \qquad\qquad , \qquad q^I = 1-\theta, \tag{13.113}$$

where y_0 is evaluated from Eq. 13.91 with $x_A = x_0$. This is very analogous to Eqs. 13.106 and 13.107 in the cocurrent plug flow. Initial conditions for the system of differential equations, Eqs. 13.104 and 13.105 with Eq. 13.111, are given by

$$x_A = x_0, \quad y_A = y_0, \quad q^I = 1-\theta \quad \text{at } s = s_T, \tag{13.114}$$

and final conditions are

$$x_A = x_f, \quad y_A = y_p, \quad q^I = 1 \quad \text{at } s = 0. \tag{13.115}$$

Then, numerical integration of the system of equations is iterated with an initial guess of the unknown cut θ so as to find a satisfactory value of cut θ until its final value at $s = 0$ is equal to one. If the feed mole fraction x_f is specified instead of x_0 a similar scheme of the iteration method can be devised. The detailed discussion for design problems and the corresponding computor programs are also presented in Appendices B and C.

Again, Blaisdell demonstrated the superiority of countercurrent plug flow in the hollow-fiber type of membrane modules as shown in Fig. 13.33 and 13.34.

One-Side Mixing

In a hollow-fiber-type membrane module where the length of the shell is much shorter than that of the bundle of hollow fibers (see Fig. 13.32) the phase outside of the hollow fibers is likely to be completely mixed when both the longitudinal and normal velocity components are small. However, the phase inside of the hollow fibers is likely to be in plug flow. If the feed stream is fed to the inside of the hollow fibers the permeate phase becomes completely mixed and its mole fraction y_p is uniform. In this case, the integration of Eqs. 13.89 and 13.90 with a constant $y_A = y_p$ yields after some transformation

$$\frac{q^I(1-x_A)}{1-x_f} = \left(\frac{\zeta + a_4 + a_5}{\zeta_f + a_4 + a_5}\right)^{n_6} \left(\frac{\zeta + a_4 - a_5}{\zeta_f + a_4 - a_5}\right)^{n_7}, \qquad (13.116)$$

where

$$a_4 = \frac{1}{2} \frac{a^* (1-Pr\ y_p) - 1 + Pr(1-y_p)}{Pr(1-y_p)},$$

$$a_6 = \frac{a^* Pr\ y_p}{Pr(1-y_p)},$$

$$a_7 = \frac{1-Pr(1-y_p)}{Pr(1-y_p)},$$

$$a_5 = (a_6^2 + a_7)^{0.5},$$

$$n_6 = \frac{-(a_4 + a_5 + a_7)}{2a_5},$$

$$n_7 = \frac{(a_4 - a_5 + a_7)}{2a_5},$$

$$\zeta = \frac{x_A}{1-x_A}. \qquad (13.117)$$

Thus the dimensionless are a s_T becomes

$$s_T = a^* \int_{\zeta_f}^{\zeta_0} \frac{q^I(1-x_A)(\zeta - a_7)d\zeta}{(\zeta^2 + 2a_4\zeta - a_6)\left[\frac{1}{(1+\zeta)} - Pr(1-y_p)\right]}. \qquad (13.118)$$

Similarly to the case of the cocurrent plug flow, there also exists a minimum stripping concentration x_{oM} given by Eq. 13.111.

When the feed stream is fed to the outside of the hollow fibers, the differential material balance yields, similarly to the previous case,

$$q^{II}\left(\frac{dy_A}{ds}\right) = x_0 - \text{Pr } y_A - y_A\left(\frac{dq^{II}}{ds}\right), \quad s \neq 0,$$

$$\frac{dy_A}{ds} = 0, \qquad\qquad\qquad\qquad s = 0, \qquad (13.119)$$

$$\frac{dq^{II}}{ds} = x_0 - \text{Pr } y_A + \frac{1}{a^*}\left[1-x_0-\text{Pr}(1-y_A)\right] . \qquad (13.120)$$

These differential equations are integrated at a constant x_0 with initial conditions

$$y_A = y_0, \; q^{II} = 0 \quad \text{at } s = 0 \qquad (13.121)$$

to find

$$y_A = y_p, \; q^{II} = \theta \quad \text{at } s = s_T \qquad (13.122)$$

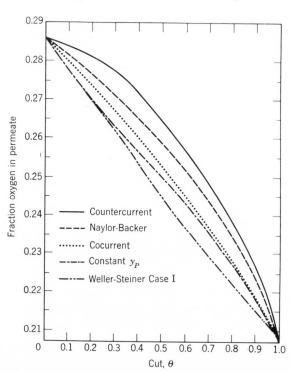

Fig. 13.35 Comparison of various cases for air separation ($a^* = 2.05$, $\text{Pr} = 0.359$, $x_f = 0.209$) [49]. Stripping section. Courtesy *Chemical Engineering Science.*

The system of differential equations is, fo course, subject to the overall material balance, Eq. 13.79. The integration yields an analytical solution similar to that of Eq. 13.117 of the previous case. However, a numerical integration is still required to evaluate the membrane area even in this solution. Therefore, a direct numerical integration with an appropriate iteration scheme is often more convenient. It should be noted that this case is the least efficient of all cases that have been considered so far. However, a relatively high pressure can be applied in such a hollow-fiber type of membrane module since a polymeric fiber can usually withstand a compression force more readily than an expansion force. Also, the pressure loss due to the hydrodynamic drag, and therefore the loss of the net driving force for permeation can be expected to be much smaller than in the cases where the high-pressure stream is fed to the inside of the hollow fibers.

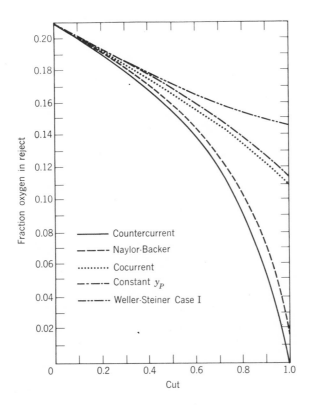

Fig. 13.36 Comparison of various cases for air separation ($a^* =. 2.05$, $Pr =. 0.359$, $x_f = 0.209$) [49]. Stripping Section. Courtesy *Chemical Enginerring Science.*

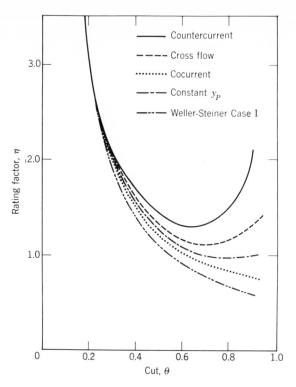

Fig. 13.37   Rating factors for various cases ($a^* = 2.05$, $Pr = 0.359$, $x_f = 0.209$).

Comparison of Various Cases and Effects of Operating Parameters

Blaisdell [49] carried out computations for oxygen enrichment of air when a silicone-rubber membrane is used as a barrier ($a^* = 2.05$). The computed results of the air separation in various cases are shown in Fig. 13.35 and 13.36. Indeed, cases with the plug-flow mode are more efficient than those with longitudinal mixing. Of all cases, the countercurrent scheme is the most advantageous. When a rating factor η is defined as the degree of separation per unit membrane area, that is,

$$\eta = \left(\frac{1}{s_T}\right)\left(\frac{y_p}{1-y_p}\right)\left(\frac{1-x_0}{x_0}\right) , \qquad (13.123)$$

the advantage of the plug-flow case can be more explicitly seen in Fig. 13.37. For different mixtures of oxygen and nitrogen as a feed stream, various cases are also compared in Fig. 13.38. This figure, in effect, represents a "stage

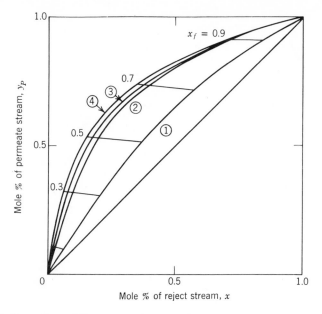

Fig. 13.38 Separation of the oxygen-nitrogen mixture for various cases ($q^* = 2.05$, Pr = 0.1, $\theta = 0.9$). 1: Weller-Steiner I; 2: cocurrent; 3: cross flow; 4: countercurrent.

equilibrium diagram," which very much resembles the equilibrium curve in stagewise distillation. Again, the figure indicates the effectiveness of the plug-flow mode.

The effects of operating parameters, pressure ratio Pr and cut θ, on the air separation are illustrated in Fig. 13.39 and 13.40 for the complete-mixing case, and in Fig. 13.41 and 13.42 for the countercurrent case. When mixtures of oxygen and nitrogen are used as feed streams, the effects of operating parameters are also shown in Fig. 13.43 and 13.44 for the complete-mixing case, and in Fig. 13.45 and 13.46 for the countercurrent case. Here the membrane is taken as silicone rubber with an ideal separation factor of $a^* = 2.05$.

Partial Mixing — Breuer Case [51]

In previous sections, several extreme cases have been considered depending on the degree of longitudinal mixing in both the permeate and the reject streams. Actually, the degree of mixing is determined largely by the extent of dominance of molecular diffusion over convective motion. In other words, when a mixture to be separated is fed into the membrane channel, the concentration of the higher permeable species A of the stream is depleted as the stream moves toward the downstream side. The resulting concentration gradient in

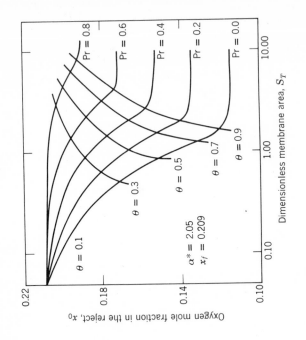

Fig. 13.40 Effects of pressure ratio Pr and cut θ on membrane area required for the air separation with complete mixing.

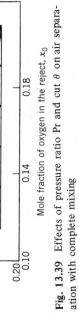

Fig. 13.39 Effects of pressure ratio Pr and cut θ on air separation with complete mixing

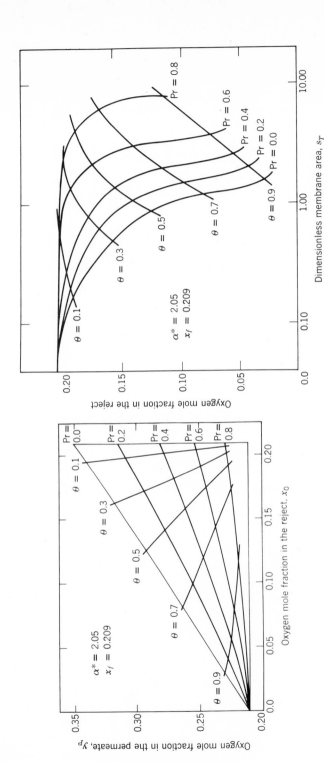

Fig. 13.41 Effects of pressure ratio Pr and cut θ on air separation with countercurrent flow.

Fig. 13.42 Effects of Pr and θ on membrane area required for the air separation with countercurrent flow.

389

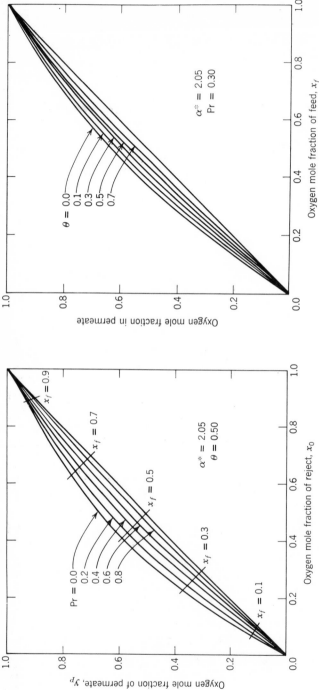

Fig. 13.43 Effect of Pr on separation of O_2-N_2 mixture with complete mixing.

Fig. 13.44 Effect of θ on separation of O_2-N_2 mixture with complete mixing.

Left figure axes and labels:

Oxygen mole fraction of permeate, y_p

Oxygen mole fraction of reject, x_o

$x_f = 0.9$
$x_f = 0.7$
$x_f = 0.5$
$x_f = 0.3$
$x_f = 0.1$

Pr = 0.0
0.2
0.4
0.6
0.8

$\alpha^* = 2.05$
$\theta = 0.50$

Right figure axes and labels:

Oxygen mole fraction in permeate

Oxygen mole fraction of feed, x_f

$\theta = 0.0$
0.1
0.3
0.5
0.7

$\alpha^* = 2.05$
Pr = 0.30

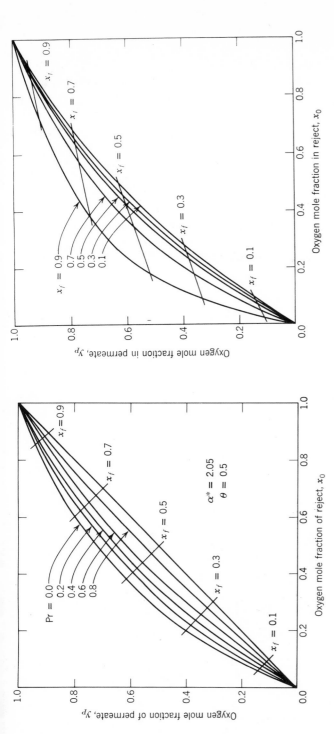

Fig. 13.45 Effects of Pr on separation of O_2-N_2 mixture with countercurrent flow.

Fig. 13.46 Effect of θ on separation of O_2-N_2 mixture with countercurrent flow.

the longitudinal direction induces selective migration of the species A by molecular diffusion towards downstream. Therefore, the concentration of the upstream is always lowered by the selective migration, resulting in reduction of the net driving force for permeation of the higher permeable species. Downstream, on the other hand, the concentration of the species is enriched, and its driving force is enhanced. However, as most of the separation occurs at the upstream location because of the higher concentration of A, the overall efficiency of the membrane module is generally reduced when longitudinal molecular diffusion occurs. In the spiral hollow-fiber-type membrane modules where the fiber length is relatively greater than that of the outer shell, the mixing is greatly enhanced by the occurrence of additional eddy motion.

In general, the longitudinal molar flow rate of the species A in a channel of cross-sectional area $\mathcal{A}$ can be expressed as the sum of convective and molecular diffusion, that is,

$$q'x_A + \left[- D_A \mathcal{A} C \frac{dx_A}{dx} \right].$$

Here C is the total molar concentration $[(g)(mole)/cm^3]$, and x is the longitudinal distance measured from the inlet of the channel. Thus, the membrane area S up to the point x becomes

$$\begin{aligned} S &= wx \qquad \text{for a single sheet membrane of width } w \\ &= 2\pi r x \qquad \text{for a single tubular membrane of radius } r. \end{aligned}$$

For multiple configurations the right-hand sides must be multiplied by the number of individual membranes. If the mixing is caused largely by eddy motion as in the hollow-fiber membrane module, the diffusivity D_A can be replaced by an effective dispersion coefficient, which depends largely on the ratio of the fiber to shell length, the flow rate q', diffusivity D_A, permeation rates, and other geometrical parameters. Therefore, the material balance over a differential membrane area dS of the sheet membrane yields

$$- d \left(q'^I x_A - w\mathcal{A}^I D_A^I C^I \frac{dx_A}{ds} \right) = \frac{D_A P'^I}{l} dS(x_A - \text{Pr } y_A)$$

$$= d \left(q'^{II} y_A - \mathcal{A}^{II} D_A^{II} C^{II} w \frac{dy_A}{dS} \right), \qquad (13.124)$$

$$-d \left[q'^I(1-x_A) + w \, \mathcal{A}^I D_A^I C^I \frac{dx_A}{dS} \right] = \frac{Q_B P'^I}{l} dS [1-x_A-\text{Pr}(1-y_A)]$$

$$= d \left[q'^{II}(1-y_A) + \mathcal{Q}^{II}D_A^{II}C^{II} \ w \ \frac{dy_A}{dS} \right] . \qquad (13.125)$$

Here parallel flow is assumed. These differential equations were first considered by Breuer [51]. In the case of cross flow, the expression for the longitudinal molar flux in the permeate (low-pressure) phase II is simply replaced by Eq. 13.91. More conveniently, Eqs. 13.124 and 125 can be transformed into dimensionless forms:

$$- d \left[q^I x_A - \frac{1}{Pm} \frac{dx_A}{ds} \right] = d \left[q^{II} y_A - \frac{1}{\beta \ Pm} \frac{dy_A}{ds} \right]$$

$$= (x_A - Pr \ y_A) \ ds , \qquad (13.126)$$

$$- d \left[q^I(1-x_A) + \frac{1}{Pm} \frac{dx_A}{ds} \right] = d \left[q^{II}(1-y_A) + \frac{1}{\beta \ Pm} \frac{dy_A}{ds} \right]$$

$$= \frac{1}{a^*} \ [1-x_A - Pr(1-y_A)] \ ds , \qquad (13.127)$$

where the dimensionless mixing parameters Pm and β are introduced as

$$Pm = \frac{l(q_f')^2}{w \ \mathcal{Q}^I D_A^I C^I Q_A P^I} \qquad (13.128)$$

and

$$\beta = \frac{\mathcal{Q}^I D_A^I C^I}{\mathcal{Q}^{II} D_A^{II} C^{II}} . \qquad (13.129)$$

The boundary conditions for the cocurrent flow are

$$q^I = 1, \quad q^{II} = 0 ,$$

$$x_f = x_A - \left(\frac{1}{Pm} \right) \left(\frac{dx_A}{ds} \right) ; \quad \frac{dy_A}{ds} = 0 \quad \text{at } s = 0, \quad (13.130)$$

$$\frac{dx_A}{ds} = \frac{dy_A}{ds} = 0 \qquad \text{at } s = s_T. \quad (13.131)$$

In countercurrent flow, the boundary condition, $q^{II} = 0$ at $s = 0$, is replaced by the condition $q^{II} = 0$ at $s = s_T$. The conditions for quantities in the phase II

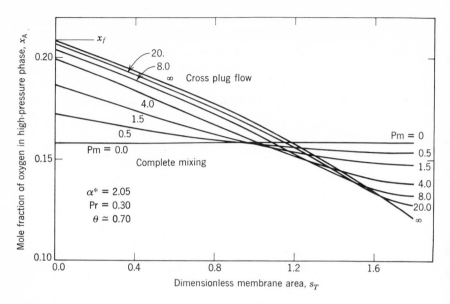

Fig. 13.47 Effect of Pm on concentration profile in high-pressure phase of cross flow.

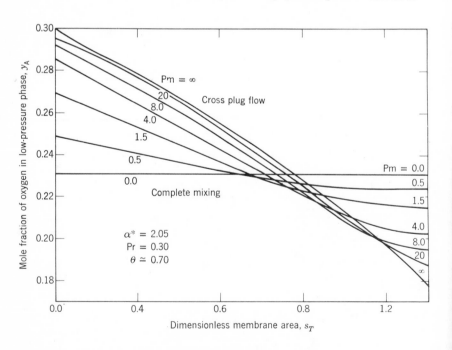

Fig. 13.48 Effect of Pm on concentration profile in low-pressure phase of cross flow.

are, of course, not required in the cross flow. Actually, Breuer assumed that the concentration gradient is zero at the entrance of the channel and the initial concentration x_A at $s = {}^+0$ is generally different from the valus of x_f. However, the initial condition, Eq. 13.130, is more rigorous.

It should be noted that the mixing parameter Pm can be physically interpreted as

$$Pm = \frac{(\text{longitudinal convective diffusion})^2}{(\text{longitudinal molecular diffusion})(\text{permeation rate})}.$$

Thus, a large value of Pm indicates the dominance of convective diffusion, that is, plug flow, in the reject (high pressure) phase I, with a small value of Pm implying a high degree of mixing in the phase. Similarly, β Pm is a measure of the degree of mixing in the permeate (low-pressure) phase II.

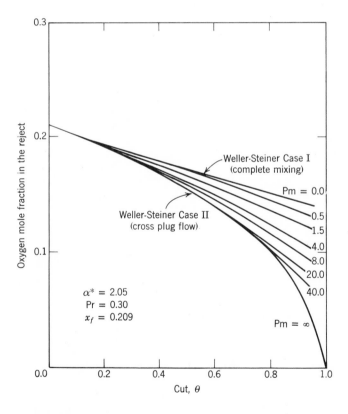

Fig. 13.49 Effect of Pm on reject mole fraction in cross flow.

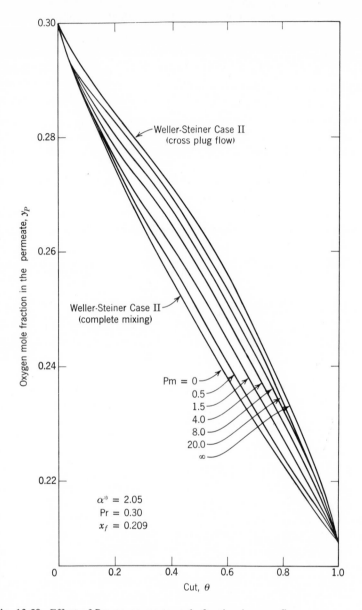

F ig. 13.50 Effect of Pm on permeate mole fraction in cross flow.

In cross flow, the system of differential equations, Eqs. 13.126 and 13.127, were solved numerically; and computed concentration profiles in the high- and low-pressure phases are presented in Fig. 13.47 and 13.48. As the mixing parameter Pm decreases, that is, the degree of the longitudinal mixing increases, the upstream concentrations are indeed reduced, but the downstream concentrations are enhanced by the longitudinal mixing. Consequently, the overall efficiency of the membrane module decreases with the decrease of the value of the mixing parameter Pm as shown in Fig. 13.49 and 13.50. Thus, the membrane module in cross flow finally reaches the complete mixing state for a very small value of Pm. As indicated in these figures, cross *plug* flow may be assumed in a practical membrane module for values of Pm greater than 50, whereas complete mixing prevails for values smaller than 0.1. A similar situation may also be expected in the counter- and cocurrent flows.

Obviously, the less-permeable species B in the high-pressure phase becomes enriched as the reject stream is stripped of the higher-permeable species. The downstream concentration of the less-permeable species can consequently become higher than the upstream one, and migration of species B by molecular diffusion due to the resulting concentration gradient occurs in the direction opposite to the total stream flow. Therefore, this mixing phenomenon is often called backdiffusion.

Pressure Loss Along Membrane Channel

In previous sections, pressures in the high- and low-pressure phases have been assumed to be uniform along the membrane channel. However, when a thin channel or a high flow rate is used to ensure the advantages of plug flow, pressures decrease along the channel, and the module efficiency is correspondingly reduced. The hydrodynamic drag at the wall is mainly responsible for the pressure loss. In addition, permeation of molecular species from the high- to the low-pressure phase further augments the pressure loss in the high-pressure phase, although increasing the pressure in the low-pressure phase. Thus, the driving force for permeation is reduced, and the module efficiency is lowered. The effect of pressure loss becomes important, particularly when a high cut is desired with a low pressure ratio Pr, or when a thin and long membrane channel is used such as a hollow fiber. In an extremely long channel, the pressure loss can be so great that the pressures in the reject and permeate streams become equal. Therefore, there exists a maximum attainable cut in an actual membrane module beyond which selective permeation will cease.

Consider a bundle of n hollow fibers of inside radius r. When the stream inside of hollow fiber is in laminar flow, the pressure gradient along the fiber can be expressed by the Poiseuille equation in *dimensional* units as

$$\frac{dP'}{dx} = -\frac{q'}{P'} \frac{8\mu RT}{n\pi r^4} \tag{13.132}$$

or in *dimensionless* units

$$\frac{dP}{ds} = -A_p \frac{q}{P},$$ (13.133)

where

$$A_p = \frac{8\mu R T}{n\pi r^4} \frac{q_f'}{(P_f')^2} \frac{1}{2\pi n r} \frac{l\, q_f'}{Q_A P_f'}.$$ (13.134)

Here the dimensionless pressure P is defined as

$$P = \frac{P'}{P_f'},$$ (13.135)

when P_f' is the pressure of the feed stream. The parameter A_p may be physically interpreted as the pressure loss along the channel of the membrane area,

$$s^* = \frac{l\, q_f'}{Q_A P_f'},$$

when no permeation occurs. Rigorously speaking, permeation of molecular species influences the velocity field inside of the hollow fiber, and the pressure gradient is different from that given by Eq. 13.132. However, the Berman perturbation solution [25] indicates that Eq. 13.132 is a good approximation when the permeation rate is as low as it normally is in gaseous diffusion.

When a gas mixture is fed to the inside of hollow fibers where the pressure loss is very significant, the differential material balances, Eqs. 13.89 and 13.90, must be modified. That is, the dimensionless pressure $P^I = P'^I/P_f'$ must be multiplied on the right-hand sides of the two equations. The pressure ratio is then a function of the dimensionless membrane area s. Thus, it is required to solve the system of differential equations that consist of the two modified material balances and the momentum equation, Eq. 13.133. In cross flow, where the pressure of the permeate stream outside of the hollow fiber is uniform, the system of equations is integrated; the results are presented in Fig. 13.51. The dotted line in the figure shows a locus of maximum attainable cuts for given values of parameter A_p. As indicated in the figure, the pressure loss is not significant in the system where the parameter A_p is smaller than 0.1.

Numerical Example – Problem Statement

It is desired to prepare a nitrogen-enriched stream containing at least 91% nitrogen from air. The hourly production is to be 750 ft^3 (STP). The membrane is made of silicone rubber sheet with thickness of 0.0367 cm. The maximum operating pressure for this material is 40 psig. It is therefore recommended

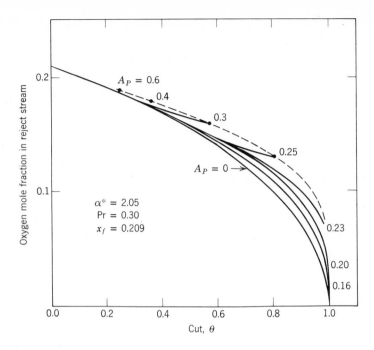

Fig. 13.51 Effect of pressure loss on separation in cross flow.

that the membrane module be operated at the safe pressure of 30 psig on the high-pressure side and atmospheric pressure on the low-pressure side. The ideal separation factor α^* is 2.00, and the permeabilities of oxygen and nitrogen are 6.0×10^{-8} and 3.0×10^{-8} cc(STP) cm/(cm^2)(cm Hg)(sec), respectively. The reject stream, therefore, will be the product.

Weller-Steiner Case I

The desired enrichment of nitrogen cannot be achieved with a single module when designed and operated so as to attain complete mixing in the high- and low-pressure phases. If a cascade system without recycle is used as described in Fig. 13.52, at least three stages of modules are required. This can be seen from the following McCabe-Thiele-type method. The minimum stripping concentration is given by Eq. 13.85 for a given feed concentration. When a McCabe-Thiele type of diagram is constructed using this equation as shown in Fig. 13.53, the minimum number of stages required for the enrichment is three. Therefore, three stages of the simple cascade system will be used. It should be noted that in this case two operating parameters, say, cuts of the first and second stages, are free. Here, the operating parameters are decided such that the overall membrane area will assume a minimum value. The calculated

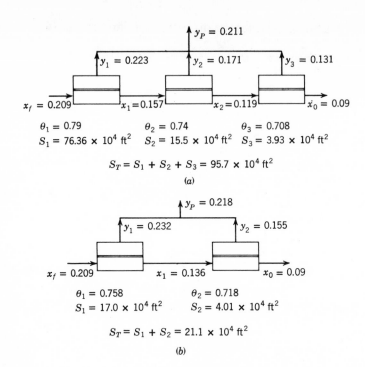

Fig. 13.52 Simple cascade system for Weller-Steiner Case I (a) and for cocurrent flow (b).

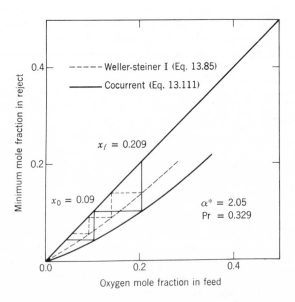

Fig. 13.53 Minimum number of stages in Weller-Steiner Case I and cocurrent flow.

Table 13.4 Numerical Example[a]

Cases	Membrane area (ft^2) $S_T \times 10^{-4}$	O_2 Conc. of Permeate (y_p)	Cut θ	Degree of separation a	Rating factor $\eta \times 10^6$
Weller-Steiner Case I	95.7	0.211		2.70	2.82
Cocurrent	21.1	0.218		2.82	13.36
Cross flow	9.69	0.228	0.863	2.99	20.86
Countercurrent	5.86	0.240	0.793	3.19	54.55

[a] Simple cascade systems are used for the Weller-Steiner Case I and the cocurrent case. Degree of separation a is defined as

$$a = \frac{y_p}{1-y_p} \frac{1-x_0}{x_0},$$

and the rating factor is defined as $\eta = \dfrac{a}{S_T}$.

results are presented in Fig. 13.52*a*. The required membrane area can, of course, be reduced with an appropriate recycle and also with more stages. Indeed, the area could be reduced to that of the Weller-Steiner Case II if an infinite number of stages were used in such a system.

Weller-Steiner Case II

The results of calculation with a single stage are shown in Table 13.4. In the *cocurrent* flow scheme the minimum number of stages required in this case is two, since a minimum stripping concentration exists according to Eq. 13.111 (see Fig. 13.53). Also, one operating parameter, for example, the cut of the first stage, is determined such that the overall membrane area will be a minimum. The calculated results are presented in Fig. 13.52*b*.

The results of calculation in *countercurrent* flow with a single stage are also shown in Table 13.4. The superior performance in countercurrent flow is obvious, and the performance of cross flow is also excellent. In a practical design of a cross-flow module, for example, the design parameter Pm must be larger than 50 in order to ensure plug flow, and also the parameter A_p must be smaller than 0.1 so that the pressure loss does not influence the separation significantly.

8 DESIGN OF LIQUID-PHASE MODULES – OSMOTIC PROCESSES

The most important osmotic process from an industrial viewpoint are reverse osmosis and ultrafiltration. Consequently, these two processes will be discussed in some detail. Design considerations for other liquid-phase processes, except for electromembrane phenomena, are very much like those of reverse osmosis and ultrafiltration. Some pertinent comments were, therefore, included in the treatment of ordinary dialysis and Donnan dialysis in Chapter VIII, and further discussion is not needed.

The basic features in the design of gaseous-diffusion modules hold in principle also for liquid-phase processing. Effects of longitudinal mixing, pressure loss along the channel, and cross or parallel flow are thus equally important. However, the main difference lies in the order of magnitude of the respective diffusivities. The diffusivity in the liquid phase is usually much smaller by a factor of 10^4 than in the gas phase. Therefore, the concentration distribution in the liquid phase is much sharper than in the gas phase. The significant effect of the concentration polarization near the membrane wall has already been discussed. Also, the longitudinal mixing by molecular diffusion is generally negligible in the liquid phase and the liquid stream assumes plug flow. When mechanical stirring devices or other turbulence promoters are used in order to reduce the concentration polarization, the longitudinal mixing by eddy diffusion becomes, however, significant.

The four types of commercial modules in use have already been discussed in Chapter VIII. To repeat, they are the plate-and-frame, the spiral-wound, the tube, and the hollow-fiber types. Also, it was pointed out that these commercial modules can be classified according to the degree of longitudinal mixing, and the two extreme cases, complete mixing and nonmixing plug flow, were discussed in general terms.

When the reject and/or permeate streams are in plug flow, the module can, in principle, be designed in countercurrent, cocurrent, or cross plug flow, as described in Fig. 13.1, or more specifically in Fig. 13.28 for gaseous diffusion. Since a highly selective membrane is used in most practical osmotic processes, the concentration of the solute to be separated is extremely small and almost uniform along the channel in the permeate stream. Thus, the Raleigh relationship, Eq. 13.88, is approximately valid. In other words, the osmotic module behaves like that of cross flow in gaseous diffusion, even when the module is designed in counter- or cocurrent plug flow under a uniform external pressure. Therefore, only the case of plug flow, where the Raleigh relationship is valid, will be treated in detail.

In the hollow-fiber type of modules, on the other hand, the concentration polarization is negligible when the reject stream flows inside of the fibers. Gill and Bansal [52] showed that the polarization effect is practically absent, even when the reject stream flows outside the fibers, which are closely packed in a shell. However, the external pressure loss along the fibers due to hydrodynamic friction significantly influences the separation in this pressure-driven osmotic process. The effect of the pressure loss, therefore, requires special attention.

Complete Mixing

When a solution containing solvent B and a single solute A is to be separated, the overall material balance over an osmotic module, as shown in Fig. 13.54, can be written as follows, using Eqs. 8.11, 8.14, and 13.32:

$$q'_p = \frac{Q_B S_T}{l\, C_B} (\Delta P' - \pi'_W + \pi'_P) \tag{13.136}$$

$$= k S_T \ln \left(\frac{C_{AW} - C_{Ap}}{C_{Ao} - C_{Ap}} \right), \tag{13.137}$$

$$q'_p\, C_{Ap} = \frac{Q'_A S_T}{l} (C_{AW} - C_{Ap}), \tag{13.138}$$

$$q'_f = q'_p + q'_0 \tag{13.139}$$

$$q'_f\, C_{Af} = q'_p\, C_{Ap} + q'_0\, C_{Ao}. \tag{13.140}$$

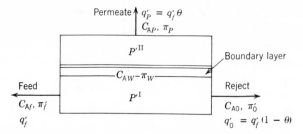

Fig. 13.54 Osmotic modules of complete mixing.

Here it is assumed that the solution is so dilute that the flow rate of the solute is approximately a product of its concentration C_A [(g)(mole)/cm^3] and the solvent volumetric flow q' (cm^3/sec). The hydraulic pressure P' is assumed to be uniform throughout the module. Similarly to a module of gaseous diffusion, it is convenient to express Eqs. 13.136 through 13.140 in the following dimensionless forms:

$$\theta = s_T \left[1 - \pi_f (\Gamma_W - \Gamma_p) \right] \tag{13.141}$$

$$= a_q \, s_T \, \ln\left(\frac{\Gamma_W - \Gamma_p}{\Gamma_0 - \Gamma_p} \right), \tag{13.142}$$

$$\theta \Gamma_p = a_A s_T \, (\Gamma_W - \Gamma_p) , \tag{13.143}$$

$$1 = \theta \Gamma_p + (1-\theta)\Gamma_0 , \tag{13.144}$$

where dimensionless quantities are defined as

$$\theta = \frac{q'_p}{q_f}, \qquad \Gamma = \frac{C_A}{C_{Af}}, \qquad s_T = \frac{Q_B \Delta P'}{q'_f C_B l} S_T, \qquad \pi_f = \frac{\pi'_f}{\Delta P'},$$

$$a_q = \frac{k C_B l}{Q_B \Delta P'}, \qquad a_A = \frac{Q'_A}{Q_A} \frac{C_B}{\Delta P'} . \tag{13.145}$$

Here the van't Hoff equation for the osmotic pressure is assumed to be valid such that

$$\frac{\pi'}{\pi'_f} = \frac{C_A}{C_{Af}} = \Gamma .$$

The physical meanings of the dimensionless quantities are self-evident. The solute separation may be defined as $(1 - \Gamma_p)$ and the fractional product recovery by θ. The dimensionless membrane area s_T is very analogous to that in gaseous diffusion if the "transfer unit" or "capacity factor" s^* (see Eq. 13.82) is defined as

$$s^* = \frac{q_f' \; C_B l}{Q_B \; \Delta P'} . \tag{13.146}$$

Then the actual membrane area is given by Eq. 13.83.

An osmotic membrane is often characterized by the solute rejection SR rather than by a_A. The parameter SR is usually measured in a well-agitated batch osmotic cell, that is, $a_l \to \infty$. In this case, the solute rejection SR becomes, from Eqs. 8.16, 13.141, and 13.143,

$$SR = \frac{1}{1 + a_A + \pi \Gamma_p} .$$

Since the parameter SR depends on operating variables $\pi \Gamma_p$, the osmotic membrane cannot be satisfactorily characterized by SR. However, SR becomes an almost constant value of $1/(1 + a_A)$ when the external pressure is much larger than the osmotic pressure and also a highly selective osmotic membrane is used.

In ultrafiltration, the boundary-layer resistance is often controlling and the solute concentration C_{AW} at the membrane wall becomes a constant gel point C_{Ag}. Also, the solute concentration C_{Ap} in the permeate stream is negligible. In this case, the dimensionless membrane area may be defined as

$$s_T = \frac{k}{q_f'} \; S_T . \tag{13.147}$$

Then the material balances, Eqs. 13.141 to 13.144, reduce to

$$\theta = s_T \; \ln \frac{\Gamma_g}{\Gamma_o} , \tag{13.148}$$

$$1 = (1 - \theta) \; \Gamma_o , \tag{13.149}$$

where

$$\Gamma_g = \frac{C_{Ag}}{C_{Af}} . \tag{13.150}$$

Thus, from definition of s_T, the capacity factor becomes

$$s^* = \frac{q_f'}{k}. \tag{13.151}$$

Plug Flow

Similarly to the case of complete mixing, the material balance over a differential membrane area dS as shown in Fig. 13.55 yields in dimensionless forms

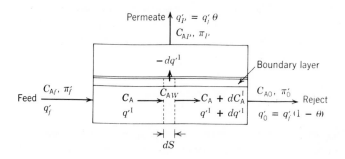

Fig. 13.55 Osmotic module of plug flow.

$$-dq^I = ds \ [1 - \pi(\Gamma_W - \Gamma^{II})] \tag{13.152}$$

$$= a_l \ ds \ \ln\left(\frac{\Gamma_W - \Gamma^{II}}{\Gamma^I - \Gamma^{II}}\right), \tag{13.153}$$

$$-d(q \ \Gamma^I) = a_A \ ds \ (\Gamma_W - \Gamma^{II}), \tag{13.154}$$

$$d(q^I \ \Gamma^I) = \Gamma^{II} \ dq^I, \tag{13.155}$$

and the overall material balance, Eq. 13.144. Here, the dimensionless volumetric rate q^I of the solvent B in the reject stream and the dimensionless area s are introduced as

$$q^I = \frac{q'^I}{q_f'}, \qquad s = \frac{Q_B \ \Delta P'}{q_f' C_B l} \ S. \tag{13.156}$$

It should be noted that the parameter a_l, which expresses the boundary-layer resistance, is generally dependent on s and the local flow rate q^I in the entrance region of the channel where the boundary layer is developing. Also, Eq. 13.155 is the Raleigh relationship. The system of differential equations is thus solved with the initial condition

$$\Gamma^I = 1, \quad q^I = 1 \quad \text{at } s = 0 \tag{13.157}$$

to find

$$\Gamma^I = \Gamma_o, \quad q^I = 1-\theta \quad \text{at } s = s_T . \tag{13.158}$$

Then, the concentration Γ_p in the permeate stream can be evaluated from the overall material balance Eq. 13.144. When the parameter a_l is not sensitive to the local variables s and q^I and also the external pressure is much higher than the osmotic pressure, the integration simply yields

$$\Gamma^I (q^I)^n = 1 ,$$

$$s = 1 - q^I ,$$

$$\Gamma^{II} = \frac{a_A e^{1/a_l}}{1 + a_A e^{1/a_l}} \Gamma^I ,$$

$$\Gamma_W = \frac{(1+a_A)e^{1/a_l}}{1+a_A e^{1/a_l}} \Gamma^I , \tag{13.159}$$

where

$$a = \frac{1 - a_A e^{1/a_l}}{1 + a_A e^{1/a_l}} . \tag{13.160}$$

Thus, the fractional product recovery and the concentration of the product shown can be obtained from Eq. 13.144 as

$$(1 - \theta)^{a-1} (1 - \theta\Gamma_p) = 1 . \tag{13.161}$$

Here, the pressure loss along the membrane in either phase is assumed to be negligible

Hollow-Fiber Module — Effect of Pressure Loss

Since a polymeric fiber can usually withstand a higher compression force than an internal expansion force, the reject (high-pressure) stream is assumed to flow outside of the fibers. Thus, the pressure P'^I of the reject stream is almost uniform, but the pressure P'^{II} of the permeate stream is subject to variation due to hydrodynamic friction. Also, in the system of hollow fibers under consideration, the concentration polarization is negligible, and both the reject and permeate streams are in plug flow. In this case, the mass conservation over a differential area dS (see Fig. 13.56) yields in dimensionless forms

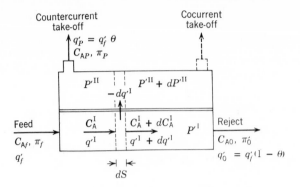

Fig. 13.56 Osmotic module with a pressure loss along the channel of the permeate phase.

$$-dq^I = dq^{II} = ds\ [P - \pi(\Gamma^I - \Gamma^{II})]\ , \qquad (13.161)$$

$$-d(q^I\Gamma^I) = \Gamma^{II}\ dq^{II} = a_A\ ds\ (\Gamma^I - \Gamma^{II})\ , \qquad (13.163)$$

and the momentum conservation becomes

$$\frac{dP}{ds} = A_p q^{II}\ . \qquad (13.164)$$

Here the dimensionless pressure drop across the membrane is defined as

$$P = \frac{P'^I - P'^{II}}{\Delta P'_f}\ , \qquad (13.165)$$

where

$$\Delta P'_f = P'_f - P'_p\ . \qquad (13.166)$$

The parameter A_p is introduced similarly to gaseous diffusion as

$$A_p = \frac{8\mu q'_f}{\pi r^4 n \Delta P'_f}\ \frac{1}{2\pi m}\ \frac{q'_f C_B l}{Q_A \Delta P'_f}\ . \qquad (13.167)$$

The perturbation of the radial permeation on the velocity field is usually so small that the simple Poiseuille equation is again assumed to be valid. Also, the approximation of the Raleigh relationship is used in Eq. 13.163.

The system of differential equations can then be integrated with boundary conditions for the *cocurrent* flow

$$\Gamma^I = 1, \quad q^I = 1, \quad q^{II} = 0 \quad \text{at } s = 0$$

$$P = 1 \qquad\qquad\qquad s = s_T, \qquad (13.168)$$

and for the *countercurrent* flow

$$\Gamma^I = 1, \quad q^I = 1, \quad P = 1 \quad \text{at } s = 0$$

$$q^{II} = 0, \qquad\qquad\qquad \text{at } s = s_T, \qquad (13.169)$$

When the external pressure is much larger than the osmotic pressure, the integration becomes for the cocurrent flow

$$P = \frac{\cosh\ (A_p^{\frac{1}{2}}s)}{\cosh\ (A_p^{\frac{1}{2}}s_T)}, \qquad (13.170)$$

$$q^I = 1 - \frac{\sinh\ (A_p^{\frac{1}{2}}s)}{A_p^{\frac{1}{2}}\ \cosh\ (A_p^{\frac{1}{2}}s_T)}, \qquad (13.171)$$

and for the countercurrent flow

$$P = \frac{\cosh\ [A_p^{\frac{1}{2}}(s_T - s)]}{\cosh\ (A_p^{\frac{1}{2}}s_T)}, \qquad (13.172)$$

$$q^I = 1 + \frac{\sinh\ [A_p^{\frac{1}{2}}(s_T - s)]}{A_p^{\frac{1}{2}}\ \cosh\ (A_p^{\frac{1}{2}}s_T)} - \frac{1}{A_p^{\frac{1}{2}}}\ \tanh\ (A_p^{\frac{1}{2}}s_T). \qquad (13.173)$$

Thus, the fractional product recovery (cut) θ for both flows is

$$\theta = \frac{1}{A_p^{\frac{1}{2}}}\ \tanh\ (A_p^{\frac{1}{2}}s_T). \qquad (13.174)$$

Equation 13.174 was first derived by Hermans [53]. The more rigorous perturbation solution including the effect of the pressure variation on the shell side was extensively discussed by Gill and Bansal [52]. However, they characterized the osmotic membrane with the solute rejection SR instead of the parameter a_A.

Numerical Examples

Computer programs for the above-mentioned cases are given in Appendix C. These programs essentially compute the permeate concentration and the membrane area required for a given fractional recovery and other operating parameters. In order to elucidate the main features encountered in various cases, some numerical examples are given in this section.

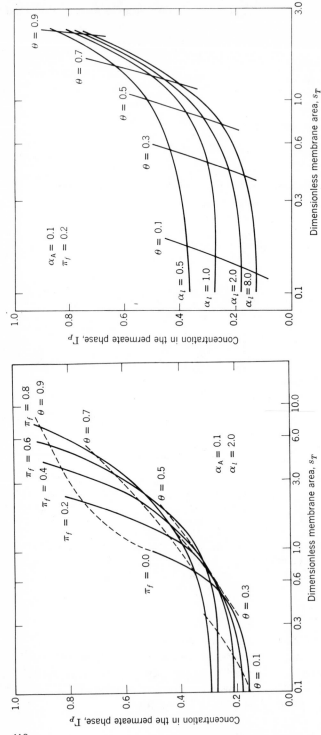

Fig. 13.57 Effect of π_f on osmotic separation with complete mixing.

Fig. 13.58 Effect of concentration polarization on osmotic separation with complete mixing.

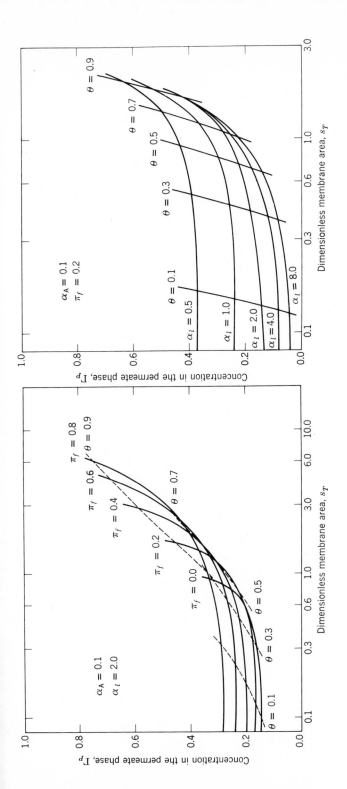

Fig. 13.59 Effect of π_f on osmotic separation with plug flow.

Fig. 13.60 Effect of α_l on osmotic separation with plug flow.

411

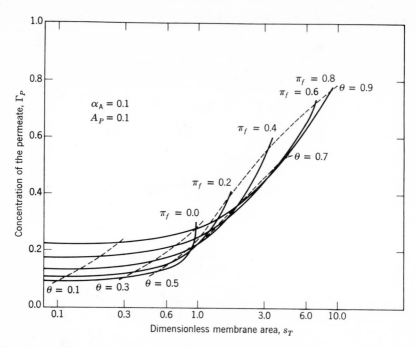

Fig. 13.61 Effect of π_f on osmotic separation in cocurrent flow with pressure loss along the permeate channel.

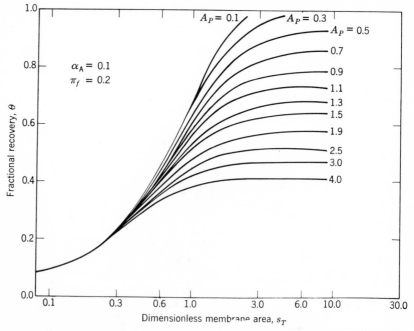

Fig. 13.62 Effect of A_p on fractional recovery in cocurrent osmotic process.

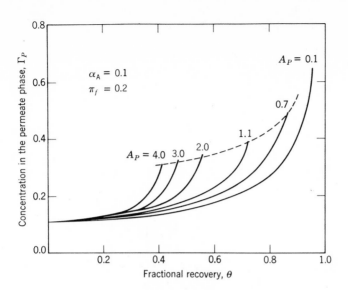

Fig. 13.63 Permeate concentration versus fractional recovery in the cocurrent osmotic process (dotted line represents the maximum attainable separation).

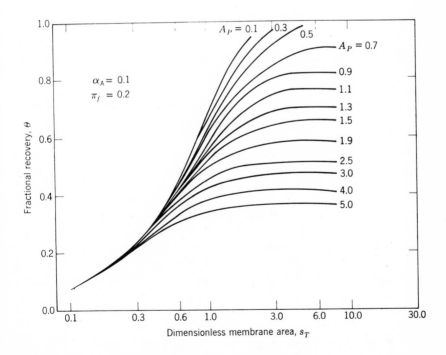

Fig. 13.64 Effect of A_p on fractional recovery in the countercurrent osmotic process with a pressure loss along the permeate channel.

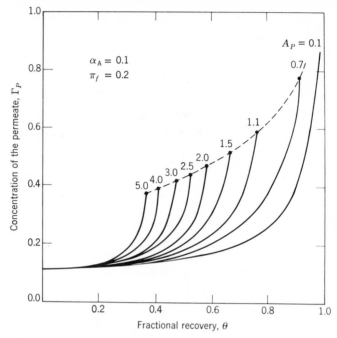

Fig. 13.65 Permeate concentration versus fractional recovery in the countercurrent osmotic process (dotted line represents the maximum attainable separation).

Effects of parameters π_f and α_l on the osmotic separation are shown in Fig. 13.57 and 13.58 for the complete-mixing case. In the plug-flow case, where the boundary layer is fully developed, these effects are presented in Fig. 13.59 and 13.60. Indeed, the advantage of plug flow is obvious. Comparing Fig. 13.58 and 13.60, it can, however, be seen that the effect of longitudinal mixing is secondary to that of concentration polarization. Figure 13.61 illustrates the effect of π_f for the cocurrent case. The influence of the parameter A_p on the fractional recovery and the permeate concentration is shown in Fig. 13.62 and 13.63 for cocurrent flow, and in Fig. 13.64 and 13.65 for countercurrent flow. The maximum attainable recovery for a given A_p in the latter case is obviously higher than in the former case. It should be emphasized that the validity of the Raleigh relationship is assumed in these computations. The advantage of the actual countercurrent case may, therefore, be more pronounced when rigorous material-balance equations are used instead of the simple Raleigh relationship.

It is evident that in an actual osmotic module the above-mentioned three phenomena influence the separation altogether. Mathematical modeling of such cases and optimum criteria for the design and operation were discussed by Fan et al. [54]. and also by Murkes and Bohman [55].

9 ELECTROMEMBRANE PROCESSES

The phenomena of concentration polarization and longitudinal mixing are also very important in these processes. The adverse effect of concentration polarization on the electric transference of an ionic species was extensively discussed in terms of the limiting current densities in Chapter IX. As indicated in Eq. 9.87, the limiting values depend on the mass-transfer coefficient and the difference of the transference number of the membrane phase and that of the solution phase. When such limiting current densities are specified, the actual current density can be evaluated from Eq. 9.86 for a given external electric potential drop across the repeating unit of the membrane stack. Then, the electrically transferred amount of the ionic species can be calculated from Eqs. 9.75 to 9.78, which represent the effects of concentration polarization and diffusional leakage. The corresponding solvent flux is also determined by Eq. 9.54. When both the dialysate and the diffusate streams are in complete mixing, the overall material balance can thus be set up to obtain the total amount of the ionic species that is electrically transferred across the electromembrane. If both streams are in plug flow, the differential material balance can also be integrated. Such procedures are very analogous to those that have been discussed in the previous section.

Nomenclature

Symbol	Meaning
A_p	Parameter defined by Eqs. 13.134 or 13.167
$\mathcal{C}$	Cross-sectional area of a membrane channel
B	Withdrawal rate of cascade bottom stream
C	Total molar concentration
C_A	Solute molar concentration
C_B	Solvent molar concentration
C_{Ag}	Gel point of solute
D	Withdrawal rate of cascade overhead stream
D_A	Diffusion coefficient of solute in the solution phase
F	Feed rate of cascade system
$\mathcal{F}$	Faraday constant
Gr	Grashof number defined by Eq. 13.44
I	Electric current density
I_l	Limiting electric current density defined by Eq. 9.87
L	Interstage downflow rate in a cascade system or the length of a membrane channel
N	Number of total stages of enriching section

Symbol	Meaning
N_A	Solute molar flux
N_B	Solvent molar flux
P	Dimensionless pressure drop defined by Eq. 13.165
Pe_L	Longitudinal Peclet number (uL/D_A)
Pe_W	Wall Peclet number defined by Eqs. 13.50 or 13.53
Pm	Mixing parameter defined by Eq. 13.128
Pr	Pressure ratio defined by Eq. 13.73
P'	Pressure
$\Delta P'$	Pressure drop across a membrane
Q_A	Permeability of more-permeable species
Q_A'	Permeability of solute defined by Eq. 8.14
Q_B	Permeability of less-permeable species or solvent in the liquid-phase process
R	Gas constant
Re	Reynolds number defined by Eq. 13.41
S	Membrane area
Sc	Schmidt number defined by Eq. 13.42
Sh	Sherwood number defined by Eq. 13.38
$\overline{Sh}$	Average Sherwood number defined by Eq. 13.62
SR	Solute rejection defined by Eq. 8.16
S_T	Total membrane area
T	Temperature
V	Interstage upflow rate in a cascade system
V'	Volume of permeation equipment
a	Constant defined by Eq. 13.160
a_1, a_2, a_3, a_4	Constants defined by Eq. 13.95
b	Breadth of a sheet membrane channel
f	Function defined by Eq. 13.21
g	Function defined by Eq. 13.22, or gravitational constant
k	Mass-transfer coefficient defined by Eq. 13.30
$\overline{k}$	Average mass-trnasfer coeficient by Eq. 13.61
l	Membrane thickness
n	Number of hollow fibers in a membrane module
$n_1, ..., n_7$	Constants defined by Eqs. 13.40, 13.45, 13.95, or 13.102
q	Dimensionless total flow rate defined by Eqs. 13.92 or 13.156
q'	Total flow rate
r	Inside radius of a hollow fiber or tube
r_h	Equivalent hydraulic radius (see Eq. 13.38)

s	Dimensionless membrane area defined by Eqs. 13.93 or 13.156
s_T	Dimensionless total membrane area defined by Eqs. 13.80 or 13.145
s^*	Capacity factor defined by Eqs. 13.82, 13.146, or 13.151
t	Variable defined in Eq. 13.95
u	Longitudinal velocity in a membrane channel
v	Velocity component normal to the membrane surface
w	Width of a sheet membrane channel
x	Longitudinal distance
x_A, x_n	Mole fraction of the reject stream or downflow stream in a cascade system
x_{oM}	Minimum reject concentration
y_A, y_n	Mole fraction of the permeate stream or upflow stream in a cascade system
z	Distance normal to the membrane surface
Γ	Relative concentration of solute defined by Eq. 13.145
Γ_g	Dimensionless gel point defined by Eq. 13.150
a	Separation factor defined by Eq. 13.78
a^*	Ideal separation factor defined by Eq. 13.77
a_A	Solute separation fractor defined by Eq. 13.145
a_l	Relative mass-transfer coefficient defined by Eq. 13.145
γ	$= (1 - \theta)/\theta$
β	Parameter defined by Eq. 13.129
δ	Boundary-layer thickness
ρ	Density of a solution
ν	Kinematic viscosity
π	$= 3.14159$
π'	Osmotic pressure
π_f	Osmotic pressure of the feed stream in dimensionless form (see Eq. 13.145)
θ	Cut or fractional recovery
ξ	Variable defined by Eqs. 13.49 or 13.52
ζ	Variable defined by Eq. 13.95
η	Rating factor defined by Eq. 13.123

Subscripts

A	Refers to more-permeable species or solute in the liquid-phase process

Symbol	Meaning
B	Refers to less-permeable species or solvent, or to bottoms in a cascade system
D	Refers to overhead in a cascade system
F	Refers to feed in a cascade system
N	Refers to feed stage
b	Refers to the bulk phase in a solution
i,j	Refer to any stage
f	Refers to the feed stream in a single module
m,n	Refer to any stage in stripping and enriching sections
o	Refers to the reject stream
P	Refers to the permeate stream

Superscripts

| I | Refers to the reject stream |
| II | Refers to the permeate stream |

References

1. D. R. Paul, *Ind. Eng. Chem. Process Des. Develop,* **10,** 375 (1971).
2. B. D. Smith, *Design of Equilibrium Stage Processes,* McGraw-Hill, New York, 1963.
3. E. J. Henley and H. K. Staffin, *Stagewise Process Design,* Wiley, New York, (1963).
4. M. Benedict and R. L. Pigford, *Nuclear Chemical Engineering,* McGraw-Hill, New York, 1957.
5. K. Cohen, *The Theory of Isotope Separation,* McGraw-Hill, New York, 1951.
6. A. M. Rozen, "Theory of Isotope Separation in Columns" (U.S.S.R.). Translation available from O.T.S. (JPRS: 11213; CSO: 6667-N).
7. P. C. Wankat, *Separat. Sci.,* 7, 233 (1972).
8. S-T. Hwang and K. Kammermeyer, *Can. J. Chem. Eng.,* **43,** 36 (1965).
9. S. Weller and W. A. Steiner, *J. Appl. Phys.,* **21,** 279 (1950); *Chem. Eng. Progr.,* **46,** 585 (1950).
10. R. W. Naylor and P. O. Backer, *A.I.Ch.E. J.,* 1, 95 (1955).
11. S. A. Stern, T. F. Sinclair, P. J. Gaveis, N. P. Vahldieck, and P. H. Mohr, *Ind. Eng. Chem.,* **57,** 49 (1965).
12. S. Blumkin, Union Carbide corp., Nuclear Div., Rept. K-OA-1552 (Nov. 1967); obtainable from NTIS, 5285 Port Royal Rd., Springfield, Va. 22151.
13. R. H. Rainey, W. L. Carter, and S. Blumkin, ORNL-4522; obtainable from NTIS as in Ref. 12.

14. R. E. Lacey and S. Loeb, Eds., *Industrial Processing with Membranes*, New York, 1972.
15. K. Higashi, H. Doi, and T. Saito, *Energ. Nucl.*, **17**, 98 (1970).
16. L. M. Litz and G. E. Smith, Union Carbide Research Institute, Rept. UCRI-701 (Sept. 1972).
17. S. Sourirajan, *Reverse Osmosis*, Academic, New York, 1970.
18. W. B. McCormack, U.S. Patent 3,246,764 (1966).
19. M. S. Mintz, *Ind. Eng. Chem.*, **55**, (6) 18 (1963).
20. E. D. Sloan, Jr. and R. C. Harshman, "The Combined Effects of Natural and Forced Convection in a Dual Membrane, Horizontal, Parallel Plate Ultrafilter," paper presented at the AIChE Meeting, New Orleans, March 1973.
21. K. Ramanadhan and W. N. Gill, *A.I.Ch.E. J.*, **15**, 872 (1969).
22. J. S. Newman, *Electrochemical System*, Prentice-Hall, N. J., 1973.
23. L. Dresner, "Boundary Layer Build-up in the Demineralization of Salt Water by Reverse Osmosis," Oak Ridge Natl. Lab. Rep., p. 3621, May 1964.
24. W. N. Gill, "Convective Diffusion in Laminar and Turbulent Hyperfiltration (Reverse Osmosis) Systems," in *Surface and Colloid Sciences*, Vol. IV, E. Wiley, New York, 1971.
25. A. S. Berman, *J. Appl. Phys.*, **24**, 1232 (1953).
26. S. W. Yuan and A. B. Finkelstein, *Trans. A.S.M.E.*, **78**, 719 (1936).
27. T. K. Sherwood, P. L. T. Brian, and R. E. Fisher, *MIT Desalination Res. Lab. Rep.*, 295-1 (1963); also, Sherwood et al., *I&BC Fund.*, **4**, 113 (1965).
28. R. E. Fisher, T. K. Sherwood, and P. L. T. Brian, *MIT Desalination Res. Lab. Rep.*, 295-5 (1964).
29. W. N. Gill, D. W. Zeh, and C. Tien, *A.I.ChE. J.*, **12**, 1141 (1966).
30. C. Mastromonico, "Reverse Osmosis in Laminar Flow in Annuli," M.S. thesis, Clarkson College of Technology, Potsdam, N.Y., 1968.
31. P. L. T. Brian, *Ind. Eng. Chem. Fund.*, **4**, 439 (1965).
32. W. N. Gill, C. Tien, and D. W. Zeh, *Intern. J. Heat Mass Trnasfer*, **9**, 907 (1966).
33. J. G. Knudsen and D. L. Katz, *Fluid Dynamics and Heat Transfer*, McGraw-Hill, New York, 1958.
34. W. H. Linton, Jr. and T. K. Sherwood, *Chem. Eng. Progr.*, **46**, 258 (1950).
35. W. N. Gill and M. Sher, *A.I.Ch.E. J.*, **7**, 61 (1961).
36. F. W. Dittus and L. M. K. Boelter, *Univ. of Calif. (Berkeley) Publ., Eng.*, **2**, 443 (1930).
37. A. S. Michaels, *Chem. Eng. Progr.*, **64**, 31 (1968).
38. M. C. Porter and A. S. Michaels, *Chem. Tech.*, **1**, 56 (1971).
39. M. C. Porter and A. S. Michaels, *Chem. Tech.*, **1**, 440 (1971).
40. M. C. Porter, *IEC Prod. Res. Dev.*, **11**, 234 (1972).
41. S. I. Rubinow and J. B. Keller, *J. Fluid Mech.*, **11**, 447 (1961).
42. P. G. Staffman, *J. Fluid Mech.*, **1**, 540 (1956).
43. H. J. Bixler and J. C. Rappe, U.S. Patent 3,541,006 (1970).
44. A. A. Sonin and R. F. Probstein, *Desalination*, **5**, 293 (1968).

45. W. F. Savage, in *Saline Water Conversion Report 1970-1971*, Section 2, U.S. Dept. of the Interior, Office of Saline Water.
46. W. G. B. Mandersloot and R. E. Hicks, *IEC Proc. Des. Dev.*, **4**, 304 (1965).
47. G. Belfort and G. A. Guter, *Desalination*, **10**, 221 (1972).
48. S. T. Hwang, *Separat. Sci.*, **4**, 167 (1969).
49. C. T. Blaisdell and K. Kammermeyer, *Chem. Eng. Sci.*, **28**, 1249 (1973).
50. J. Oishi, Y. Matsumura, K. Higashi, and C. Ike, *J. Atom. Energ. Soc. (Japan)*, **3**, 923 (1961).
51. M. E. Breuer and K. Kammermeyer, *Separat. Sci.*, **2**, 319 (1967).
52. W. N. Gill and B. Bansal, *A.I.Ch.E.J.*, **19**, 823 (1973).
53. J. J. Hermans, cited in J. P. Davis and T. A. Orofino, "Mass Transport in Hollow Fiber Assemblies," presented in the 163 ACS Meeting, Boston, Mass., April 9, 1972.
54. L. T. Fan, C. Y. Cheng, L. Y. S. Ho, C. L. Hwang, and L. E. Erickson, *Desalination*, **5**, 237 (1968).
55. J. Burkes and H. Bohman, *Desalination*, **11**, 269 (1972).

Chapter **XIV**

MEMBRANES AND THEIR PREPARATION

421

Membrane preparation has developed into a rather sophisticated science mixed with a good deal of art. The subject is so extensive that a host of books and articles have been written, and thus only an overview is presented. The discussion of techniques must differentiate between microporous membranes and nonporous media.

The techniques required to prepare the two types of membranes are quite distinct. In microporous membranes the aim must be to create a structure with adequately small pores to permit separative flow as opposed to bulk flow normally encountered in the usual porous medium. Such techniques involve the application of either powder metallurgical procedures, or the leaching out of a discretely distributed constituent. On the other hand, the creation of a nonporous structure comes down to the preparation of as thin a film of material as possible and to avoid the presence of pinholes.

1 MICROPOROUS MEMBRANES

The question whether a porous medium is microporous or macroporous in nature is really dependent on its intended use. In general, when pore sizes are 10μ or less, one is likely to use the term *microporous*. Still, this is a somewhat arbitrary cutoff point. Although pore sizes of a few microns will certainly act as a separation medium in many liquid systems, any pore structure as large as 1μ, that is, 10,000 Å, will be too large to act as a micropore in gas separation. Considering that the mean free paths of gases at standard conditions are in the range of 560 Å for CO_2 to 2500 Å for He, it becomes evident that a microporous medium for gas separation must have pore sizes in the range of 50 to 100 Å. Although just a short time ago only porous Vycor glass—and surely the AEC barrier—were known to meet this requirement, there have been some startling developments in the last few years so that several additional structures are now available.

There are only two ways to make a porous barrier: One can start with a solid matrix and make holes, or start with a composition having large openings and proceed to make them smaller. Holes can be created by penetration, such as the nuclear-track registration and chemical-etching method, or by incorporating a soluble phase in the matrix and leaching. For instance, the latter procedure is used in the preparation of microporous Vycor glass, resulting in a structure having an average pore diameter of about 50 Å.

Ceramic Compositions

The ceramic materials that have been used to prepare microporous membranes comprise porous siliceous and clay-type matrices, with perhaps the addition of porous carbon structures, which, although not ceramic in nature, behave like the ceramic creations.

Microporous Vycor Glass

This material has been one of the most useful aids for conducting studies in gaseous diffusion. For this purpose it must be used in the form of a membrane, flat or tubular. The preparation of the porous material was first described by Hood and Nordberg in 1938 [1]. Briefly, a glass composition is used that results in the separation of an acid-soluble phase, and this phase can be removed by acid leaching. Thus, a microporous structure is obtained as an intermediate stage in the manufacture of Vycor-brand glass.

The behavior and structural parameters of the microporous stage were described in some detail by Nordberg in 1944 [2]. Early studies in gaseous diffusion, such as those of Huckins [3] and Russell [4], were possible because of the availability of this unique microporous medium, as its average pore diameter of about 50 Å permits attainment of free molecular diffusion at atmospheric pressure.

The material was available in the shape of flat disks, test tubes, and tubing. With the test-tube shape a multigraded seal had to be used to provide a small-bore glass tube connection. The tubing shape should, in principle, be suitable for multitubing exchange equipment. However, when a tubular exchanger was constructed, the rather brittle nature of the microporous glass gave rise to excessive fracturing of the tubes mounted in a steel shell. Even relatively small temperature differences were enough to induce expansion and contraction stresses that the fragile glass did not tolerate [5]. Thus, flexible end mounting will be necessary with such tubes, and this will limit their use to appropriately low temperatures.

For many years the microporous glass was available only with the 50-Å pore structure. Recently, however, a "controlled pore glass" material has been reported by Haller [6]. The range of average pore diameters available in the market is from 240 to 2000 Å, as stated by Corning Glass Works. Haller [7-9] has shown that the controlled pore glass is capable of selectively separating many chemical compounds. The material is primarily available in the form of granules or powders. Thus, it falls into the category of column packing for chromatography, and in this manner it does not act as a membrane.

When microporous Vycor glass is heated to elevated temperatures above about 300°C for long periods, or to much higher temperatures for shorter periods, it loses its microstructure and becomes the well-known solid Vycor glass material.

Microporous Porcelain

Selas Flotronics is marketing finely porous porcelain shapes primarily for fluid filtration purposes. These materials are available in disk, candle, and fine-bore tubular shapes. The finest grades possess a fair number of small enough pores so that they are marginally suitable for gas separations. The separating action of the material could be enhanced if the separation process were to be carried out at subatmospheric pressures or elevated temperatures where the mean free path of gases would become sufficiently large to ensure a high proportion of molecular flow.

Compressed powders of alumina, silica, and graphite are suitable for membrane preparation. In fact, disks and plugs of such materials formed the basis of many of Barrer's elegant experiments [11]. The compaction needed to attain microporous structures is accomplished by appropriately high pressures only, without use of elevated temperatures. Of course, this procedure requires that the particle size of the original powders be fine enough to permit the creation of pores in the proper angstrom range.

Metallic Compositions

The preparation of microporous membranes from metallic materials is generally carried out by a powder-metallurgical process. Here, one starts with a powdered material and reduces large openings to smaller ones through sintering and usually also pressure compaction. The attainment of the final desired pore size is then accomplished by further controlled sintering and judicious use of pressure. Presently, however, there is no structure available that possesses the necessary pore sizes for effective gas separation. However, those available are of sufficient importance to deserve discussion. Generally such structure will permit operation at elevated temperatures. As the mean free path of gases is directly proportional to the absolute temperature, some of these structures will then become more efficient if used accordingly.

Tungsten Membrane

A novel technique was developed by the Hughes Research Laboratories to prepare porous matrices of tungsten, iridium, and other metals [12, 13]. Very-small-diameter tungsten spheres were warm pressed and sintered at a high vacuum. The surface, which would be bumpy from the tungsten spheres alone, can be machined to smoothness after impregnation with copper. Finally, the copper is vaporized to recreate the porous structure.

Silver Membrane

Selas Flotronics [10] developed a very thin silver membrane that is primarily intended as a filter for fine cleanup of liquids. The membranes are available in nominal pore sizes of 0.2 to 5 μ, and circular diameters of 13 to 293 mm, with a thickness of 2 to 4 mils. Thus, for gas separations at atmospheric pressure their pore structure would be too large, and one would have to operate at a substantial vacuum for effective use.

It is not surprising that some of the structures created by powder compaction and sintering will be sensitive to high-temperature exposures. This is particularly the case with a structure composed of finely divided silver particles. Thus exposure to high temperature alone, or to moderately high temperature for a fairly long time, is likely to result in structural changes.

Other Configurations

An interesting procedure was used by early investigators in the study of Knudsen flow (reference not known). It consisted of tightly winding thin wire on a porous tube, and the very fine slits between coils of wire were sufficiently small to give openings of microporous consistency.

An experimentally produced microporous structure has been reported by Desorbo and Cline [14], who were able to prepare uniform submicron-size pores from directionally solidified eutectics of NiAl-Mo and NiAl-Cr, by subsequent etching. They also prepared a microporous gold film by using a plastic replica of the eutectic structure and vacuum deposition of gold on the replica. The pore diameters were in the $0.5\text{-}\mu$ range.

Applications

Most of the strictly microporous media have been considered mainly for gas separations. However, there is no real cutoff point from the standpoint of pore size for their use in liquid-phase processes. This situation is well illustrated in some of the large-scale examples presented in Chapter XV. There the Du Pont Permasep hollow fiber and the asymmetric cellulose acetate membrane have been used for the separation of gas mixtures, even though they had originally been developed for the reverse-osmosis process.

At present the commercially available effective microporous media are still very limited, as shown in Table 14.1. No detailed information is available on pore size and porosity of the U.S. Atomic Energy Commission's barrier for UF_6 separation. The table also includes microporous membranes prepared from polymeric materials, and these are treated in the following section.

Polymeric Compositions

Only a limited number of porous and microporous polymeric materials are available. In most instances the procedures used to create micropores are not divulged by the manufacturer.

Polymer Filters

The *Millipore Filters* [15] have been well known for some time. They represent a microporous plastic composition with straight-through pores. The manufacturing procedure is, however, not known. Membranes of a similar nature are the *DIAFLO ultrafilters* [16]. A comparative study by Preusser

Table 14.1 Commercially Available Microporous Media

Medium	Material	Pore Diameter	Ref.
Porous Vycor Glass Corning Glass Works	Mainly silica Glass	About 50 Å average fairly sharp distribution	2
Porous Porcelain Selas Corp. of America	Porcelain	Smallest max. pore diameter 0.3 μ	10
Flotronic Membrane Selas Flotronics	Silver metal	Max. pore size: five sizes 0.2-5.0 μ	10
Porous Tungsten Hughes Research Labs (also iridium and other metals)	Tungsten metal	Avg. pore diam. 1.4-6.9 μ	12
Millipore Filters Millipore Corp.	Cellulosic	Millipore VF: avg. 0.05 μ Pellicon: avg. 0.03 μ	15

Diaflo Membrane Filters Amicon Corp.	Hydrated Polymer	XM-50: avg. 0.02 μ XM-100: avg. 0.03 μ	16
Nuclepore Membrane Filters Nuclepore Corp.	Polycarbonate	Nominal sizes 0.03 to 8.0 μ Max. effective pore diam. 0.21 to 10.3 μ	18-21
Seprane Membrane Sepco, Los Altos, Cal.	Teflon base	About 0.05 μ	23
Celgard Celanese Plastics Cor.	Polypropylene	Nominal size, pore width less than 0.1 μ	22
Metricel Membrane Filters Gelman Instr. Co.	Cellulosics	5 μ 75 Å	
Selectron Membrane Filters Schleicher & Schuell, Inc.	Cellulosics	8 μ 50 Å	

427

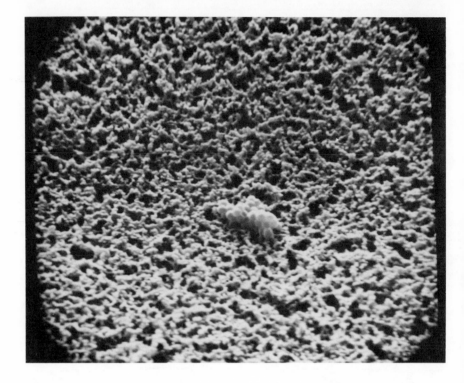

Fig. 14.1   Surface structure of Millipore Filter. (Courtesy Millipore Corporation.)

[17] indicates that the average pore size of the DIAFLO membranes is about one-half of that of Millipore filters, but about the same as the Pellicon membrane manufactured by Millipore Corporation. Figure 14.1 pictures the surface structure at 4500X of type-GS Millipore Filter, 0.22 μ (with a staphylococcus microcolony on the surface).

Nuclepore Membranes

These structures represent a microporous medium with a nominal pore size range of 0.03 to 8 μ. The pores are essentially straight-through holes with a frequency of 100 million per square centimeter of membrane surface [18]. Although the 100 million number amounts to only 10% porosity, compared to an 80% porosity for some cellulosic membranes, the lower wall thickness of the Nuclepore material, about 10 μ against some 120 μ for cellulosics, off-sets the lower porosity in terms of flow rates; also, the Nuclepore path is straight against a normally tortuous configuration in other microporous media.

The difference between a 120-μ-thick cellulosic and a 10-μ-thick Nuclepore structure is strikingly illustrated in Fig. 14.2 [18]. The interwoven and tangled

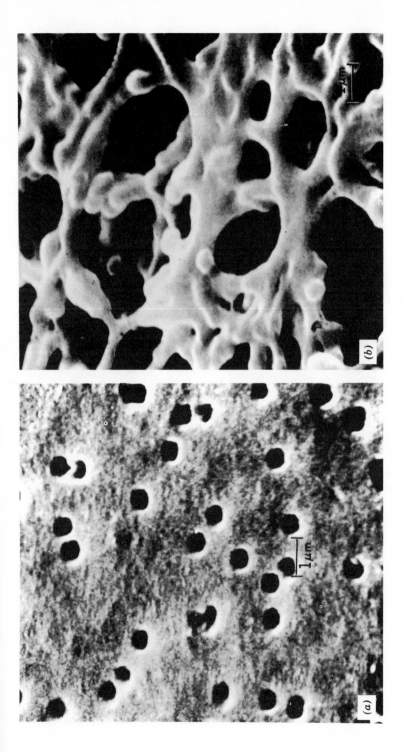

Fig. 14.2 Scanning electron micrographs. (a) Nuclepore structure of a membrane filter with 0.4 μ nominal size pores. (b) Cellulose membrane filter of nominal 0.45-μ pore size. (Courtesy Nuclepore Corporation.)

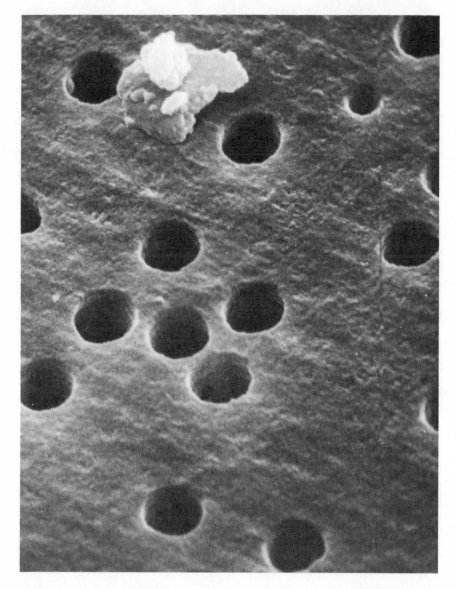

Fig. 14.3 Electron micrograph of nominal 1-μ pore size Nuclepore membrane. (Courtesy Nuclepore Corporation.)

appearance of the cellulosic matrix is in evident contrast to the definite hole structure of the Nuclepore membrane. Another view of the Nuclepore structure is of interest, as it gives a clear picture of the sharpness with which the pores are etched into the polymer and the smoothness of the membrane. This is illustrated in Fig. 14.3, where the pore diameter is 1 μ. The particle on the membrane is not identified; prior to photographing the membrane was used as an in-line filter in a gas stream.

The normal base material is polycarbonate film. The "particle-track-etching" process [19-21] is used to create the pores. The polymer film, f.i. poly(bis-phenol-A-carbonate), Kimfol, is irradiated in a collimated beam of ^{235}U fission fragments. The tracks that are created alter the polymer structure, so that subsequent leaching with warm sodium hydroxide opens the pores. Size of the pore opening becomes a function of etching time. An interesting point is that the locations of the pores cannot be seen even with an electron microscope after exposure to radiation but before etching.

Present emphasis on applications appears to be mainly as a unique filter medium, ranging from use in diagnostic cytology, through liquid filtration with bacterial removal, to air filtration. For convenient use several housing modules have been developed, such as a Stak-Plate configuration and a disposable pleated cartridge. It would seem reasonable that these membranes will find useful applications in a variety of membrane-separation processes.

Celgard Membranes

This structure is described as a thin polypropylene film that contains uniformly dispersed pores [22]. The film thickness is nominally 1 mil, and the pore width is less than 0.1 μ (1000 Å). As can be seen from the electron micrograph in Fig. 14.4, the "pores" are actually discrete, somewhat tortuous channels thus giving a slitwidth of 0.1 μ and appearing rather uniform in configuration. The film is opaque in the dry state but becomes transparent when the pores are impregnated with a liquid. Porosity is in the 35% range, and pressure required to initiate water penetration is 600 psi.

The 0.1-μ pore size is marginal for gas separations, but is quite in the range for liquid-phase separations such as ultrafiltration. Because of its chemical composition the film is quite resistant to acids and alkali, is rather tough, and can be used at temperatures of -80 to $160°F$. If the film is mounted so that it is constrained, it is servicable up to $250°F$ without loss of porosity.

Flow-rate curves for water, solvents, and gases [21] show that the flow of both liquids and gases is essentially laminar, that is, a straight-line increase in rate with increasing pressure. This behavior is as expected, because the pore width of about 0.1 μ would not yield any separative flow at atmospheric and superatmospheric pressure.

Seprane Membranes

The most recent development of a truly microporous membrane is exemplified by the Seprane types of the Sepco Company [23]. Permeabiltiy data supplied by the manufacturer are plotted against molecular weight in Fig. 14.5, and the

Fig. 14.4 Electron micrographs of Celgard membrane and comparisons with some other membrane structures. Upper left represents a membrane filter. (Courtesy The Celanese Plastics Company.)

corresponding values for a microporous Vycor glass membrane are given for comparison. This graph is very instructive. It shows that the Seprane permeabil-

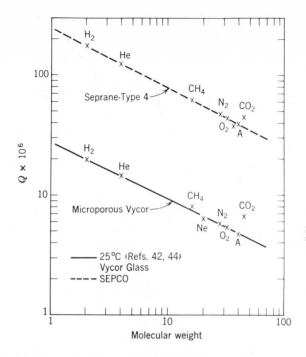

Fig. 14.5 Permeability versus molecular-weight data for Seprane membrane compared with microporous Vycor glass.

ities are almost ten times as great as the Vycor data, so that a corresponding tenfold reduction in transfer area is achieved. The slopes of the lines are also the same and closely correspond to the $M^{1/2}$ relationship. This indicates that both membranes give almost pure Knudsen flow, that is, molecular diffusion. The fact that CO_2 does not fall in line with the other gases is not disturbing. It represents the usual behavior of CO_2 and shows that this compound is adsorbed to a fair extent and so gives rise to surface flow.

These Seprane membranes are discribed as being Teflon derivatives with included reinforcing fibers. They are made in thicknesses of 5 to 10 mils and possess a porosity of 30 to 32%, which is quite similar to the porosity of microporous Vycor glass. Pore size can be varied to a limited extent and present manufacture yields an apparent average pore size of 0.05 μ, that is, 500 Å. As one would expect from the composition, the membranes are quite solvent, and corrosion resistant and can be used up to about 400°F. With water the membranes inhibit flow below about 175 psig, and then begin to show a very low flux.

2   NONPOROUS MEMBRANES

Nonporous membranes are normally understood as being of polymeric origin. However, as described in Chapter VI, a number of such inorganic membranes have found important applications. Some commercial uses of metallic membranes are also presented in Chapter XV.

In the present section the treatment will be limited to *polymeric* membranes. Although modern polymer chemistry is highly proficient in tailoring polymers to specific aims, the processing operations involved in making the different shapes introduce a variety of factors that can exert strong effects on ultimate barrier behavior. In principle, it is a well-established fact that mass transfer will be greater in the amorphous phases of a nonporous organic matrix. Thus, crystallization and orientation are to be avoided as much as possible. However, the strength and physical properties of the polymer may then be adversely affected, and the final product will represent a compromise between necessary strength properties and desirable mass-transfer rates. A further complication is our inability to predict mass-transfer behavior in relation to polymer structure. At present we do not have any really satisfactory correlations between barrier properties and mass transfer.

The principal aim is to create as thin a barrier as possible, consistent with minimum strength porperties and freedom from pinholes. There are two basic configurations: flat sheeting or tubular shapes, be they composed of polymers or metals. The preparation of polymer films is covered authoritatively in an extensive two-volume treatise edited by Sweeting [24]. Film shapes can be prepared by casting from solution or melt, extrusion, blow molding, and press molding. Tubular shapes, including hollow fibers are made by extrusion with central gas injection. Melt extrusion or bath spinning is suitable. Also, wire coating and stripping can be carried out, but this procedure is more likely to be used only for a laboratory procedure.

3   MICROPOROUS MEMBRANES – LIQUID PHASE

One could object to the designation of microporous membranes as discussed in this section. Certainly, some parts of these membranes possess a definite porous structure. However, whether the so-called active portion of the membranes is acting because of microporosity or as a very thin nonporous skin is open to question. This statement really brings up the whole matter of the possible mechanism of separative flow in liquid-phase separations—is it flow through micropores, or it is diffusive solubility flow, so that pore flow if any, is incidental? We do not wish to open this Pandora's box as it would lead to lengthy polemics. But we think that the reader should form his own opinion.

Semipermeable Membranes

From the pioneering studies of Reid [25] on reverse osmosis to the sophisticated structure of present-day membranes, there must have been numerous attempts to reach the capabilities of the so-called asymmetric or composite structures now in use. However, this is a rather arbitrary distinction. The membranes used in liquid-phase processes are commonly referred to as semipermeable membranes. As reported in Chapter XV, such membranes have also been found of value in a gaseous-diffusion separation of helium from natural gas [26].

The effectiveness of reverse osmosis as an industrial operation largely depends on the performance of the membrane, just as in other membrane processes. High flux, high selectivity, and the physical and chemical stability for a reasonable period are important requirements for an economically viable membrane. As mentioned previously, cellulose acetate, especially in its anisotropic form, so far has been known to best meet those requirements in desalination and also has been successfully and widely employed on a large scale. It has been recognized that the performance of a membrane can be improved to a great extent by preparing and tailoring the membrane in a suitable way, as well as by designing and operating the osmosis unit under optimum conditions.

For many years now, the most commonly used membrane has been cellulose acetate. A statement contained in an Eastman Chemical Products bulletin [27] expresses the state of the art as recently as 1971: "Despite many years of work in many locations, no alternate has been found which matches the overall qualifications of cellulose acetate." This statement was based on the contention that cellulose acetate's inherent salt-rejection capability and availability in thin pinhole-free films placed it in a most favored position.

As successful a process as reverse osmosis, however, will generate extensive investigational activity, and it is no wonder that a strong competitor has arisen in the form of Du Pont's Permasep membrane. The membrane itself is described as an asymmetric polyamide structure and is primarily used in the form of hollow fibers [28]. Some of the details concerning these hollow fibers will be discussed later in this chapter. Suffice it to say here that its compactness, its strength that allows it to withstand a high pressure gradient, and its chemical stability make it very attractive for use in reverse osmosis and other membrane separations.

Anisotropic Cellulose Acetate Membrane

Cellulose acetate at this time still is and is likely to remain one of the mainstays for reverse-osmosis membranes. It is therefore appropriate to discuss the membrane development and preparation in some detail. Excellent discussions of membrane preparation and of performance characteristics are given in the books of Merten [29], Sourirajan [30], and Lacey and Loeb [31]. The following material represents a summary presentation based on these references.

The Loeb-Sourirajan type of the anisotropic cellulose acetate membrane essentially consists of a very thin, dense surface layer and a rather porous, thick substrate, all in the same configuration. The permeant solution molecules experience the flow resistance and separation primarily in the thin layer. The flux is therefore considerably higher than that through an unmodified cellulose acetate membrane of the same thickness, yet the salt rejection is almost the same. Electron micrographs and other studies of its transport properties show that the thin layer is about 0.25 μ thick out of a total thickness of 100 μ, and that its pore size is in the order of 100 Å. The pore size of the substrate is in the range of 0.1 to 0.4 μ. Transport properties of the thin layer are very similar to the unmodified membrane, in spite of the finely porous character of the skin layer.

An especially informative picture of the structure of a Loeb-Sourirajan modified cellulose acetate membrane is shown in Fig. 14.6, which represents a 30-μ cross section: the details of structural tapering are clearly evident. Supporting views of structural details are shown in Fig. 14.7. Here, Views 1 and 2 are cross-sectional pictures of top and bottom surface, when the membrane was dried in air. View 3, of the top surface, shows no evidence of pores down to about the 100-Å level. View 4 pictures a microwrinkled structure of the bottom surface, which is characteristic of this membrane. Finally, View 5 exemplifies the highly porous substructure of the membrane.

These facinating photographs were obtained by Riley and co-workers through a replication electron microscopy scheme. The scheme and photographs have been published in Refs. 32 to 34.

An improved procedure of membrane preparation has been developed by Kesting. The description of the procedure is quoted directly from the 1971-1972 Saline Water Conversion Summary Report [37]:

"A new process for fabricating reverse osmosis membranes has been developed by Chemical Systems Inc. The Kesting process, named after its inventor, is a dry process that eliminates the necessity for immersion and annealing steps, and appears to offer very significant advantages as compared to conventionally prepared membranes with regard to quality control, wet-dry reversibility, shipping and storage (can be shipped and stored dry, in which condition they are lighter, temperature insensitive, and not biodegradable). The ability of the membrane to retain its dimensional stability on drying is attributed to the relatively large pores in the substructure. An interesting variation was made with a casting solution containing both cellulose diacetate and cellulose triacetate.

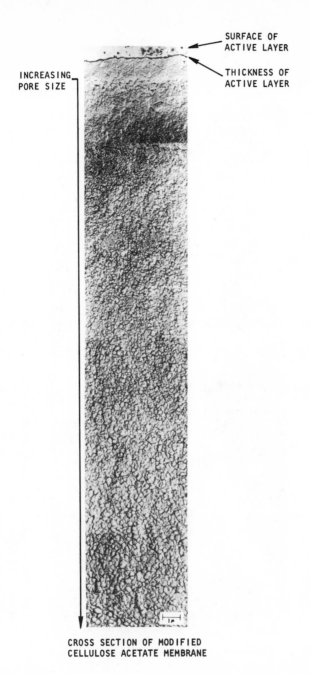

SURFACE OF
ACTIVE LAYER

THICKNESS OF
ACTIVE LAYER

INCREASING
PORE SIZE

CROSS SECTION OF MODIFIED
CELLULOSE ACETATE MEMBRANE

Fig. 14.6 Structure of Loeb-Sourirajan anisotropic cellulose acetate membrane. (Courtesy Gulf Environmental Systems Company.)

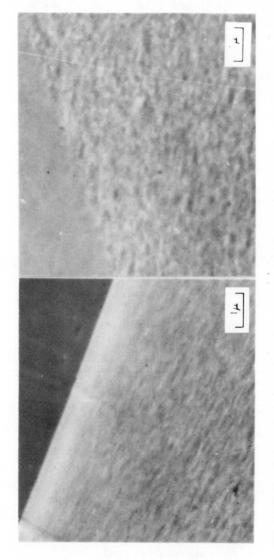

Fig. 14.7 Detail views of portions of membrane structure in Fig. 14.6. (Courtesy Gulf Environmental Systems Company.) (a) Ultramicrotomed cross sections of the air-dried and bottom surfaces of a modified reverse-osmosis membrane. Transmission electron micrographs (10,000X).

Fig. 14.7 (*b*) Top, bottom, and cross section of a freeze-dried modified reverse-somosis membrane. Electron micrographs of preshadowed carbon replicas (20,000X).

The triacetate was found to be concentrated in the skin, and the diacetate in the substructure; with great care the skin could actually be 'peeled' off of the substructure. This offers the very exciting possibility of making a composite membrane from two different polymers in a single casting solution. Another virtue of this new process was ability to control the skin thickness (and thereby flux) by variations in the casting solution."

Considerable details on this development have been published very recently by Kesting [73].

Membrane Preparation and Its Effect on Performance

The procedure of membrane preparation involves the dissolving of the polymer in a solvent to get a casting solution (dope), casting on a smooth glass plate, evaporation of the solvent at a low temperature, gelation in ice water, and finally annealing in hot water.

A typical casting solution is composed of the following components [30]:

Cellulose acetate	22.2% by weight
Mg $(ClO_4)_2$	1.1% by weight
Water	10.0% by weight
Acetone	66.7% by weight
	100.0% by weight

The average molecular weight and the distribution of molecular weight of the raw powder of cellulose acetate are not so important. However, the salt rejection can be greatly improved with a high acetyl content within the limit of the solubility of cellulose acetate in acetone. The 2.5-acetate (approximately 39.8% acetyl content) was usually recommended.

The solution is cast on a very smooth glass plate at a temperature between -5 and $-10°C$, and the solvent is allowed to evaporate in the air at that temperature for 3 min. The thin, dense layer is apparently formed in this step. The cast is then immersed into ice water for about an hour to leach out the electrolyte and to permit gelation. At this step, the cast membrane is too porous and exhibits low mechanical strength and low salt rejection, but high flux. However, if the cast membrane is annealed in hot water, the pores of the membrane are shrunk, and the thin layer grows in thickness. The salt rejection and mechanical strength are therefore greatly improved. As the annealing temperature increases, the flux becomes reduced, but the salt rejection increases and then levels off above a certain temperature. A high annealing temperature also reduces the pressure dependence of the membrane parameter Q_B/l in Eq. 8.12 of Chapter VIII and influences the activation energy of the molecular diffusion through the membrane. Hence, one may utilize this annealing step at an optimum temperature to fit a given requirement. This tailorability is the most distinctive and useful feature of the membrane separation. In desalination, for instance, the annealing is effectively conducted at a temperature between 65 and 85°C for 5 min.

The anisotropic cellulose acetate membrane prepared in this way possesses a flux of $N_B = 3 \times 10^{-4}$ g/(cm^2)(sec), and a salt rejection greater than 99% at a pressure differential of 102 atm.

Later, Manjikian et al. [35] attempted to prepare cellulose acetate membranes from different casting solutions that did not contain electrolytes. They found that the performance of a membrane made from the following casting solution was equal to or better than the Loeb-Sourirajan-type membrane:

Cellulose acetate	25% by weight
Formamide	30% by weight
Acetone	45% by weight

The fabrication of the membrane is comparatively simpler, and the casting can be conveniently done at room temperature.

Other Membrane Research

Not surprisingly, much work is still under way in the search for imporved membranes. The general aim of such research is the development of hydrophilic polymers that possess properties equal or superior to cellulose acetate. This means performance in respect to salt rejection, as well as mechanical ruggedness for satisfactory long-time, in-plant operation. The literature dealing with this subject is so extensive that it is impossible to cover it in detail within the scope of the present treatise. At the danger of being accused of omissions, only a limited number of references have been cited. Actually, much of the technology is described in patents, and these need to be consulted for essential details.

Although Eastman Kodak has given up the marketing of reverse-osmosis membranes, it should interest the reader to study the technology described in U.S. Patent 3,525,335 [36]. Other pertinent references are Refs. 29, 31, and 37.

Of the large number of materials that have been investigated as to their suitability for reverse-osmosis applications, many have been found to be of marginal value and more to be of no value at all. Reports indicate that ethyl cellulose has received a fair amount of attention [29, 30, 38], but it does not seem to give membranes comparable to the acetates in overall performance. Staude [39] mentions polyvinyl pyrollidone, polyvinyl alcohol, polyvinyl chloride, and polystyrene, all of which were considered of inferior quality. As a matter of fact Reid's early paper [25] shows that quite a number of such polymers had been screened in his investigations. Recent studies by Sachs and Lonsdale [40], however, report that they have found a polyelectrolyte membrane to be competitive with cellulose acetate for desalination of dilute solutions. The membrane was prepared by casting an aqueous solution of polyacrylic acid on a porous support membrane.

An intriguing approach has been followed by Stille [41, 42], who developed the idea of preparing block polymers of the type poly(methacrylic acid)-*b*-poly-(2-vinylpyridine) to see if complementary selectivity characteristics could be attained. Such a copolymer, containing anionic and cationic blocks, could exhibit a more uniform rejection of all salts and offer the polymer compatibility that is present in block polymers, but often missing in homopolymer mixtures. In addition, charge-mosaic membranes that contain microdomains should exhibit piezodialysis. Stille reasons that it should not be too difficult to prepare such pressure dialysis membranes, and that they would have improved strength characteristics and carry microelectric fields with minimal distances between charges.

The salt-rejection behavior of block polymers prepared from methacrylic acid and 2-vinylpyridine shows that reasonable rejection could be attained [42] in the 75-to-80% range, with a sharp minimum spike at the isoelectric point as shown in Fig. 14.8.

A rather informative treatise on all aspects of membrane research is contained in an extensive OSW report by Riley et al. [43]. It covers the preparation of thin-film membranes, especially attempts to deposit thin films of cellulose acetate on a finely porous substrate of a cellulose acetate - cellulose nitrate combination. This report is an excellent example of the comprehensive activity in this field of endeavor. Similarly, Lonsdale and Podall's book on *Reverse Osmosis Membrane Research* [44] is a most helpful reference text.

The reader who is particularly interested in this phase should continue to pursue the publications of the Office of Saline Water, as the most up-to-date information is likely to be found there, rather than in the normal literature channels. Some specific bulletins should be cited. Thus, a listing of OSW Research and Development Progress Reports is available. It contains the titles, order number, and price of 780 documents as of October 1972. The most recent issue of the yearly reports is for 1972-1973, and there additional available documents are listed up to No. 785 [37]. This listing, of course, covers all methods, processes, and equipment used in the demineralization of waters, but a fair number are concerned with membrane processes. Another compilation is the *Desalting Plants Inventory*. The latest issue available at this time is Report No. 3, and an updated version is under way.

Intense current interest in development of membranes for reverse osmosis is evident from a survey by Walch [45, 46]. His summaries of noncellulosic-type membranes are particularly pertinent and most instructive. The data represent a good collection of transfer coefficients and thus are presented in Chapter XI.

What appears to be a new concept is described in the 1971-1972 Saline Water Conversion Summary Report as follows:

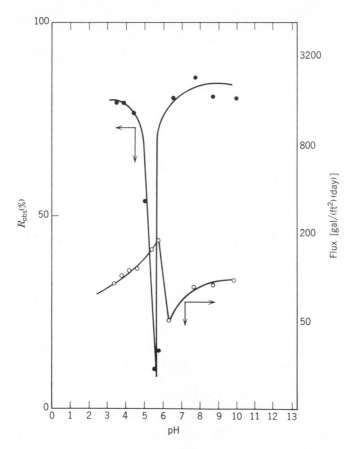

Fig. 14.8 Salt rejection and flux of block polymer versus pH: pressure 950 psi [42].

"A novel desalination technique is under development, whereby an external electrical circuit regulates the charge density around pores in a metallic membrane containing pores of *ca.* 100 Ångstrom units in diameter. Electrochemical theory indicates that in this way it should be possible to have complete salt rejection along with high water flux, in a membrane system far more durable than any other available."

The 1972-1973 Saline Water Conversion Summary Report (May 1973) presents considerable new and interesting material in progress with membrane processes, particularly in reverse osmosis [37]. Highlights of membrane research and technology are quoted. Thus, in composite *membrane technology:*

"The R & D studies on the cellulose acetate, polyacrylic acid, PPO-SO$_3$H, and most recently the NS-1 composite membrane (thin-film, non-cellulosic) have now clearly shown the viability of the thin-film composite membrane approach. The key factors responsible for the success obtained to date on this approach have been: (1) The fineness of the surface porosity of the support film and (2) the method of forming the barrier on the microporous support.

"Other studies include: (1) Radiation grafting of polyvinylpyridine and polystyrene sulfonic acid to microporous polypropylene supports for selective separations, (2) deposition of thin polyelectrolyte layers on partially annealed cellulose acetate membrane for selective separations, (3) ultraviolet surface photopolymerization of vinylpyrrolidone and other hydrophilic vinyl monomers on microporous supports, and (4) formation of Blodgett multilayers from spread polymer films on microporous polysulfone to control the structure of the permselective barrier."

Progress in *noncellulosic* membranes is reported to deal with noncelluosic polymers as candidate reverse-osmosis membranes. Compositions investigated include: aromatic polyamide-imides; polyimidazopyrrolone precursors; chemically modified polysulfones; ethylene oxide grafted nylon-6, and polyoxetane amides. The key attribute of these polymers is their greater chemical stability compared to presently used commercial membranes.

Also, particularly for reverse-osmosis uses, new membranes from polyamides, polyamidehydrazides, and polybenzimidazole look very promising. Technology has been developed for making and utilizing "large fibers" and/or "small tubules," which retain the high-flux advantages of tubular membranes and the pressure-containment characteristics of hollow fibers. The hydrazide composition is one of those developed by Du Pont and shows real promise.

4 MEMBRANES FOR OTHER LIQUID-PHASE PROCESSES

The preceding material is essentially a description of membranes developed for use in reverse osmosis. However, these membranes and their variations are equally suitable for certain applications in ultrafiltration and dialysis.

The details of preparation of most ultrafiltration membranes are not published. Only the polyelectrolyte membranes (the UM series of Amicon Company) are reportedly known to be cast from a shielding solution (e.g., $CaBr_2$ solution) containing a polyion complex. The complex is formed by reaction between a polycation like poly(vinylbenzyltrimethyl ammonium chloride) and a polyanion such as sodium poly(styrenesulfonate) [49, 50].

5 TUBULAR SHAPES AND HOLLOW FIBERS

The distinction between tubes and hollow fibers, obviously, is on the basis of the diameter. In a manner of speaking, the cutoff point is rather arbitrary, and certainly there is a region of considerable overlap where one may apply either description equally well.

Usually, if the internal diameter is in the fractional millimeter range as for fine-bore tubing, say, from ¼ mm on up (~10 mils), the material might be labeled tubular. The Du Pont Permasep confirmation has an i.d. of about $40\,\mu$, which is $\frac{1}{25}$ of a millimeter, and it is called a hollow fiber.

The matter of producing hollow cylindrical shapes calls for two possible approaches. One procedure would be to make unsupported tubes, such as fine-bore silicone-rubber tubes (see Chapter V) and the Permasep hollow fiber. Evidently, this calls for a most skillful operation of fiber spinning with the creation of a hollow core. The other approach is the deposition of the polymer matrix on a tubular support. Both methods will be exemplified in teh following sections.

It may be surprising to the reader that the earliest hollow fibers were produced prior to World War II in what is now East Germany [51]. A very sophisticated fiber technology had been developed in the eastern part of Germany in the late 1920s, and a heavy concentration of fiber industry grew up there. These early fibers were nylon, and the cross section of the hollow core was somewhat triangular in shape.

Fig. 14.9 Du Pont Permasep hollow-fiber configuration; 750 X. (Courtesy Du Pont Company.)

Some of the early fine tubes produced in the United States had a pearlike shape, but present manufacturing techniques produce well-rounded contours Fig. 14.9 of Du Pont's hollow fibers at 750X.

Hollow-Fiber Polyamide Membrane

The only polymer membrane in extensive use, besides cellulose acetate, is the asymmetric *polyamide* membrane of Du Pont Permasep devices. There are numerous patents in existence that cover composition as well as processing and design features of equipment. A detailed coverage would be too lengthy, and the most pertinent patents are listed in Ref. 52. Also some aspects of membrane preparation are covered in the book by Lonsdale and Podell [44]. This membrane is employed only in tubular form. Considerable details are given in Ref. 28, and a worthwhile generalized description is contained in the article covering the Kirkpatrick Chemical Engineering Achievement Award in 1971 to the Du Pont Company [53]. Figure 14.10 is a schematic from the Du Pont bulletin giving nominal dimensions of the fiber.

B-9 Fiber

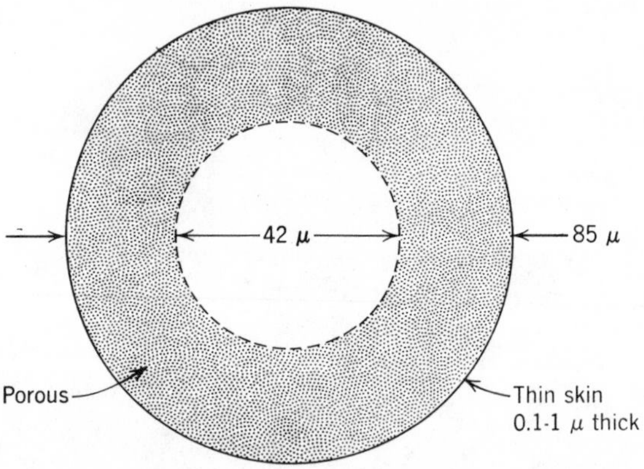

1μ = 1 micron = 0.00004 in.
        = 4/100,000 in.

42μ = 0.0016 in.
85μ = 0.0033 in.

Asymmetric
Aromatic polyamide

Fig. 14.10 Dimensions of Permasep hollow fiber.

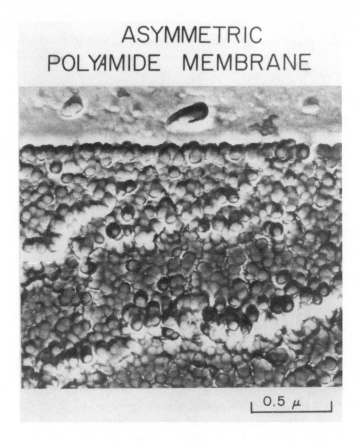

ASYMMETRIC
POLYAMIDE MEMBRANE

0.5 µ

Fig. 14.11   Photomicrograph of Du Pont Permasep membrane strucutre. (Courtesy Du Pont Company)

A photomicrograph of the so-called B-9 polyamide membrane structure is shown in Fig. 14.11. It clearly pictures the difference in microstructure of the active skin (top) and the porous substrate. A great advantage of this configuration is the inherent strength of the material, so that it does not require a mechanical backing support. The flux is rather low, that is, about 1×10^{-4} g/(cm^2)(sec), and the salt rejection is only moderately good. But the ruggedness of this material and its ability to withstand high pressure differentials give it great operational advantages. A view of the assembly is given in Chapter XV.

Hollow-Fiber Cellulose Acetate Membrane

A very promising configuration has been developed by the Monsanto Research Corporation under OSW sponsorship. The product is a true hollow fiber of cellulose acetate having the asymmetric structure of the flat membranes. The fiber has outside and inside diameters of 300 and 100 μ, respectively, At an operating pressure of 250 psig the flux tested out at 2.3×10^{-4} g/(cm^2)(sec), with about 95% salt rejection, when charging a 3000-ppm NaCl solution.

An OSW Report [54] and several company technical reports [55-57] give considerable insight into the technology of hollow-fiber preparation, techniques of attaining asymmetric structures, effects of different polymer composition, and hydraulic considerations in operation of hollow-fiber systems.

It seems that these fibers have been investigated thoroughly in Field Service installations in capacities up to 2000 gal/day. Larger units are in progress: some operational details are presented in Chapter XV.

Tubular Cellulose Acetate

The tubular form of the cellulose acetate membrane has attracted much attention due to its distinctive operational advantages. Mahon [58] was able to fabricate very fine tubes of cellulose triacetate, and Loeb [29] prepared tubes of a somewhat larger size from the formamide casting solution, wrapped with Nylon cloth. As a general rule the tubular form of cellulose acetate requires some mechanical structure as a support to allow the relatively high pressures used in the liquid-phase processes. Thus, porous tubes of varying construction and composition are in use, and the membrane can be placed either on the inside or the outside of the tubular support. A specific design calls for an enclosing porous tube of Fibergalss-reinforced plastic made by Havens [59]; thus device is now available commercially.

A rather novel approach to provide strong tubular support by means of so-called resin-bonded sand logs is being investigated by Westinghouse [60]. Here the porous matrix serves as the flow path for the liquid on the low-pressure side of the separating element. A schematic view of the tubular unit is shown in Fig. 14.12.

Other Configurations

Many of the tubular elements in use are based on the spiral-wound design, similar to that proposed by Osburn for gaseous-diffusion cells (see Chapter XII). A description of the commercial design is given in Chapter XV. A new version of a hollow-fiber design is described by McCormack [61] in a U.S. Patent. It concerns the coating of a nonporous organic polymer film onto hollow porous glass filaments. The filaments are made of Corning's microporous Vycor glass and serve as a strong support for the thin polymer membrane. The use of small bore tubing has already been covered in Chapter V, where

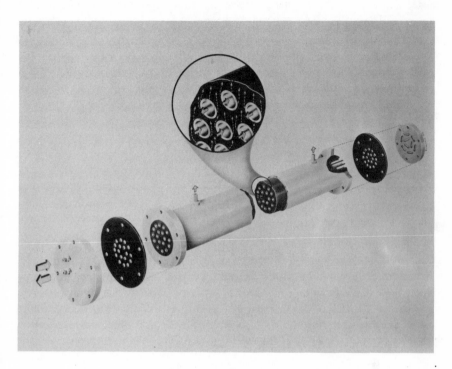

Fig. 14.12 Schematic of resin-bonded sand log construction. (Courtesy Office of Saline Water.)

Blaisdell's studies on nonstretching and expandable silicone-rubber tubing were illustrated. In addition Du Pont's Permasep hollow-fiber design was covered in Chapters V and VII. Silicone-rubber tubing in varying sizes is available from Dow Corning [62] and the General Electric Company [63] and also from a number of plastics processing operators. A capillary-type tubing consisting of a silicone-polycarbonate block polymer designated as MEM-213 was available from General Electric, but encountered pinhole difficulties in extrusion.

Flat film shapes of all types of compositions are available from a very large number of processors. There are several instances of specialty-type preparations where particularly thin films have been prepared. Thus, flat films of dimethyl silicone and copolymer of 1 mil thickness or less, pinhole free, are available from General Electric's Development Operation.

Successful preparation of ultrathin membranes was reported by Stancell and Spencer [64]. These investigators pursued the problem in two directions. One aim was to reduce the thickness of the membrane well below the mil thickness range, actually into the micron range, by film deposition on a supporting substrate. The well-known method of Loeb and Sourirajan [65] for preparing an asymmetric membrane is a fine example of this technique. The other objective was to modify such thin membranes structurally to attain imporvement in permselectivity. Stancell and Spencer were able to prepare membranes as thin as 0.5 μ with simultaneous improvement in permselectivity of hydrogen relative to methane. This was accomplished by deposition, from a plasma, of a highly cross-linked coating onto permeable polymer film substrates. The latter were poly(-phenylene oxide), silicone-carbonate copolymer, and cellulose acetate butyrate. Plasma depositions covered a host of organic compounds, and improvements in permeability ratios for hydrogen-methane mixtures were, indeed, remarkable. However, experiments with butene-1 and isobutene did not result in any improvement in the permselectivity of plasma deposit films for these two compounds. The authors do not indicate if these membranes are suitable for liquid-phase separations.

A most unusual approach to membrane preparation was disclosed by Correns et al. [66], who studied the formation of ultrafiltration membranes by growing cellulose-forming microorganisms. Incorporation of other components, such as fibers, led to modifications in the bacterial membranes. The morphological structure, in contrast to filter membranes, is that of a fibrile network. Water permeability and ultrafiltration results with aqueous electrolyte and nonelectrolyte solutions were reported at the meeting but not listed in the abstract.

6   ION-EXCHANGE MEMBRANES

A synthetic ion-exchange membrane consists of a matrix of hydrocarbon chains carrying fixed ionic groups. The area resistance of the membrane is usually less than 20 $(\Omega)(cm^2)$ in a univalent electrolyte. A membrane is called a cation-exchange membrane if the fixed ionic group is acidic, and an anion-exchange membrane if the group is basic. In an amphoteric membrane, on the other hand, the fixed acidic groups are intermingled with the fixed basic groups, and both groups are so uniformly distributed within the membrane that they are physically indistinguishable. Typical examples are the UM-series of ultrafiltration membranes, which have already been discussed.

Ion-exchange membranes can be prepared essentially in the following three ways. Ionic monomers, or mixtures of ionic and nonionic film-forming monomers, are directly polymerized to form a film. Also, nonionic monomers are polymerized, and then fixed ionic groups are introduced in the film, or the fixed ionic groups are introduced during polymerization of the nonionic monomers. The general features in the preparation of these membranes are briefly reviewed here. The earlier methods of preparation have been outlined in some detail by Sollner [71], and recent reviews by Helfferich [67] and by

Friedlander and Rickles [68] are worthwhile references. However, the detailed methods of preparation of commercial ion-exchange membranes are usually not disclosed.

Fixed Ionic Groups

The most common fixed ionic groups in *cation*-exchange membranes are the sulfonic group of a strong acid and the carboxylic group of a weak acid. Of course, the acidic strength also depends on the nature and configuration of the hydrocarbon matrix. The phosphoric, phosphinic, arsonic, and selenonic groups are often found to be useful. In *anion*-exchange membranes, the fixed ionic groups are usually quaternary ammonium, quaternary phosphonium, and tertiary sulfonium for strong bases, whereas the amino group is commonly used for a weak base. The basic groups are generally known to be inherently less stable than the acidic group, and therefore a cation-exchange membrane is usually more stable than an anion-exchange membrane.

It can easily be postulated that an ion-exchange membrane would prefer counterions that tend to associate to form a complex with its fixed ionic groups or with its matrix. Therefore, a suitable membrane for a given counterion can be prepared by incorporating these materials so that a high selectivity can be achieved. For instance, a membrane carrying weak acidic groups shows a high selectivity for a hydronium ion, since the acidic groups easily associate with the ion. However, the mobility of the counterion in such membranes is generally small.

Factors Affecting Membrane Performance

High flux, high selectivity, and good physical and chemical stability are as desirable for an efficient ion-exchange membrane as for any other membrane structure. Such properties depend largely on the nature and composition of the fixed ionic group and the matrix backbone; the balance of the hydrophilic and the hydrophobic content; degrees of cross-linking and entanglement; porosity and asymmetry, if any; other physical and chemical posttreatments such as annealing, compression, and high-energy irradiation. In particular, the degree of cross-linking in the polymeric membrane is very important, as well as the nature and composition of its components. In a highly cross-linked membrane, the Donnan exclusion is effective and mechanical stability is excellent. But this membrane is hardly swollen and the mobility of the counterion is unattractively small. On the other hand, a linear, non-cross-linked membrane is easily swollen and the ionic mobility is high, but this membrane is too soft. Thus, an appropriate degree of cross-linking can be controlled, to some extent, by using a polyfunctional substance in polycondensation, or by introducing a cross-linking agent in addition polymerization. The typical example in the former case is adding an unsubstituted phenol in the polycondensation of *p*-substituted phenol with formaldehyde. In addition polymerization of styrene, divinylbenzene has been extensively used as a cross-linking agent. Radiation cross-linking has already been discussed in the preparation of membranes for gaseous diffusion and pervaporation.

Homogeneous Membranes

Ion-exchange membranes can be prepared simply by polycondensation of monomers carrying the fixed ionic group on acid- or base-resistant film-casting plates. Alternatively, a precondensed or partially condensed mixture can be heated between glass plates to form a thin film. Phenolsulfonic acid or salicylic acid is polycondensed with formaldehyde, for example, to produce a cation-exchange membrane. Aromatic amine or aliphatic polyamine is condensed with formaldehyde to form an anion-exchange membrane. The membranes prepared in these ways are usually homogeneous in the scale of the wavelength of visible light. However, such membranes are often too weak for industrial applications. Therefore, reinforced membranes are generally employed in practice.

Furthermore, a homogeneous membrane can be prepared by addition polymerization of nonionic monomers with a cross-linking agent and immersion in an acidic solution. Such membranes are generally more stable, chemically and thermally, than the above-mentioned condensed polymeric membranes. The degree of cross-linking and the pore size are also more readily controlled. A styrene type of cross-linked membrane is commonly used for this purpose. This membrane is then immersed in a concentrated sulfuric or chlorosulfuric solution, for example, to give a strong cation-exchange membrane. For an anion-exchange membrane, the base group can be introduced in such a styrene type of membrane by chloromethylation and subsequent treatment with ammonia or alkylamines.

Similarly, a fixed ionic group can be introduced into a graft-copolymer membrane to produce an ion-exchange membrane that generally possesses good mechanical and electrochemical properties. A polyethylene film is, for example, impregnated with a mixture of styrene and divinylbenzene and then exposed to γ irradiation such that the styrene and divinylbenzene are grafted onto the polyethylene base. Sulfonation of this graft-copolymer membrane gives a cation-exchange membrane of strongly acid character, whereas the basic group can be introduced by amination. Also, an anion-exchange membrane can be prepared by amination of a graft copolymer of polyethylene and vinylpyridine.

If a microporous nonionic or weakly ionic membrane strongly adsorbs a polyelectrolyte, the membrane can be directly impregnated with a solution containing the polyelectrolyte. A microporous collodion membrane is, for example, immersed in polystyrene-sulfonic acid to give a strong cation-exchange membrane. These membranes are, however, chemically less stable, and thus their capacities diminish slowly as time elapses.

In contrast to the above-mentioned methods, an ion-exchange membrane can be prepared by evaporating a casting solution containing a linear polyelectrolyte and a linear inert polymer. This casting technique has already been discussed in preparation of the cellulose acetate membrane for reverse osmosis and the UM-series of ultrafiltration membranes. Such membranes are generally not soluble in aqueous solutions, since the linear polymers are intermingled with one another and thus the hydrophilic polyelectrolyte can hardly be leached out. The solvent must possess an intermediate cohesive energy or be a mixture of polar and nonpolar components such that it can dissolve the hydrophilic polyelectrolyte as well as the hydrophobic inert polymer. Butyrolacetone, dimethyl formamide, and dimethylsulfoxide are typical examples. For such solvents, one may, for example, use a copolymer of vinylchloride and acrylonitrile as an inert polymer and polystyrenesulfonic acid for an acidic polyelectrolyte, or polyvinylimidazol quaternized with methyliodide for a basic polyelectrolyte. The collodion membrane prepared by dissolving collodion in ether-alcohol mixtures is a classical example.

Also, a polymer possessing a high thermal stability, such as perfluorocarbon sulfonyl fluoride copolymer, can be directly melted and fabricated into a film by conventional plastics-processing methods. Subsequent saponification with hot caustic gives a sodium salt that can be converted with acid, such as nitric acid, to an ion-exchange membrane in acid form. The *Nafion* perfluorosulfonic acid membrane is manufactured by Du Pont Company in this way. In such preparation, it is easy to achieve good electrochemical properties and high mechanical strength by tailoring the polymer structure and also by special techniques of fabricating and reinforcing membranes [72].

Heterogeneous Membranes

These membranes consist of fine colloidal ion-exchange particles embedded in an inert binder such as polyethylene or phenolic resins. Such a membrane can be prepared simply by calendering ion-exchange particles onto an inert plastic film. Another procedure is dry molding of inert film-forming particles and ion-exchange particles and then milling the mold stock. Also, ion-exchange particles can be dispersed in a solution containing a film-forming binder, and then the solvent is evaporated to give an ion-exchange membrane. Similarly, ion-exchange particles are dispersed in a partially polymerized binder polymer, and then the polymerization is completed. Most heterogeneous membranes that possess adequate mechanical strength generally show poor electrochemical properties. On the other hand, a membrane that contains ion-exchange particles large enough to show an adequate electrochemical performance exhibits poor mechanical strength. Therefore, homogeneous membranes are ussally more attractive.

References

1. H. P. Hood and M. E. Nordberg, U.S. Patent 2,106,744 (1938).
2. M. E. Nordberg, *J. Amer. Ceramic Soc.*, **27**, 299 (1944).
3. H. E. Huckins and K. Kammermeyer, *Chem. Eng. Progr.*, **49**, 180, 294, 517 (1953).
4. E. R. Gilliland, R. F. Baddour, and J. L. Russel, *A.I.Ch.E. J.*, **4**, 90 (1958).
5. Departmental Res. Rept., Chemical Engineering, University of Iowa.
6. W. Haller, *J. Chem. Phys.*, **42**, 686 (1965).
7. W. Haller, *J. Chromatog.*, **32**, 676 (1968).
8. W. Haller, *Nature*, **206**, 693 (1965).
9. W. Haller, *Virology*, **33**, 740 (1967).
10. Selas Flotronics, Bulletins.
11. R. M. Barrer, *Diffusion in and through Solids*, Cambridge University Press, 1951.
12. Hughes Research Labs Bulletin.
13. R. R. Turk, "Tungsten Ionizers with Controlled Porosity for Cesium Ion Engines," in *Modern Developments in Powder Metallurgy*, Vol. 2, H. H. Hausner, Ed., Plenum, New York, 1966.
14. W. Desorbo and H. E. Cline, *J. Appl. Phys.*, **41**, 2099 (1970).
15. Millipore Corp., company catalogues.
16. Amicon Corp., company literature.
17. H. J. Preusser, *Kolloid Z.*, **250**, 133 (1972).
18. M. C. Porter and H. J. Schneider, "Nuclepore Membranes for Air and Liquid Filtration," *Filtration Engineering*, Jan./Febr. (1973).
19. R. L. Fleischer, P. B. Price, and R. M. Walker, *Science*, **149**, 383 (1965); *Sci. Amer.*, **220**, (6) 30 (1969).
20. R. L. Fleischer, H. W. Alter, S. C. Furman, P. B. Price, and R. M. Walker, *Science*, **178**, 255 (1972).
21. Nuclepore Corp., BUlletins.
22. H. S. Bierenbaum, R. B. Isaacson, M. L. Druin, and S. G. Plovan, *Ind. Eng. Chem. Prod. Res. Develop.*, **13** (1), 2 (1974); Celgard bulletin, Celanese Plastics Co.
23. Sepco Co., Bulletins.
24. O. J. Sweeting, in *The Science and Technology of Polymer Films*, O. J. Sweeting, Ed., Interscience, New York, 1968.
25. C. E. Reid and E. J. Breton, *J. Appl. Polymer Sci.*, **1**, 133 (1959).
26. L. M. Litz and G. .E. Smith, Union Carbide Res. Inst. Rept. UCR-J-701 (1972).
27. Eastman Chemical Products, Kingsport, Tenn., Publication No. PM-4A (1971); now withdrawn from market.
28. E. I. du Pont de Nemours & Co. (Inc.), "Permasep" products.
29. S. Loeb, "Preparation and Performance of High Flux Cellulose Acetate Desalination Membranes," in *Desalination by Reverse Osmosis*, U. Merten, Ed., M.I.T. Press, Cambridge, Mass., 1966, Chap. 3, p. 55.

30. S. Sourirajan, *Reverse Osmosis,* Academic, New York, 1970.
31. R. E. Lacey and S. Loeb, *Industrial Processing with Membranes,* Wiley-Interscience, New York, 1972.
32. R. L. Riley, J. O. Gardner, and U. Merten, *Science,* **143,** 801 (1964).
33. R. L. Riley, U. Merten, and J. O. Gardner, *Desalination,* **1,** 30 (1966).
34. R. L. Riley, G. R. Hightower, and C. R. Lyons, *Gulf Environmental Systems Rept.* EN-412293 (1972).
35. S. Manjikian, S. Loeb, and J. W. McCutchan, *Proc. First Intn'l Desalination Symp.,* Paper SWD/12, Washington, D.C., Oct. 3-9, 1965.
36. M. E. Rowley, U.S. Patent 3,522,335 (1970).
37. Office of Saline Water, U.S. Dept. Int., "Saline Water Conversion Report." Yearly reports and individual progress reports since 1952; also Commerce Dept. NTIS Rpt. AD-769-208/TA (1974).
38. E. Klein and J. K. Smith, *IEC Prod. Res Dev.,* **11,** 207 (1972).
39. E. Staude, *Chem. Z.,* **96,** (1/2) 27 (1972).
40. S. B. Sachs and H. K. Lonsdale, *J. Appl. Polymer Sci.,* **15,** 797 (1971).
41. M. Kamachi, M. Kuribara, and J. K. Stille, *Macromolecules,* **5,** 161 (1972).
42. M. Kuribara, M. Kamachi, and J. K. Stille, *J. Poly. Sci., Poly. Chem. Ed.,* **11,** 587 (1973).
43. R. L. Riley, C. E. Milstead, H. K. Lonsdale, and K. J. Mysels, Gulf Gen. Atomic, R&D Prog. Rept. No. 729, U.S. Dept. Int. (Dec. 1971).
44. H. K. Lonsdale and H. E. Podall, Eds., *Reverse Osmosis Membrane Research,* Plenum, New York, 1972.
45. A. Walch, *CZ Chem.-Tech.,* **2,** 7 (1973).
46. A. Walch, *Batt. Inform.,* **14,** 20 (1972).
47. D. L. Hoernschemeyer, R. W. Lawrence, C. W. Saltenstall, Jr., and O. S. Schaeffler, *Polymer Reprints,* **12,** 284 (1971).
48. L. T. Rozelle, J. E. Cadotte, and D. J. McClure, Office of Saline Water, Res. Dev. Progr. Rept. No. 531 (June 1970).
49. A. S. Michaels, *Ind. Eng. Chem.,* **57,** 32 (1965).
50. A. S. Michaels, U.S. Patent 3,276,598 (1967).
51. Personal communication.
52. Selected *Du Pont patents* concerning development of Permasep equipment: Selection of the polymeric material, U.S. 3,172,741 (Jolley); Design of the spinnerette, U.S. 3,313,000 (Hays), U.S. 3,397,427 (Burke, Hawkins); Process for spinning hollow fibers, U.S. 3,259,674 (Scott); Permeation apparatus, U.S. 3,198,335 (Lewis, Rogers), U.S. 3,246,764 (McCormack), U.S. 3,339,341 (Maxwell, Moore, Rego); Process for manufacture of permeation apparatus, U.S. 3,339,341 (Maxwell, Moore, Rego), U.S. 3,442,002 (Geary, Harsch, Maxwell, Rego); Encapsulation of hollow fibers, U.S. 3,492,698 (Geary, Harsch, Maxwell, Rego); Providing uniform openness of hollow fiber ends, U.S. 3,503,288 (Swartling), U.S. 3,507,175 (Swartling).
53. *Chem. Eng.,* **78,** (27) 53 (1971).
54. T. A. Orofino, W. F. Savage, and K. C. Channabasappa, "Development of

Hollow Filament Technology for Reverse Osmosis Desalination Systems," Res. Devel. Progr. Rept. No. 549, Office of Saline Water (May 1970).

55. T. A. Orofino, "Cellulose Acetate Hollow Fibers for Reverse Osmosis Desalination," presented at 69th Mtg, A.I.Ch.E., Cincinnati, Ohio May 1971 (MRC, Chemstrand Res. Center).

56. J. P. Davis and T. A. Orofino, "Mass Transport in Hollow Fiber Assemblies," presented at 163rd Mtg, Amer. Chem. Soc., Boston, Mass, Oct. 1972 (MRC, Chemstrand Res. Center).

57. J. K. Lawson, J. D. Bashaw, and T. A. Orofino, "Hollow Fiber Technology for Advanced Waste Treatment," presented at Environmental Systems Conf., San Diego, Calif., July 1973 (MRC, Chemstrand Res. Center).

58. H. I. Mahon, "Hollow Fibers as Membranes for Reverse Osmosis," *Natl. Acad. of Sci., Nat'l Res. Council* Publ. No. 942, 345 (1961).

59. Sea Water Conversion, Brochure of Havens Industries, Spring 1964.

60. News Item, *Chem. Eng.*, **79**, (13) 42 (1972); also News Release by OSW, May 3, 1972.

61. W. B. McCormack, U.S. Patent 3,246,764 (1966).

62. Dow Corning, Bulletin 14-416, Medical Products.

63. General Electric Co., Chemical and Medical Division, Membrane Products Bulletins.

64. A. F. Stancell and A. T. Spencer, *J. Appl. Polymer Sci.*, **16**, 1505 (1972).

65. S. Loeb and S. Sourirajan, *Advan. Chem. Ser.*, **38**, 117 (1962).

66. E. Correns, J. Purz, and H.-H. Schwarz, abstract of a paper presented at 10th Magdeburg Conf. on Separation Technoloyg, in *Filtration and Separation* (Jan.-Feb. 1973).

67. F. Helfferich, *Ion Exchange*, McGraw-Hill, New York, 1962.

68. H. Z. Friedlander and R. N. Rickles, *Anal. Chem.*, **37**, (8) 27A (1965).

69. R. E. Lacey, in *Industrial Processing with Membranes*, R. E. Lacey and S. Loeb, Eds., Wiley-Interscience, New York, 1972, Chap. 1.

70. R. N. Rickles, *Membranes, Technology and Economics*, Noyes Devel. Corp., Park Ridge, N. J., 1967.

71. K. Sollner, *Ann. N.Y. Acad. Sci.*, **57**, 177 (1953).

72. D. J. Vaughan, *du Pont Innovation*, **4**, (3) 10 (1973).

73. R. E. Kesting, *J. Appl. Polymer Sci.*, **17**, 1771 (1973).

Chapter **XV**

LARGE-SCALE AND INDUSTRIAL APPLICATIONS

The development of large-scale applications of membrane-separation processes presents a spotty picture. Some processes such as electrodialysis have been used for some 20 years, and more, in reasonably large-scale practice, Gaseous diffusion, of course, goes back to the AEC installations of World War II, Reverse osmosis was used occasionally many years ago but, if anything, only in a small-scale fashion. Dialysis as such, shows an on-and-off history in usage. What happens, of course, is that some process will finally come into its own through a particular set of circumstances. The most recent example is that of reverse osmosis as an active competitor in desalination of waters. Usually, some specific breakthrough occurs, and in the case of reverse ofmosis this happened with the creation of high-flux cellulose acetate membranes, and most recently the developments in hollow-fiber technology.

As one might expect, the general interest in applications of membrane processes has stimulated publications dealing with medium- to large-scale accomplishments. As yet, looking at the membrane process field as a whole, there are not very many installations in operation, but the trend towards utilization of these processes is definitely upward, and a full-blown boom seems to be developing.

A look into the recent past and a forecast of what the near future may bring can be attained by perusing such reviews as "Membrane Processes for Industry" [1] and Membranes, Technology and Economics [2]. The former is a symposium report with papers by individual contributors, and the latter is an exceptionally good treatise by Rickles with emphasis on industrial aspects. Also, Merten's book on reverse osmosis [3] contains pertinent material. More recent texts by Sourirajan [4], and by Lacey and Loeb [5] indicate the accelerating pace of membrane process development.

Considerable progress has been made in modeling of multistage separation processes, especially in the gaseous-diffusion field: There are numerous computational examples to be found in the literature, and they are, of course, of some value in bringing the potentialities of stagewise diffusional operation into proper perspective. However, they are like a kiss over the telephone: it's a kiss all right, but not the real thing.

1 GASEOUS DIFFUSION — POLYMERIC MEMBRANES

For many years now, promising large-scale applications of gaseous diffusion as a separation process have been just around the corner.

In 1968 the status of large-scale use of the gaseous-diffusion process was summarized [6]. The essential conclusion was that, except for the Atomic Energy Commission's gaseous-diffusion plants and Union Carbide's hydrogen purification with palladium alloy foils, no other industrial applications had materialized.

At that time the technology of preparing hollow fibers was not sufficiently advanced to make them readily available. Today, however, the experience gained with hollow fibers in large-scale reverse-osmosis installations can be utilized to design and construct gaseous-diffusion cells with an effective area-to-volume ratio, and capable of operation at substantial pressure differentials across the fiber wall.

Du Pont Gaseous-Diffusion Plants

The Du Pont Company has extended its Permasep design to gaseous-diffusion applications. In 1958 Jolley [7] filed a U.S. Patent application covering films of polyacrylonitrile and polyethylene terephthalate for separating hydrogen or helium from gas mixtures. A later patent, issued in 1965 to Lewis et al. [8], describes in considerable detail the design and construction of a permeation apparatus for the separation of fluids, gas or liquid, by using hollow filaments. Thus, it covers gaseous-diffusion as well as reverse-osmosis applications. Figure 15.1 shows a diagrammatic picture of the Permasep consturction. The permeator is said to contain 50 Million hollow polyester fibers, 50 μ o.d. with a surface area of 5 acres. A listing of pertinent patents dealing with the various aspects of equipment development is given in the reference section of Chapter XIV. In the design illustrated, the high-pressure gas is fed to the inside of the fibers.

The first commercial installation for the recovery of hydrogen from refinery gases has been in operation since about 1969, operating at a temperature of 100°F and a high side pressure of 550 psig. A diagrammatic flow sheet of a hydrogen recovery unit that was put on stream in October 1972 is demonstrated in Fig. 15.2 [9].

The hollow fibers used in the permeators are of Dacron polyester type, 18 μ i.d. and 32 to 38 μ o.d. They are encased in a steel shell rated at 650 or 1500 psi, depending on which operating pressure is needed. The overall length of each permeator is 18 ft, with a 12-in. i.d. for the 650-psi unit. The permeabilities of the particular polymer for various components are listed in Table 15.1. Magnitudewise, the permeabilities are very low, but the mechanical ruggedness

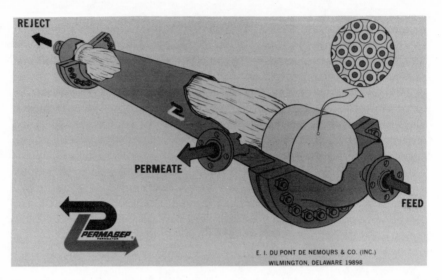

Fig. 15.1 Du Pont Permasep module construction for gas separation. (Courtesy Du Pont Company.)

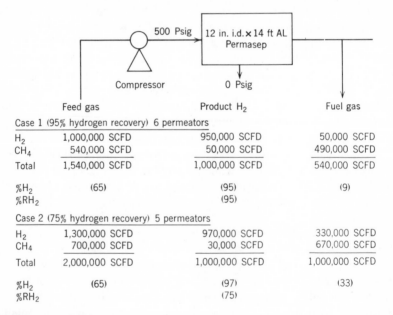

	Feed gas	Product H$_2$	Fuel gas
Case 1 (95% hydrogen recovery) 6 permeators			
H$_2$	1,000,000 SCFD	950,000 SCFD	50,000 SCFD
CH$_4$	540,000 SCFD	50,000 SCFD	490,000 SCFD
Total	1,540,000 SCFD	1,000,000 SCFD	540,000 SCFD
%H$_2$	(65)	(95)	(9)
%RH$_2$		(95)	
Case 2 (75% hydrogen recovery) 5 permeators			
H$_2$	1,300,000 SCFD	970,000 SCFD	330,000 SCFD
CH$_4$	700,000 SCFD	30,000 SCFD	670,000 SCFD
Total	2,000,000 SCFD	1,000,000 SCFD	1,000,000 SCFD
%H$_2$	(65)	(97)	(33)
%RH$_2$		(75)	

Fig. 15.2 Flow diagram for off-gas hydrogen recovery [9]. Minimum requirements: 1,000,000 SCFD, 95% H$_2$.

460

of the hollow fibers allows the use of rather higher pressures that make it possible to obtain economical flow rates.

Table 15.1   Permeability — Du Pont Fiber

	Permeability $\frac{(cc)(cm) \times 10^{12}}{(sec)(cm^2)(cm\ Hg)}$	Relative Rate
H_2	165.0	61.1
CO_2	31.0	11.5
A	7.5	2.8
CH_4	3.5	1.3
N_2	3.1	1.2
CO	2.7	1.0

Union Carbide—Helium Recovery

Union Carbide has conducted a large-scale test on helium recovery from natural gas. Considerable details on technology and operating data were reported by Litz and Smith [10]. The test was carried out at the Navajo Helium Facility in Shiprock, New Mexico, where a natural-gas supply was available with about 5% helium.

Figure 15.3 is a schematic of the flow sheet for the two-stage operation. The first stage bank of permeators was operated at 850 psig pressure, and the second stage at 950 psig. The downstream pressures in each of the stages were near atmospheric, so that appreciable interstage compression was necessary. It should be noted that the process arrangement is somewhat unusual, in that the feed streams to the modules in the stages are in series, whereas the product take-off streams are in parallel.

The membranes used for the separation were Eastman Chemical Company's KP-98 cellulose acetate films previously used in reverse-osmosis desalination.* These asymmetric-type membranes contain a very thin layer of dense film, about 0.00001 in. thickness, with a rather porous sublayer about 0.003 in. thick. The permeator design was that developed by Stern and his co-workers [11]. It is a flat-sheet arrangement with spacers between successive layers of the membrane as illustrated in Fig. 15.4. A sandwich type of structure is enclosed in a cylindrical steel shell, so that about 200 ft^2 of active permeation area are available in a module approximately 10 in. in diameter and 5 ft long.

*Eastman Kodak no longer markets such membranes.

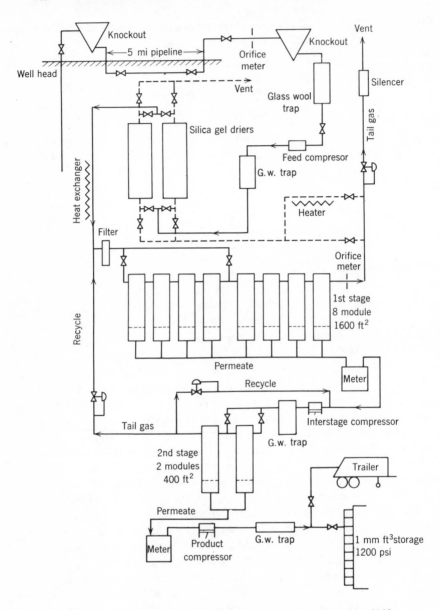

Fig. 15.3 Flow diagram for two-stage helium recovery form natural gas [10].

The magnitude of the operation is expressed by the following quantities of gas flow (average values):

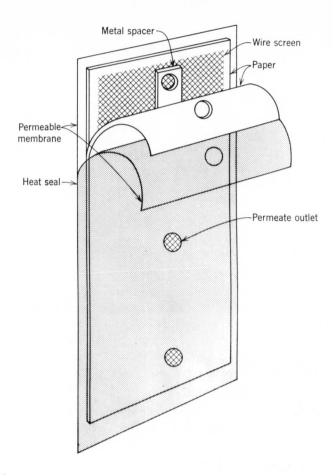

Fig. 15.4 Structure of Union Carbide gaseous-diffusion membrane [10].

To 1st Stage
Fresh gas to 1st stage: 16,300 std ft^3/hr
Helium content: 5.7%
Additional recycle of tail gas from 2nd stage
Helium content: 11.5%
(amount depends on operational splits)
From 1st Stage:
Tail gas: 14,200 std ft^3/hr
Helium content: 2.3%

To 2nd Stage:
All of porduct gas from 1st stage
Some tail gas from 2nd stage as recycle
(amounts depend on operational splits)
From 2nd Stage:
Product gas: 645 std ft^3/hr
Helium content: 82.5%
Total gas: 2880 std ft^3/hr
Helium content: 11.5%
Recovery: 62% of helium in fresh feed
Temperatures: 116°F in 1st stage
78°F in 2nd stage

Pending Developments

An indication of further growing interest in the gaseous-diffusion process is a news item from *C&EN* [12], as follows:

"A cost-cutting helium recovery process has been developed by Teijin, Ltd. The Japanese method, based on selective permeability of a synthetic hollow-fiber material, operates at room temperature and about 425 psig. Compared to conventional helium recovery/purification methods—removal of other gases by liquefaction or by adsorption—the Teijin process features a 30% saving in operating cost at about the same initial investment. It's also usable to purify hydrogen streams in refineries."

No further details are available at present. However, one would infer that the process parallels the Permasep deisgn of the Du Pont Company.

2 METALLIC MEMBRANES

The only known large-scale process is the separation or purification of hydrogen by means of palladium alloy membranes. The status of this process was reviewed in 1968 [6], where the pertinent literature was surveyed. For the sake of convenience a few essential items should be repeated here.

Union Carbide Hydrogen Purification

The most informative description of the process still is that of McBride and McKinley [13] reported in 1965. Operating conditions require feed pressures of 500 psi and temperatures of 300 to 400°C or higher. The potentially deleterious effect of gas impurities, such as CH_4, C_2H_4, CO, and H_2S, especially at higher temperatures, imposes a practical upper temperature limit. Economic evaluations are presented and show that profitable operation is attainable with plants that produce about 4 million ft^3 of high-purity hydrogen per day (greater than 99.99% H_2). Sulfur compounds and carbon monoxide must be removed from the feed stock to as large an extent as possible, as they are particularly prone to poison the barrier in forming sulfides and carbides.

Information supplied in 1972 by R. B. McBride of the Union Carbide Corporation lists the present status as follows:

"We are still operating this process to recover and purify hydrogen from our olefins plants by-product fuel gas at Whiting, Indiana; Texas City, Texas; and Ponce, Puerto Rico. The unit at Puerto Rico has recently been expanded in conjunction with the installation of our new petrochemical complex on the island. We have found that these palladium diffusion units yield high-purity by-product hydrogen at low cost for sale or captive use at hydrogen consuming units in our chemical plant complexes."

"Although hydrogen purification through very thin palladium barriers of large area is a more delicate unit operation than we are normally accustomed to encounter in petrochemical plant operation, our experience has demonstrated that barrier deterioration can be held to a minimum if proper operating procedures and protective instrumentations are provided. The diffusion unit in Puerto Rico completed five years of operation in early 1972, with only two interruptions to normal operation for barrier repair in that 60-month time span."

"We are not using the process for the production of ultra-high purity hydrogen, and are content with product purities in excess of 99.0 per cent. The hot hydrogen product effluent from the diffusers is passed through a catalyst bed to reduce the carbon monoxide content to a non-detectable level in those instances when hydrogen consuming units contain catalyst systems that are susceptible to CO poisoning."

Exxon Carbon Monoxide Enrichment

Another equally interesting account of a commercial installation is given by Meinhold [14]. It concerns a unit operated by the former Humble Oil and Refining Company. In this instance, the main purpose is the removal of hydrogen from a gas mixture to prepare a relatively pure carbon monoxide synthesis stream. Operating conditions are stated to be 700°F (370°C) and about 450 psi pressure drop, with a daily throughput of nearly 150,000 std ft^3 of gas. The process is conducted in 24 diffusion cells requiring a space of only 6.5 X 13 X 6 ft. The membrane is made of a Pd-Ag alloy, which may be the reason why the quoted operating temperature can be used with the carbon monoxide.

Matthey Bishop Hydrogen Purification

Laboratory- and semiscale hydrogen purification units are being manufactured and marketed by Matthey Bishop, Inc. A U.S. Patent by Hunter [15] appears to be the basis of commercial hydrogen diffusion today. The patent describes the use of Ag-Pd alloys, containing from 10 to 50% of silver, with 20 to 40% as a preferred range.

Typical permeability values are presented in Table 15.2.

Table 15.2 Permeability of Hydrogen through Pd and Alloy Membranes[a]

Pressure (psig)	Temperature			
	450°C		500°C	
	Pd	Ag-Pd	Pd	Ag-Pd
100	84	144	128	167
200	145	228	219	274
300	198	302	286	353

[a]Thickness of membrane: 0.001 in.; alloy composition: 27% Ag; permeability units: $ft^2/(hr)(ft^2)$.

It is evident that the alloy composition permits a higher permeation rate. However, the main reason for using an alloy is that pure palladium undergoes a structural change from α-β-phase recycling with increasing and decreasing temperatures. Consequently, there is danger of dimensional variation and ultimate destruction of a membrane [16].

Modern construction of the diffusion cells uses tubular membranes with a considerable improvement of membrane area-to-volume ratio [17]. A rather informative description of the process and equipment was published by Roberts [18]. The presently available units are marketed in capacities of 10, 25, and 50 standard ft^3/hr. Custom-made installations of larger sizes are also available.

The 50-SCFH model has dimensions of 11 × 11 × 8½ in., operating temperature 750°F, inlet pressure 250 psig, and maximum outlet pressure 150 psig. G. P. Matlak states that:

"By far the greatest commercial use of the principle of hydrogen purification by diffusion is on the laboratory or small industrial scale typified by our HP Series of Hydrogen Purifiers. They are used in semiconductor work, epitaxial crystal growth, for analytical purposes such as chromatography and micro-coulometry and other critical applications where a consistent and reliable source of ultrapure hydrogen is required. The larger units such as the 1500 SCFH unit are not in great demand becuase of the readily available economical sources of bottled and liquid hydrogen that usually have sufficiently high purity for most large-scale operations."

A rather interesting combination unit is available that produces hydrogen by decomposition of ammonia and simultaneous membrane separation of the hydrogen from the reaction gases [19]. The flow diagram for this process is

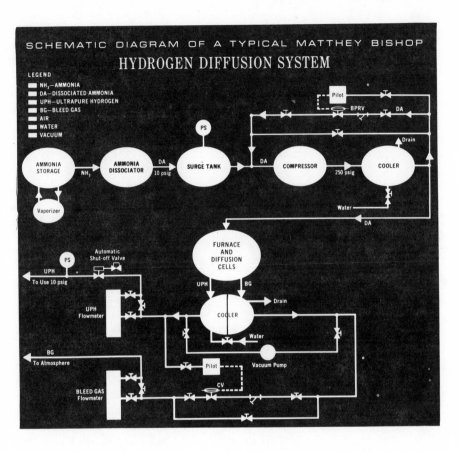

Fig. 15.5 Flow diagram for production of hydrogen from ammonia with simultaneous membrane purification [19].

shown in Fig. 15.5 and a good description is presented in a U.S. Patent by Hunter [20].

Potential Applications

There is one other application that should be mentioned, even though it has not found any commercial use as yet. This is the possibility of oxygen enrichment from air by means of a silver or silver alloy membrane. The basic concepts were described in a U.S. Patent by Mullhaupt [21]. As with other nonporous metal membranes, a fairly high operating temperature is required, that is, 500 to 600°C. Interest in this work has been revived from time to time, but there is no evidence that it has been used successfully in either laboratory- or large-scale operation.

3 GLASSY MEMBRANES

In general, one is not likely to think of a glass or glasslike composition as being a suitable barrier for separations. Obviously, the structure is so "tight" that whatever flow takes place is quite low. Thus, if such a material were to be used as a barrier, it must show some particular advantages. This happens to be the case with the helium-quartz system.

Actually, the fact that some gases may diffuse through glass has been known for a long time. However, one must credit Norton's [22] publications with creating a definite interest in the quartzlike and glassy materials for helium separation. The advantage that makes the process possible is that the separation of helium from other gases approaches what one might call the ideal process, that is, essentially a "go, no-go" operation or an almost complete one-step separation.

The discussions of the process by McAfee [23], and of Melnyk and Habgood [24] are so comprehensive that only a few items should be mentioned here. To get an idea of the flow rates, a round numbers comparison is of interest. Quoted in standard permeabilities, the flow of helium through silica is about 2×10^{-9}, through microporous Vycor glass 15×10^{-6}, and through polyethylene, roughly 1×10^{-9}, However, in order to quote realistic values, the silica figure is for a temperature of 700°C, whereas the other two values are for room temperature. This points up one of the difficult processing requirements for glasses. The temperature must be quite high, otherwise the permeation rates are simply too low for serious consideration. Some years ago, a considerable amount of effort went into possible designs for using quartz tubing in heat-exchange arrangement. The design called for an operating temperature of 900°C and high pressure differentials. There is reason to believe that a successful design of workable header construction could not be accomplished, and thus the effort was shelved [25].

4 POLYMERIC MEMBRANES IN LIQUID PHASE

Although Hagerbaumer's early studies with microporous Vycor glass [26] indicated that such media may find applications in liquid-phase membrane separation, the major thrust has been in the use of polymeric structures. Especially the asymmetric types of polymer membranes have been quite successful in bringing liquid-phase separations to economically competitive status.

Office of Saline Water Activities

The Office of Saline Water has now been in existence somewhat more than 20 years. The overall objective was to develop recovery of fresh water from saline and brackish sources by whatever means possible, with heavy emphasis on economics. To do this on a national scale, it was realized that the program would require a heavy infusion of federal funds.

As one of the more promising areas of technology, membrane-separation processes were considered from the very beginning. Although the primary emphasis was directed toward the reverse-osmosis process, it was expected that any successful developments in membrane technology were likely to benefit all other membrane processes. This has certainly been the case and has shown up particularly well in regard to electrodialysis. Additionally, sophisticated membrane preparation has had an impact on ultrafiltration and dialysis performance. Thus, a review of the present state of development of the OSW activities is of considerable interest as it highlights what has been accomplished in the last two decades and indicates what can be expected in the near future.

The most important large-scale use of polymer membranes has been in reverse osmosis as applied to the desalination of waters. While production of fresh water from sea water (3.5% salts) is one of the major aims, the treatment of brackish waters, or waters that have an unpleasantly high mineral content, has reached a high level of development. Thus, most of the available information is in the area of water cleanup to a satisfactory level, rather than extensive reduction of salt content.

The relative importance of large-scale installations using the reverse-osmosis process for desalting of waters can best be judged from the publications of the Office of Saline Water. A paper containing guidelines was published by Miller [28]. It contains informative statistics and material on comparative equipment configurations. Operating conditions and economic value judgments are presented in the 1971-1972 and 1972-73 OSW reports [27].

A rather important consideration is the relatively high pressure level needed in reverse osmosis. Therefore considerable effort is under way in the R & D category to develop membranes that can operate at substantially lower pressures than the current 600-to-1000-psi level. A comparison of test data is given in Tables 15.3 and 15.4, showing that operating levels of 150 to 300 psi are possible [27].

It seems that membrane desalination has made the greatest progress in the conditioning of brackish waters. This is indicated by a summary type of graph on cost figures presented in the 1972-1973 Summary Report of OSW [27]. This chart, Fig. 15.6, presents expected improvements in cost reduction as a function of plant capacity in million gallons per day (MGD). Presumably, all types of membrane processes are considered. It is, of course, difficult to judge if the quoted future costs are optimistic or realistic. A similar chart is shown in the OSW publication for sea-water desalting based on distillation technology. One can infer that for sea water, that is, 35,000 ppm solids, the distillation process is considered to be more favorable in the present outlook.

Table 15.3 Summary of Field Test Data for Hollow Fiber Units Roswell, N. Mexico[a,b]

Design Product Water Production (gal/day 25°C)	Hours of Operation	Operating Pressure (psig)	Flux [gal/(ft²)(day) at 25°C]	Salt Rejection (% based on feed)	Recovery (%)	Measured Production (gal/day at 25°C)
20,000	60	300	3.8	91	40.6	28,100
	1674	300	3.45	85	34.0	25,600
30,000	100	600	2.0	95	40.8	32,200
	1957	600	1.9	91	35.4	30,600

[a] Feed water, 3690 ppm TDS.
[b] Reference 27.

Table 15.4 Status of Low-Pressure-Process Brackish Water Membrane Development[a]

Manufacturer	Membrane Type	Permeability [gal/(ft²)(day) at 100 psi]	NaCl Rejection (%)	Operating Pressure (psi)	Flux [gal/(ft²)(day)]
Achieved under current programs (lab bench-scale units):					
GESCO	Diacetate (sheet)	10.0	56	200	20.0
Monsanto	Diacetate (HFF)	6.0	75	150-200	9.0
Dow	Triacetate (HFF)	3.0	70	200-300	6.0

[a] Reference 27.

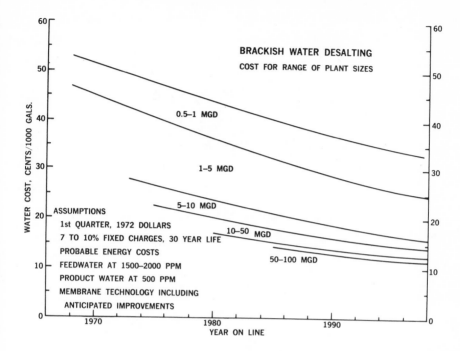

Fig. 15.6 Cost estimates for desalination of brackish water with membrane technology. (Courtesy Office of Saline Water.)

Significantly, the costs for the distillation technique in sea-water treatment are roughly three times those for brackish water.

Configurations of Commercial Membranes

Before presenting examples of large-scale installations it will be of interest to take a look at the commercial types of membranes that are available.

There are four basic types of configurations that are manufactured on a commercial scale. They are

> Plate-and-frame construction
> Spiral-wound construction
> Tubular systems
> Hollow-fiber systems

A detailed discussion of all commercial variations would be lengthy and for the most part not pertinent. Only a limited number of devices will be discussed in order to present the most important features of design and construction. The designs used for reverse osmosis and ultrafiltration are mostly of the same

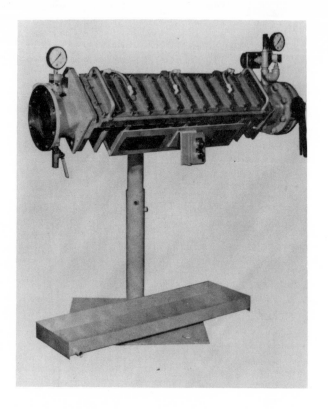

Fig. 15.7 IOPOR module of Dorr-Oliver, Inc. (Courtesy Dorr-Oliver, Inc.)

type and differ primarily in the magnitude of pressure that the modules will permit.

Plate-and-Frame Systems

The unit developed by *Aerojet-General Corporation* uses a parallel-sheet construction. The membranes, on both sides of spacer plates, are sealed to the plates at the edges. The feed stream flows alternately toward the center and to the periphery of the membranes. Permeate flows through a porous paper layer or grooved plates to the product takeoff. The assembly is enclosed in a suitable pressure container.

A filter-press type of construction is built by *A/S DeDanske Sukkerfabrikker, Copenhagen*. The plate stack is built of polyacetal injection-molded parts, and metal press flanges with central compression screw provide sufficient compression so that a pressure container is not needed. The permeate flows toward the edges of the membranes, and thus individual control over each membrane unit is possible. The largest pack that is manufactured contains 28 m² of membrane area nad is 40 cm thick.

The *Iopor* System of *Dorr-Oliver* is a block type that contains 30 parallel plates, spaced about 1/8 in. apart. Each membrane unit consists of one or several series-connected housings built of glass-fiber-enforced polyester composition. The housings contain three exchangeable membrane blocks with a total surface of 60 ft². The module is intended primarily for ultrafiltration with operating pressures below about 60 psig [29]. A view of the commercial module is shown in Fig. 15.7. This unit is used as a building block to compose large-size installations.

Spiral-Wound Systems

Gulf Environmental Systems developed the sandwich-type construction [30] where two cellulose acetate membranes are separated by a semipermeable rigid spacer. Figure 15.8 illustrates the construction. The rolled-up assemblies are inserted in series in a pressure tube. The larger module is about 4 in. in diameter. Called ROGA modules, they are designed for pressures up to 800 psi. A common operating pressure is 600 psi, which represents the intermediate pressure range for reverse osmosis. An interesting relationship is that experienced in flux decline, even without fouling, as a function of operating time. Some information on the flux performance of the ROGA modules is provided in

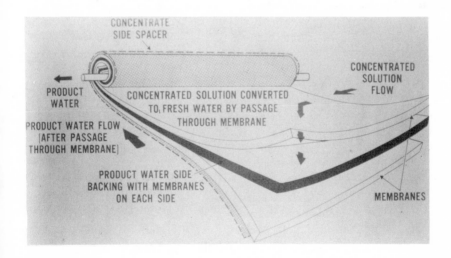

Fig. 15.8 Spiral-type construction of membrane module for reverse osmosis. (Courtesy Office of Saline Water.)

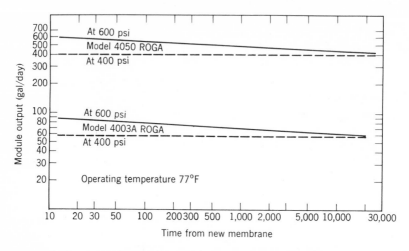

Fig. 15.9 Plot of flux decline of ROGA module (spiral wound [30]).

Fig. 15.9. There, the daily production per unit is plotted against the time elapsed from the installation of a new module. The decrease is essentially linear on a log-log scale, for the 600-psi pressure level. As indicated, no decrease was observed at the 400-psi level, even after 25,000 hr (2.85 yr).

Tubular Systems

The *Havens Industries* design uses porous glass-fiber-enforced plastic tubes for supporting the membrane that is inserted into the tube. The tubes have a wall thickness of 0.03 in., inside diameter of 0.5 in., and are 96 in. long. To promote turbulent flow and reduce boundary-layer thickness, the tubes contain rod-shaped diffusers. Tubes are arranged in bundles to form modules. Pressure operation is at about 600 psig.

Abcor Modular membranes [31] are arranged to have seven tubular modules in a pressure shell as shown in Fig. 15.10. The feed stock flows inside of the tubes, which are 10 ft long and have 2.2 ft² of membrane surface. The outer housing is 4.5 in. in diameter and is built of high-strength plastic or stainless steel. Inlet pressure is in the 50-psig range and pressure drop over the membrane is from 20 to 40 psi. The manufacturer stresses the point that this equipment is designed on a "clean-in-place" manner, which is a particularly important factor in food processing.

Romicon's module [32] is designed for ultrafiltration use. The cartridge contains 60 membrane tubes. Each tubular membrane is overbraided with Dacron, and the tubes are installed in a clear plastic jacket that serves as a permeate collector. The flow scheme is illustrated in Fig. 15.11. This is a basic module, 43 in. long with 14 ft² of active membrane area. It is supplied in two different channel sizes, resulting in capacities of 35 to 45 and 100 to 120

Fig. 15.10 Abcor tubular-module design for ultrafiltration. (Courtesy Abcor, Inc.)

gal/min, respectively. Operating pressure is 100 psi, and backpressure at the cartridge outlet is 15 psi. Maximum operating temperature is 140°F. Because this module is the one used for large-scale installations, it is also marketed in a portalbe test stand combination. Thus, full-scale operating conditions can readily be tested out, and the final scaled-up version simply consists of a multiplicity of these modules.

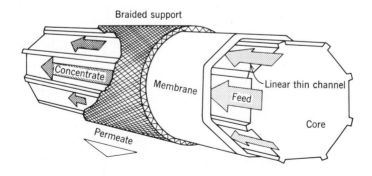

Fig. 15.11 Flow scheme of Romicon tubular module [32].

Nonporous tubular supports consisting of aluminum or fiber-inforced plastic and containing a multiplicity of fine holes are employed in the *Neger Institute-University California* design. Here the membrane also is placed inside the tube. The tubes are mounted in a vertical position.

A modification that has the membrane on the outside of the tube is manufactured by *Société Degrémont* (France). Operation also uses a vertical flow scheme.

Hollow-Fiber Systems

The attractive properties of hollow fibers have stimulated great interest, so that considerable development work is under way. A description of Du Pont's hollow-fiber configuration was given in the section on Gas Separation. There, the available information on the hollow-fiber material and the fiber-containing permeation unit were covered. Pertinent information on the nature of the fiber was presented in Chapter XIV. The Permasep equipment, however, was originally developed for reverse-osmosis applications, and the separator arrangement differs from that of the gas permeator. An essential difference is that the high-pressure feed stream goes to the outside of the hollow fibers. Thus, a cut-away view of RO assembly is shown in Fig. 15.12. It is evident that hollow-fiber assemblies, in general, will be variations of this design.

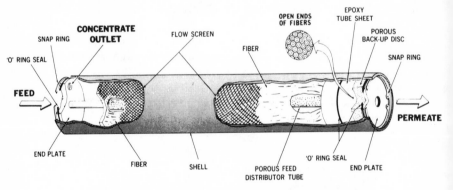

CUT AWAY DRAWING OF PERMASEP* PERMEATOR

Fig. 15.12 Schematic of Du Pont hollow-fiber assembly, Permasep unit, for reverse-osmosis use. (Courtesy Du Pont Company.)

The development work on cellulose acetate (CA) hollow fibers by Monsanto also was covered in Chapter XIV. As mentioned previously, cellulose acetate compositions are subject to hydrolytic deterioration. However, Monsanto's experience shows that a CA fiber unit operating at Webster, South Dakota, has already given satisfactory performance for over 40 months. A definite advantage of CA is that it permits a higher flux rate, that is, about 3 gal/(ft^2) (day) compared to about 1.5 for the B-9 fiber in existing installations. This differential shows up in the operating pressures. Thus, the CA modules are operable at 250 to 350 psi, whereas the polyamide units require close to 400 psi to attain the same flux capacity. OSW research results [27] indicate, however, that the aromatic amide membranes may reach flux capacities in the 4-to-5-gal/(ft^2)(day) range when operating at 1500 psi. A description of a fairly large-scale CA module is presented subsequently with the specific examples.

Module Sizes

Considerable effort is under way to improve the effectiveness of membrane modules so that a fewer number of cartridges can be used in commercial installations. A subsequent figure, Fig. 15.14, shows that a large number of cartridges has to be installed for even a moderate capacity. Thus, present and future designs are looking for an appreciable degree of scaling-up in the cartridge size.

A similar development is being accomplished with the hollow-fiber configuration. The cellulose acetate hollow-fiber modules, which have been under test so far, have risen to about 2500 gal/day capacity. Other commercial-size modules have capacities in the 1500-to-2000-gal/day range, and some as high as 7500 gal/day. A prototype that has been installed in a test facility is a CA fiber system with a nominal capacity of 20,000 gal/day. This is a substantial degree of scaling-up, and the instllation is discussed in some detail later in this chapter.

5 LARGE-SCALE OPERATIONS

The status of a process is best expressed by the attention it receives for large-scale utilization. The membrane processes that are being operated at respectable pilot size and as large installations are reverse osmosis, electrodialysis, dialysis, and ultrafiltration. Representative examples of such installations are presented herewith.

Reverse Osmosis

Most of the reverse-osmosis plants in existence are operating as municipal water-supply systems. The field of municipal water supply is an area where the reverse-osmosis process is already well accepted, and considerable growth is to be expected. The process lends itself to use on a relatively small scale, for moderate-size communities, and is easily scaled-up to even very large installations by simply increasing the number of modules. The specific examples selected are plants using a Permasep hollow fiber, a tubular cellulose acetate membrane, and a spiral-wound cellulose acetate configuration. A large-capacity module with cellulose acetate hollow fibers is also described. Although the latter represents only a field test installation, it has important implications from an engineering view point.

Du Pont Permasep Installation

An example of a reverse-osmosis plant that supplies the needs of a municipality is the 150,000-gal/day Permasep installation at Greenfield, Iowa: population about 2200 people. Operating results for the first four months of 1972 were reported by Moor [33]. Since such installations, through not of industrial scale, will be of interest to many communities it is worthwhile to quote some of the observed results.

Table 15.5 Operating Details of RO Desalting Plant, Greenfield, Iowa
(January 1 - April 30, (1972)

Hours operated	1,578
Total elapsed hours	2,904
% of time operated	54.3
Gallons of product	10,005,210
Gallons of feed	14,831,870
Gallons of reject	4,826,660
Guaranteed capacity	8,630,082
Guarantee exceeded by	15.9%
Conversion	67.5%
Feed-water temperature	87°F or 30.6°C

Thus, actual operating data reflecting plant capacity and solids removal are listed in Tables 15.5 and 15.6. The guaranteed total dissolved solids content of 300 ppm shown in Table 15.6 is not unusually high and actually representative of many municipal waters. Total costs, operating amortization for a 54.3% load factor (the unit is designed as an auxiliary to regular water plant) were experienced as 77¢/1000 gal. Obviously, operation at 100% load factor will almost halve this cost.

The flow sheet for the Greenfield installation is pictured in Fig. 15.13. The complete installation occupies about 500 ft^2 of floor space. The compactness of the multitubular arrangement, allowing for interconnected Permasep modules, can be seen from Fig. 15.14. The capacity of such reverse-osmosis plants is handled very simply by providing the required number of modules. In the present example the module size is the 4-in. permeator, but this size can also be adjusted. For instance, an 8-in. module is available with a design capacity of 7500 gal product water/day compared to the 2500-gal/day capacity of the 4-in. module. In general then, capacity of installation is attained by multiple

Table 15.6 Greenfield, Iowa Water Analysis — Dissolved Solids (ppm)

	Pretreated Feed	Actual Product	Guaranteed Product
Calcium	150	1.4	4.5
Magnesium	45	0.4	1.5
Sodium	474	38	96
Potassium	24	1.8	3.0
Bicarbonate	81	65	67.5
Sulfate	1125	17.5	105.5
Chloride	335	17.2	30
Fluoride	1	0.2	0.9
Nitrate	0.9	0.06	.37
Silica	7	0.05	3.0
Iron	2.2	0.05	0.05
Manganese	0.02	0.02	——
Phosphate	16.2	0.18	——
Conductance	3000	220	——
pH	5.7	7.5	——
Hardness	560	5.0	18.0
TDS	2250	142	300

use of modules where the feed flows in parallel to all of the units. Obviously, if needed, a variety of flow schemes can be used as commonly practiced in separation cascades.

Permutit Installation

Some of the reverse-osmosis installations are quite substantial. Thus, the municipal water supply at *Rotonda West* in Florida is provided by a plant that processes 500,000 gal/day. The appearance of the tubular batteries installed at Rotonda West would be very much like that of the Permasep rack shown in Fig. 15.14. The membranes are of tubular configuration, Du Pont B-9.

The desalting unit was installed by the Permutit Company and initially provides potable water for 3500 people. Expectations are that population growth will result in some 20,000 residents within five years, and this will call for expansion of plant capacity to 2,000,000 gal/day [34]. The feed stock to the unit runs about 7000 ppm, and the product water is in the 250-ppm range. It is hoped that the cost of water will reach 50¢/kgal on the expanded capacity.

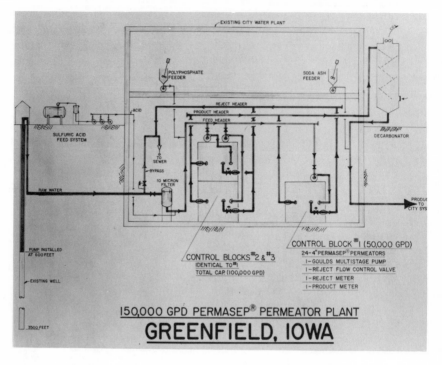

Fig. 15.13   Flow diagram of Greenfield Municipal Water Plant: reverse osmosis. (Courtesy Du Pont Company.)

Gulf Environmental Installation

Another installation of major size is the reverse-osmosis plant at the Ocean Reef Club, Key Largo, Florida. The system employs the ROGA modules of Gulf Environmental Systems using spiral-wound cellulose acetate membranes. It produces 630,000 gal/day of drinking water with 300 ppm dissolved solids from a feed stock of 6600 ppm. This capacity makes it the largest RO plant in the United States at the present time. However, what may be the largest such plant in the world is the 1,400,000-gal/day ROGA installation at Kashima, Japan.

Monsanto Cellulose Acetate Hollow-Fiber Module

What appears to be the largest single module has been installed at a field test facility in Roswell, New Mexico. It is a hollow-fiber configuration developed by the Monsanto Research Corporation and has a nominal capacity of 20,000 gal/day. Evidently, such configurations are of great interest as they will reduce

Fig. 15.14 Cartridge rack of Greenfield Reverse Osmosis Unit. (Courtesy Du Pont Company.)

the number of cartridges for a large-capacity plant. The specifications for the module are listed in Table 15.7.

Ultrafiltration

There are a wide variety of applications of this process. Of course, the principles are the same for all installations, although engineering details will vary with different specific requirements.

Table 15.7 Design and Field Data of 20,000-gal/day Unit Hollow-Fiber Module RB002, Monsanto Research Corporation Installation at Roswell, New Mexico

Hollow Fibers

Cellulose acetate, Eastman 39.4 nominal acetate content, 45 mol wt. index, dry jet-wet spun as five-filament yarn; mean dimensions in working module, 309 μ o.d., 114 μ i.d., intrinsic performance 4.9 gal/(ft^2)(day) with 94.8% rejection of 3000-ppm NaCl at 250 psig, high flow, short fiber testing at 70°F

Cartridge

Parallel fiber array with radial flow of feed through center distribution pipe; 230,000 fibers in bundle cross section, 10-ft active fiber length, 7400 ft^2 fiber area (o.d. basis)

Module

Schedule 20, nominal 8-in. coated carbon steel pipe accommodating fiber cartridge, with Victaulic end fittings; packing factor of fiber in pressure vessel is 52%; internal volume of pressure vessel-active fiber portion is 3.60 ft^3

Test Loop

Triplex positive displacement pump, cartridge filters, connecting piping; instruments and controls are assembled with module on a single skid

Performance Data

Startup was 7-14-72 on 3600-ppm mixed salts feed at Roswell, N. M. Performance at Day 40 (8-22-72) was as follows:

feed pressure	300 psig
shell pressure drop	45 psi
recovery	33%
productivity	22,570 gal/day at 70°F
rejection	89.5%
flux and rejection decline	none

Whey Processing

A substantial unit capable of processing 300,000 lb/day of cottage cheese whey was installed in 1972 by *Abcor* at the Crowley Milk Company plant in LaFargeville, New York [35, 36]. The flow sheet shown in Fig. 15.15 indicates that the liquid-phase processing portion of the plant is actually a combination ultrafiltration and reverse-osmosis procedure. The raw whey is first processed in the UF unit, where it is subjected to a 10- to 30-fold concentration. When 300,000 lb of whey liquor go to the UF unit, the UF concentrate stream amounts to some 25,000 lb. In the operating procedure where UF permeate goes to the RO unit (the scheme portrayed in the diagram), the permeate from the RO unit will become 22,000, and this results in the waste water of low

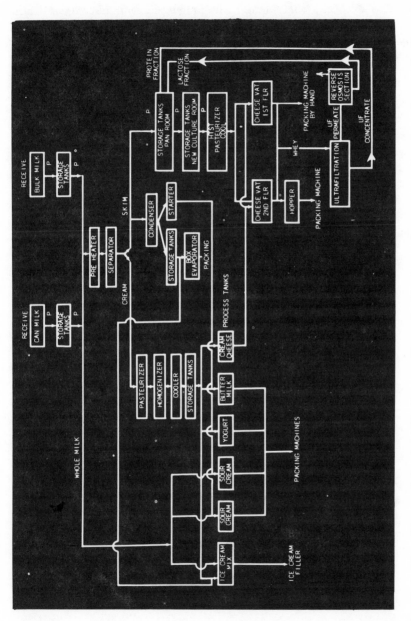

Fig. 15.15 Flow diagram of Crowley Milk Campany Plant; ultrafiltration and reverse osmosis units at lower right (Courtesy Abcor Inc.)

BOD value. The 3000 lb of concentrate will be the lactose fraction, that is recycled to the Pan Room storage. The protein concentrate contains a high proportion of proteins, 50 to 75%, and the lactose concentration depends greatly on the water removal. Thus the protein/lactose ratio changes from about 0.2 at 75% water removal to as high as 40 at 98% removal.

The ultrafiltration membrane used is cellulose based with relatively large pores and retains only the larger molecules, that is, whey protein. The composition of the membrane is not disclosed. Operation is in the 10-to-100-psi pressure range with temperature varying from 60 to 130°F depending on whey properties. The UF system is made up of six Abcor UF-480S modules. Each module contains 216 tubular membrane cores, 10 ft long and 1 in. in diameter. A battery of modules consisting of six cabinets connected in parallel flow is used in batch operation.

The UF permeate, or whole whey liquor, can be fed to the *reverse osmosis* unit, which operates at 600 to 1500 psig and at temperatures of 60 to 100°F. A similar cellulose acetate-type membrane is used in the RO unit, but it retains a high proportion of organic solutes, so that the permeate from this unit has an adequately low BOD for waste disposal.

Paint Processing in Electrocoating

A number of ultrafiltration installations are already in existence in the electrocoating industry. A rather successful example is that of a *Romicon* system

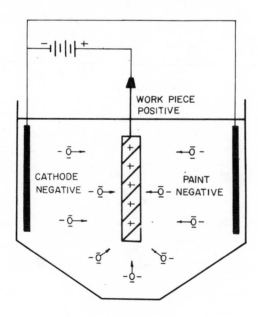

Fig 15.16 Schematic of electrocoating principle [32].

at the Equipto Corporation, Aurora, Illinois [32].

Electrocoating is a process where specially prepared water-based paints are "plated" on metallic substrates when a current is passed through the system. Typically, the metallic substrate serves as the anode, Fig. 15.16. The paint particles, negatively charged, migrate to the anode, and plate out at the surface. An accurately controlled, uniform film is deposited over all conductive surfaces. As the film deposits, it increases the electrical resistence, thereby slowing the deposition until the substrate is removed from the tank. When the substrate leaves the tank, it carries with it not only the electrodeposited paint but an additional film of adhering paint called "dragout." This highly porous coating is washed off in the rinsing sections of the electrocoating system.

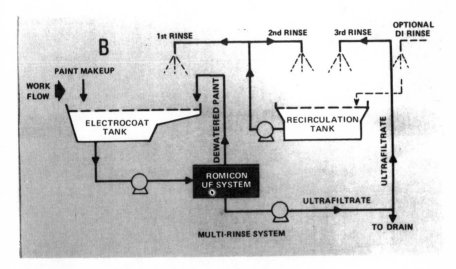

Fig. 15.17 Flow diagram of paint circulation in electrocoating [32].

Ultrafiltration, when used in an electrocoating system, performs two major tasks. Solvent used to keep the paint in suspension can build up in concentration. Since the solvent passes through the membrane, ultrafiltration can perform a kidney function on the tank. Second, the dragged-out paint washed off the plated substrates can be totally recovered using ultrafiltration in a closed-loop operation, Fig. 15.17. The paint tank is tapped, and pressurized paint solution flows through the ultrafiltration system. Dewatered paint returns to the electrocoat tank, but ultrafiltrate is used for substrate rinsing. By proper sizing of the ultrafiltration system, the paint concentration in the recirculation tank can be maintained below 0.5% paint solids. This closed-loop system only requires addition of paint solids for makeup and deionized water for evaporation losses. The ultrafiltration unit is built up from Romicon modules.

Sewage Beneficiation

An application that may have a promising future for domestic installations is exemplified by a Dorr-Oliver *Iopor* unit now in operation at a tourist facility at Pikes Peak (elevation 14,100 ft). It consists of ultrafiltration equipment that is coupled with an activated sludge sewage-treatment unit [37].

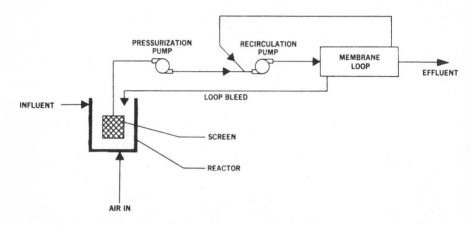

Fig. 15.18   Flow scheme for raw sewage treatment by ultrafiltration [37].

Fig. 15.19   IOPOR unit of Dorr-Oliver for sewage beneficiation. (Courtesy Dorr-Oliver.)

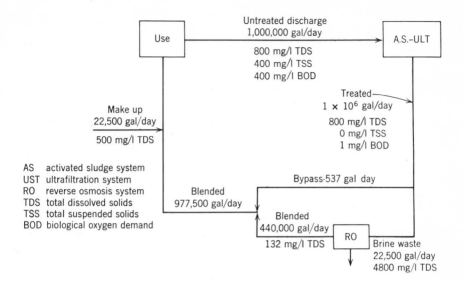

Fig. 15.20 Material-balance diagram for ultrafiltration of sewage [37]; see Fig. 15.19. AS: activated sludge system; SLT: ultrafiltration system; RO: reverse-osmosis system; TDS: total dissolved solids; TSS: total suspended solids; BOD: biological oxygen demand.

The flow diagram of Fig. 15.18 shows how screened sewage liquor is continuously pumped through a membrane loop, and the effluent becomes suitable for *nonpotable*-water recycling. The system consists of 32 modules giving a total membrane area of 1920 ft^2. Operation has been continuous through the last three summers, yielding a water flux of 8 to 11 gal/(ft^2)(day). A view of the complete UF unit is presented in Fig. 15.19. Engineering analyses have shown that the addition of a reverse-osmosis unit to the UF system would provide *potable* water. The material-balance diagram in Fig. 15.20 is based on the use of an Iopor UF system and a Du Pont Permasep RO operation in series.

Pharmaceuticals Processing

There are many uses of ultrafiltration in the pharmaceutical industry. It is difficult, however, to obtain detailed information on specific applications. Evidently, much of the processing involves valuable in-house experience and thus leads to reticence in disclosing hard-won knowledge.

This situation even seems to carry over to the customer service situation. Thus, for instance, two examples of enzyme liquor processing for customers were supplied to us by *De Danske Sukkerfabrikker* [38] where the intended plant processing and product compositions were unknown. DDS carried out pilot studies in their Model 20 type apparatus. Scaling up to manufacturing-size capacity was ultimately achieved. In both cases a substantial degree of concentration was desired. The heat sensitivity of the enzymes precluded the use of evaporation, and ultrafiltration became the preferred procedure. Specifically, *pectinase* and *amyloglycosidase* liquors were processed.

In the case of *pectinase* the starting material was a solution containing 58% dry solids. This solution was diluted to 19% dry solids and then processed by ultrafiltration. A high degree of refinement was achieved, yielding a product of excellent color and reduced extraneous solid matter. Production results were 4.9 liters of concentrate from 37 liters/(hr charge)(m² membrane surface). Dry solids concentration was raised from 19 to 35% and enzymatic activity was increased 7.5-fold.

The *amyloglycosidase* material was received as a 13.8% dry solids solution and was known to contain considerable foreign matter. One reason for substantial concentration of the liquor was the necessity of shipping the finished enzyme preparation in refrigerated containers. Much of the foreign matter and almost all of the coloring substances were eliminated by UF processing. Production results were similar to those achieved with the pectinase. Thus, 3.9 liters of concentrate containing 32.8% dry solids were produced from 37 liters of the raw liquor per square meter of membrane area. Enzymatic activity was increased by a factor of 9.4.

Table 15.8 Ultrafiltration Modules Available from De Danske Sukkerfabrikker[a]

Module Type	Effective Membrane Area (m²)	Max Operating Pressure (atm)	Height (cm)	Diameter (cm)	Approx. Weight (kilos)
40-28	28	42	204	40	510
40-7	7	42	75	40	170
20-5.4	5.4	100	194	20	110
20-1.8	1.8	100	91	20	60
20-0.36	0.36	80	64	20	55

[a]Reference 38.

The processing of such materials on a plant scale is carried out in several size modules as shown in Table 15.8. The internal arrangement of the modules is that presented in Fig. 12.13. Single modules, Fig. 15.21, are combined into multiple-unit arrangements for plant installations.

The problem of concentrating the desalting of biological preparations is, of course, a rather common problem and receives continued attention. Thus, the results of an extensive study carried out by Hancher and Ryon [39] will be of interest, even though the operation was conducted on a laboratory scale

Fig. 15.21 Production-size ultrafiltration unit Module 40-28; see Table 15.8. (Courtesy De Danske Sukkerfabrikker.)

only. The findings are reported in an ORNL publication and cover ultrafiltration experiments with transfer ribonucleic acid, bovine serum albumin, chymotrypsinogen, and catalase.

A number of examples of the commercial use of reverse osmosis and ultrafiltration are described in a recent Symposium volume entitled *Water-1972* [40].

Dialysis

A good coverage of industrial-scale dialyzers and examples of dialysis of acids and alkalies is presented in Tuwiner's book on *Diffusion and Membrane*

Technology [41]. Additionally, equipment views and a listing of applications are given in Rickles' book [2]. Also, a treatise on dialysis by Riggle [42] contains pertinent information and a brief account of caustic recovery in cellulose processing.

Most applications on the industrial scale seem to be concerned with treatment of waste streams and recovery of acids from metallurgical liquors. It is generally known, however, that the pharmaceutical industry does utilize dialysis procedures in many instances. Many such applications are probably on a rather small scale. In any case, whatever use is made of the process seems to be difficult to ascertain, and details are often not disclosed.

Dow Mini Plant Dialyzer

Information for a membrane desalting operation on the multiliter scale was supplied by The Dow Chemical Company. The equipment used was Dow's Mini Plant unit c/HFD-15, which is a hollow-fiber device of shell-and-tube design. It contains 15 ft² of nominal transfer area and uses a cellulosic membrane. If sterility in processing is needed the operation can be achieved by employing a presterilized closed system.

Table 15.9   Operating Data for Dow Mini Plant Unit in Dialysis

Conditions		
	Circulation rate through fibers	200 ml/min recycled
	Circulation rate through jacket	800 ml/min
	Pressure	10-12 psi
	Protein concentration	$\leqslant 1.0\%$
	Starting NaCl concentration	$0.5M$
	Temperature	0-4°C
	Volume	40 liters
Results		
	Final NaCl concentration	$0.003M$
	Final protein concentration	Same as starting
	Final volume	Same as starting
	Time	4.5 hr

Operating conditions for desalting of a protein solution are shown in Table 15.9, and the flow scheme is pictured in Fig. 15.22. As indicated, the dilute protein solution is circulated through the fibers, and the dialysate, which is distilled water, passes around the fibers in a single-pass operation. With the particular miniplant used, one unit of 15 ft² was connected in series for each 8 liters of product volume. This setup permitted the dialyzing of 16 to 50 liters for salt removal. For permanent plant installations, units containing up to 250 ft² are available.

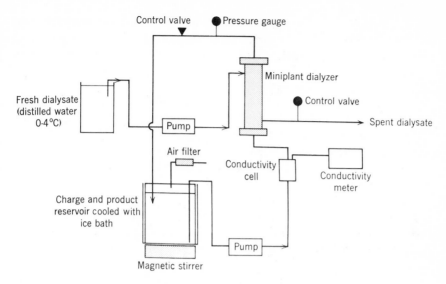

Fig. 15.22 Flow diagram for dialysis with Dow Mini Plant Dialyzer.

Some specific comments pertaining to the system are as follows. Solutions containing proteins and/or salts exert an osmotic pressure during dialysis that leads to an increase in the volume of the solution being dialyzed. With the indicated flow scheme this can be controlled by applying pressure across the fibers. However, when the volume is not critical the increase in volume may be tolerated. If the solution being dialyzed is in a closed system, the installation of an air filter to the product reservoir will prevent a vacuum (seen with decreased volume) and increased pressure (seen with an increase in volume). The utilization of a standardized in-line conductivity cell permits continuous monitoring of the salt concentration. In a closed system this eliminates breaking the system for periodic sampling. Temperature plays an important part in the efficiency of the dialysis membrane. In the example the dialysis time would be greatly decreased at room temperature.

Electromembrane Processes

It appears that the only electromembrane process that has found extensive industrial application is that of electrodialysis. Consequently, the examples will have to be limited mostly to this area. However, a description of some work with forced-flow electrophoresis is available and it is judged to be of sufficient interest to cover it also.

Electrodialysis in Desalination

The electrodialysis process also has found numerous opportunities in water treatment, especially for desalination of brackish water. In this respect it is a strong competitor to reverse osmosis. Some installations are in existence where potable water is produced from sea water, which contains 3.5% of salts, mostly NaCl. The majority of applications, however, are in treating raw water where dissolved solids are something like 10,000 ppm. Unless drinking water is loaded with about 400 to 600 ppm of *NaCl*, it will not taste "salty" to most people. Organoleptic tests [43] indicated that there was a rather wide range of threshold values; thus, the 400-to-600-ppm limits represent only an approximation. However, the aim is usually to reduce total parts per million to a value of not more than 250 ppm (recommended standard is 500 ppm maximum), and at that level no objectionable taste from dissolved solids should be encountered.

What is presently the largest electrodialysis installation in the United States for municipal water supply, is the *Siesta Key,* Florida plant having a capacity to produce 1,500,000 gal/day. The plant is an *Ionics* unit and processes brackish water. The ED equipment is contained on a concrete pad 40 by 100 ft, with a head room of 12 ft [34].

Citrus-Juice Processing

What should be an appealing and perhaps even tasty example is the use of electrodialysis to reduce acidity in orange juice or citrus juices in general. A very fine description of the process was given by Zang [44, 45] and in AMF company bulletins. The ionic flow scheme is shown in Fig. 15.23 as written for citrate ion removal. As such, the diagram is, of course, identical with Fig. 9.4 in Chapter IX, but whereas that diagram is drawn for both cation- and anion-exchange membranes, the citrate removal calls for anion membranes only.

The reason for the use of electrodialysis in this food-processing application is, of course, due to potential economic advantages gained over other alternatives. Orange juice, for instance, becomes sweeter as the growing season progresses, or, expressed in terms of acidity, the citric-acid equivalent decreases towards the end of the growing season. A high acid content is objectionable to the majority of consumers. There are, of course, other means to overcome high acidity, such as blending-off early-season high-acid material with late-season high-sugar-content juice, or the addition of sugar to reduce tartness. Clearly, blending will require storing of sweet juice, and this is often a difficult matter. Sugar addition is a simple operation, but it results in a lower market value of the commodity. So the electrodialysis process turns out to be an eminently suitable means to produce a seasonal food product at consistent quality throughout the year at an economical market value.

As is the case with many food products, certain requirements must be met in design to handle a liquid that contains a fair amount of pulp. This required some careful consideration to ensure continuous operation of what is termed a sanitary stack. The manner in which this was accomplished is presented in a concise statement prepared for the AMF bulletins by Smith et al. [45].

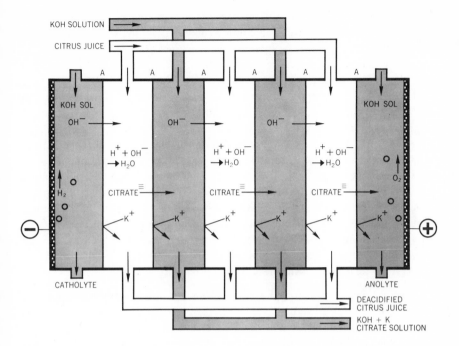

Fig. 15.23 Ionic flow scheme for citrus-juice processing. (Courtesy AMF, Inc.)

"Inasmuch as citrus juice contains pulp, it was necessary to devise a scheme whereby membranes could be kept apart, yet avoiding the use of a separator which would collect pulp and cause plugging. This was one of the more important considerations in the design for handling such suspensions."

"In the sanitary stack diagram, the flow paths of the juice must be streamlined so that pulp removal can be successfully achieved to sanitize the stack, and so that the operating stack will have to be disassembled only occasionally for complete scouring."

"To accomplish this, the electrodialysis stack was designed placing the screen-type separators in the clear alkaline streams only, leaving the juice chambers clear of all obstructions. By control of the relative pressures in the two chambers, that is, by keeping the pressure in the liquid path greater than in the separators the membranes are pushed against the screen grids."

"This leaves the juice chambers open and clear of all obstructions. This device was effective in eliminating the hang-up problem. The low rates of bacterial growth determined in this system were quite satisfactory, proving the effectiveness of the concept."

"Another requirement was that the unit be versatile enough to handle all the single strength juices normally processed and, in addition, be able to treat the various concentrates produced commercially. The process was also designed to provide simple, continuous operation with direct control of the amount of acid removal. A simple voltage control device was used to achieve this flexibility in the electrodialysis process."

"Almost all of the components in orange juice are soluble solids. In addition to the ionic character of some of the soluble solids, the pulp and fiber carry charges which can, and do, affect the operation of the anion membranes. We have observed that the resistance will double, or even quadruple, in an hour if the current flow is unidirectional. However, if a periodic current reversal is utilized, resistance changes can be kept at a minimum. Since constant voltage rectifiers are generally used (because of low cost), this is obviously an important consideration. The resistance increase, we believe, is related to ionic colloidal materials in the citrus juices. Fortunately, with a regular current reversal cycle, the resistance, and thus the power increase, is kept well under control. The accumulation of interfering ions, which tend to aggregate on the membrane surface, is eliminated when the current flow is reversed."

Obviously, the discussion dealing with fouling of the membranes and the current reversal for cleaning is a matter of concentration polarization as discussed in Chapter XIII. It represents a striking example of the importance of hardware engineering, even when process principles signify a favorable reaction scheme.

Table 15.10   Citrus Pilot-Plant Data; Material Treated: Grapefruit Juice

Feed temp	$92°F$
Feed acidity	1.52%
Product acidity	0.90%
Production rate	95 gal/hr
Cell velocity, product	0.3 ft/sec
Cell velocity, KOH	0.1 ft/sec
Voltage	167 V
Current	122 A
Current density	13 A/ft^2
Current efficiency	70%
d.c. energy comsumption	0.22 kWhr/gal

Some operating data from the pilot plant used to demonstrate the utility of the process with grapefruit juice are reported in Table 15.10. The tests were carried out at Lake Wales, Florida. (Florida Citrus Canners Corporation), and a picture of the unit is shown in Fig. 15.24. At the right is the electrodialysis stack,

Fig. 15.24 Semiplant unit for citrus-juice processing. (Courtesy AMF, Inc.)

which contains 12 cell pairs each with 9.5 ft^2 of effective membrane area. The capacity of this prototype installation was 5.3 lb of citric acid transferred per hour, which represents a throughput of 100 to 300 gal/hr of citrus juice, depending on desired acid removal.

Forced-Flow Electrophoresis

The uses of this process are presently rather limited. Bier [46] mentions that isolation and concentration of gamma globulin from serum is being conducted on a large volume scale. However, an essential part of Bier's suggested use of electrophoresis deals with its application in treating sewage to remove suspended colloidal impurities, such as algae, bacteria, and silt, as well as dissolved colloids.

Bier describes the apparatus as being "essentially a filter press with an electrical current applied across the filter elements." The application of the current prevents the clogging of the filters. The process is a great deal like electrodialysis, and the difference is primarily in the construction of the cell; see Fig. 9.1. In electrophoresis the cell spacers, which distribute the flow of liquid, are separated by an alternate sequence of membranes and filters, rather than by the ionic-type membranes used in electrodialysis. Thus, neutral cellophane membranes are most commonly used, and filter paper, microporous polymerics, as well as Millipore or Gelman filters serve as the nonmetallic filtering medium. In this arrangement the filters will readily pass water and most colloids, and the membranes allow passage of electrical current and small ions, but not of colloids.

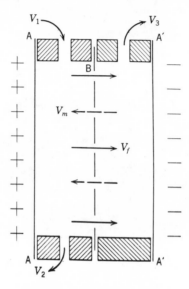

Fig. 15.25 Schematic for forced-flow electrophoresis cell [46]. A and A': membranes; B: filter.

The diagram in Fig. 15.25 illustrates the principle of operation. Shown is a cross section of a single cell, where A and A' represent membranes, with the filter B in between. Colloid-carrying solution enters at V_1 and leaves at V_2. Part of the stream V_1 is forced through the filter by an imposed pressure gradient and exits at V_3. The electrical field, perpendicular to the filter, induces negatively charged colloids to move opposite to the liquid flow V_f (solid arrow), so that an electrophoretic migration V_m (broken line) is created. Clearly, if V_m is equal to or greater than V_f, negatively charged colloids will not pass the filter, and the V_3 stream will contain only positively charged or isoelectric colloids.

It would be highly injudicious to consider the cited examples as more than a "current state of art" development. Quite likely, by the time the manuscript appears in print some examples may be on the way to obsolescence. This is particularly likely in regard to scale of operation. For instance, desalting plants are already being thought of in terms of multimillion gallons per day. This situation holds in a very general way for all liquid-phase processes.

One prediction can be made with a reasonable degree of safety. It is that polymeric nonporous membranes are likely to experience a high degree of use in gas and vapor separation. Microporous media will find application only to limited extent in special situations or in separation of gas mixtures having very large separation factors.

References

1. "Membrane Processes for Industry," Proc. Symposium May 19-20, 1966, Southern Research Institute, Birmingham, Ala.
2. R. N. Rickles, *Membranes, Technoloyg and Economics*, Noyes Dev. Corp., Park Ridge, N.J., 1967.
3. U. Merten, Ed., *Desalination by Reverse Osmosis*, M.I.T. Press, Cambridge, Mass., 1966.
4. S. Sourirajan, *Reverse Osmosis*, Academic, New York, 1970.
5. R. E. Lacey and S. Loeb, *Industrial Processing with Membranes*, Wiley-Interscience, New York, 1972.
6. K. Kammermeyer, "Gas and Vapor Separations by Means of Membranes," *Progress in Separation and Purification*, E. S. Perry, Ed., Vol. 1, Wiley-Interscience, New York, 1968, p. 335.
7. J. E. Jolley, U.S. Patent 3,172,741 (1965).
8. W. D. Lewis and S. A. Rogers, U.S. Patent 3,198,335 (1965).
9. Du Pont Co.'s Permasep Technical Bulletins 105 and 110, 1972.
10. L. M. Litz and G. E. Smith, Union Carbide Res. Inst. Rept. UCRI-701 (Sept. 1972).
11. S. A. Stern, T. F. Sinclair, J. J. Gareis, N. P. Vahldieck, and P. H. Mohr, *Ind. Eng. Chem.*, **57**, (2) 49 (1965); S. A. Stern, U.S. Patent 3,332,216 (1967).
12. News item, *C&EN*, **50**, (42) 11 (1972); *Chem. Eng.*, **79**, (24) 40 (1972).
13. R. B. McBride and D. L. McKinley, *Chem. Eng. Progr.*, **61**, (3) 81 (1965).
14. T. F. Meinhold, *Chem. Proc.*, **12**, 66 (1965).
15. J. B. Hunter, U.S. Patent 2,773,561 (1956).
16. J. B. Hunter, *Platinum Met. Rev.*, **4**, 130 (1960); also a paper presented at A.C.S. Mtg, New York (Fall 1963), Petroleum Div.
17. J. B. Hunter and G. M. Hickey, U.S. Patent 2,961,062 (1960); G. C. Pinney and H. G. Hemphill, U.S. Patent 3,226,915 (1966).
18. P. M. Roberts, *Chem. Ztg.*, **95**, 780 (1971).
19. Mathey Bishop bulletin.
20. J. B. Hunter, U.S. Patent 3,254,956 (1966).
21. J. T. Mullhaupt, U.S. Patent 3,359,705 (1967).
22. F. J. Norton, *Gen. Elec. Rev.*, September (1952); *J. Amer. Ceram. Soc.*,

36, (3) 90 (1953); (with A.U. Seybolt) *Trans. AIME,* **230**, 595 (1964).

23. K. B. McAfee, "Diffusion Separation," in *Kirk-Othmer Encyclopedia of Chemical Technology,* 2nd Suppl., A. Standen, Ed., Wiley-Interscience, New York, 1960, p. 297; *Bell Lab. Record,* **39** Oct. (1961); Staff reports; *Chem. Eng. Progr.,* **54**, (6) 106 (1958); *C&EN,* **36**, (19) 42 (1958).

24. N. Melnyk and H. W. Habgood, *Can. Mining Metall. Bulletin,* Oct., 768, (1961).

25. Personal communication.

26. K. Kammermeyer and D. H. Hagerbaumer, *A.I.Ch.E. J.,* **1**, 215 (1955).

27. Office of Saline Water, U.S. Dept. of Interior, Saline Water Conversion Reports 1971-72 and 1972-1973; yearly reports since 1952.

28. R. L. Miller, *Chem. Eng.,* **79**, (19) 24 (1972).

29. Dorr-Oliver, Inc., Bulletin Nos. 10-2 and 10-3.

30. Gulf Environmental Systems, Bulletins.

31. Abcor, Inc., Bulletins.

32. Romicon, Inc., Bulletins.

33. D. H. Moore, "Operation of a reverse osmosis desalting plant at Greenfield, Iowa," presented at 92nd Annual Conf., Amer. Water Works Assn, Chicago, Ill., June 4-8, 1972.

34. Dept. Interior News Release, Dec. 3, (1972).

35. B. S. Horton, R. L. Goldsmith, and R. R. Zall, *Food Tech.,* **26**, (2) 30 (1972).

36. R. Zelitzer, *Dairy Ice Cream Field,* **155**, (6) 34 (1972).

37. I. Bemberis, P. J. Hubbard, and F. B. Leonard (Dorr-Oliver, Inc.), "Membrane Sewage Treatment Systems Potential for Complete Wastewater Treatment," presented at 1971 Winter Mtg, Amer. Soc. Agricult. Engrs, Chicago, Ill., Dec. 1971.

38. A/S De Danske Sukkerfabrikker, Copenhagen, Denmark, "Reverse Osmosis, Ultrafiltration and Hyperfiltration," catalog.

39. C. W. Hancher and A. D. Ryon, "Concentration adn desalting of biological macromolecular by ultrafiltration," Oak Ridge National Laboratory Report ORNL-TM-3620 (1972).

40. *Water - 1972,* G. F. Bennett, Ed., *A.I.Ch.E. Symposium Series* No. 129, Vol. 69, p. 81, American Inst. Chem. Engrs., New York, 1973.

41. S. B. Tuwiner, *Diffusion and Membrane Technology,* Reinhold, London, 1962.

42. J. W. Riggle, "Dialysis," in *Chemical Engineer's Handbook,* 4th ed., McGraw-Hill, New York, 1963.

43. Departmental Res. Rept, Chem. Eng'g, Univ. Iowa, Iowa City.

44. J. A. Zang, "Sweetening Citrus Juices," in "Membrane Processes for Industry," Proc. Symp., p. 35, Southern Res. Inst., May 19-20, 1966.

45. R. N. Smith, C. T. Hicks, and R. J. Moshy, "The AMF Electrodialysis Process for Citrus Juice"; J. A. Zang, R. J. Moshy, and R. N. Smith, "The AMF Electrodialysis Process for Food."

46. M. Bier, "Forced flow electrophoresis and its biomedical applications," in "Membrane Processes for Industry," *Proc. Symp.,* Southern Res. Inst., Birmingham, Ala., May 19-20, 1966.

Chapter **XVI**

MEMBRANE SEPARATION OF ISOTOPES

The general aspects of isotope separation were discussed in Chapter V. There it was pointed out that the separation of isotopes in principle is subject to all of the phenomena inherent in the gaseous-diffusion process. Theoretical interpretations and engineering calculations are, then, the same as outlined for the various possible flow schemes.

A somewhat different situation arises, however, when the molecular weights of isotopes are very close. Under such conditions the separation factors become very small, that is, they will approach the value of 1. Thus, the previously stated molecular-weight ratio of 1.008 for the UF_6 compounds would yield an ideal separation factor of only $a^* \simeq 1.0043$. In computational procedures the term $(a^* - 1)$ is introduced. For the small differences in molecular weight of the respective isotopes this term approaches a value of zero, and this approximation is then used in the mathematical manipulations.

The subsequent material, therefore, covers the membrane separation of isotopes that exhibit a very small separation factor. In essence, the theoretical developments are based on the separation by means of a microporous membrane. As far as is known, all membrane installations for the separation of isotopes use microporous membranes.

1 FLOW THROUGH POROUS BARRIERS

Isotope separation is achieved by gaseous diffusion through a microporous membrane. This process is based on the phenomenon of molecular effusion, first described by Graham in 1846. A high-pressure vessel with fine holes contains a gas mixture of two isotopes: the lighter isotope will escape faster and more easily from the vessel than the heavier one. Thus, the remaining gas will be enriched in the heavier component. The practical separation based on this principle is important in the development of nuclear energy.

This method was first used in practice in 1920 by Aston [1] for the separation of the isotopes of neon through a porous clay tube. Consecutively, Hertz [2-4] and Woolderidge, Jenkins, and Smythe [5, 6] used the process for isotope separation of neon, hydrogen, and methane. The most outstanding applications of gas diffusion were accomplished by the U.S. Atomic Energy Commission in order to enrich $^{235}UF_6$. Partial descriptions have been given in the Symthe Report [7], and by Hogerton [8] and Keith [9]. At present, the AEC operates three plants with large separation capacity (17,000 tons/yr).

Mean Free Path

The nature of the flow through a porous medium depends on the ratio between the mean free path λ and the pore diameter (see Chapter III). The mean free path may be estimated from the fundamental kinetic theory and Newton's law of viscosity for rigid spherical molecules at low density [10]:

$$\lambda = \frac{3\mu}{2P} \left(\frac{\pi R'T}{2M'} \right)^{\frac{1}{2}} , \tag{16.1}$$

where P is the average pressure.

The values for some gases under conditions of $20°C$ and 1 atm are given in Table 16.1.

The smallness of the mean free paths necessitates the use of special barriers with extremely fine pores in order to reduce the number of intermolecular collisions.

Porous Barriers

From the point of view of their structure, barriers are characterized by their pore radius and their permeability. The theory shows that the radius msut be of at least the same order as the mean free path. Since the permeability is

Table 16.1   Mean Free Paths

Gas	$\mu(\mu P)$	λ (Å)
A	222	1017
H_2	87	1775
He	194	2809
N_2	173	947
Ne	311	2005
Q_2	203	1039
UF_6	176	279

inversely proportional to the thickness, the barriers must be as thin as their mechanical strength allows. Barriers can be classified as: simple barriers, characterized by a sturcture independent of the thickness, or composite barriers: composed of very permeable support, without separative action, and a thin layer having a small pore radius [11]. They may, for instance, be obtained by sintering of metallic powders or ceramics, or by deposition of a Teflon emulsion on a metallic grid support.

Flow Characteristics

When the mean free path is much larger than the pore diameter d, individual molecules will not collide with each other. The molecular flux F through a circular capillary is from Eq. 3.35:

$$F = \frac{4d}{3l} \frac{(P_1 - P_2)}{(2\pi M'R'T)^{1/2}}. \qquad (16.2)$$

If the conditions are such that $\lambda \ll d$ under a high pressure, the flow will be viscous and obey the well-known Poiseuille law. For circular capillaries, the laminar flow is characterized by the equation

$$F = \frac{d^2}{64\, l\, \mu\, R'T} (P_1^2 - P_2^2). \qquad (16.3)$$

In the intermediate pressure range, Present and Pollard [12] have shown that the flow may be represented as the sum of the above two equations:

$$F = \frac{a(P_1 - P_2)}{(M')^{1/2}} + \frac{b(P_1^2 - P_2^2)}{\mu}, \qquad (16.4)$$

where constants a and b can be determined by experiments. This relationship represents the flow behavior through a porous membrane in the transition pressure range from the Knudsen to the Poiseuille regime. It is important to note that separation is possible only when the flow is in the free molecular regime. Thus, as the portion of laminar flow increases, the efficiency of separation drops until complete Poiseuille flow results, in which there is no separation. A detailed discussion of the flow characteristics at low pressure has been presented by Hwang and Kammermeyer [13, 14].

Ideal Separation

When the operating pressures are relatively low, the mean free path is larger than the pore size, thus yielding Knudsen flow. Further, if the pressure on the low-pressure side is negligibly small, then there will be no backdiffusion from the downstream side to the upstream side of the barrier. These conditions result in a maximum attainable separation, and it is frequently called *ideal separation*. When the conditions of ideal separation are met, the flux equation for each component can be written as

$$F_L = \frac{aP_1 x}{(M'_L)^{\frac{1}{2}}} , \tag{16.5}$$

$$F_H = \frac{aP_1 (1-x)}{(M'_H)^{\frac{1}{2}}} , \tag{16.6}$$

where x is the mole fraction of the light component on the upstream (high-pressure) side. The ideal separation factor a^* is defined by

$$a^* = \frac{F_L/F_H}{x/(1 - x)} . \tag{16.7}$$

Substituting Eqs. 16.5 and 16.6 into Eq. 16.7, the ideal separation factor becomes

$$a^* = \left(\frac{M_H}{M_L} \right)^{\frac{1}{2}} . \tag{16.8}$$

In actual separation processes, the ideal separation conditions will not prevail. However, a fictitious quantity x^* can be defined that satisfies the following relationship for a given value of composition y on the downstream (low-pressure) side:

$$x^* = \frac{y}{y + a^*(1-y)} . \tag{16.9}$$

For mixtures of heavy isotopes, the following simplifications result:

$$a^* - 1 = \epsilon^* \simeq \frac{M'_H - M'_L}{M'_H + M'_L}, \qquad (16.10)$$

$$y - x^* \simeq \epsilon^* \, x(1-x)$$
$$\simeq \epsilon^* \, y(1-y) , \qquad (16.11)$$

where ϵ^* is a very small positive number, that is, $0 < \epsilon^* \ll 1$.

Stage Characteristics

In Chapter V, the performance characteristics of a gaseous-diffusion cell were discussed. Here, the aspects of stage efficiency for various cases will be treated. As shown in Fig. 16.1, V and L represent the flow rates and y and x represent the mole fraction of light component in the diffused and reject streams, respectively.

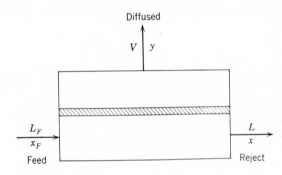

Fig. 16.1 Diffusion cell.

An abundance ratio is defined [15-17] as

$$R = \frac{y}{1-y} \quad \text{or} \quad \frac{x}{1-x}. \qquad (16.12)$$

The stage separation factor is given by

$$a = \frac{y/(1-y)}{x/(1-x)} . \qquad (16.13)$$

Since a is close to unity for most isotopes systems, we obtain a similar result to Eq. 16.11:

$$y-x \simeq \epsilon\, x(1-x) \simeq \epsilon\, y(1-y) \,, \tag{16.14}$$

where

$$\epsilon = a-1 > 0 \,.$$

Some authors [17] use a quantity, called the *heads separation factor* a', defined by

$$a' = \frac{y/(1-y)}{x_F/(1-x_F)} \,. \tag{16.15}$$

It is obvious that a' is always less than a,

$$1 < a' < a \,. \tag{16.16}$$

Frequently, the cut θ is introduced in the discussion of a gaseous-diffusion unit:

$$\theta = \frac{V}{L_F} \,. \tag{16.17}$$

Also, an important quantity to characterize the stage performance is the stage separation efficiency η, which is quite similar to the overall Murphree-plate efficiency in distillation:

$$\eta = \frac{y-x}{y-x^*} = \frac{\epsilon}{\epsilon^*} \,. \tag{16.18}$$

This stage efficiency is actually an integrated quantity of the membrane separation efficiency, which may be called the point efficiency. These two efficiencies become identical only for the perfect mixing model. The interrelationships between these two efficiencies are quite involved algebraically; therefore, they will be omitted here. For the cross-flow model, it has been shown that the stage efficiency is always greater than the point efficiency [15, 17]:

$$\eta = \eta' \, \frac{n(1-\theta)}{\theta} \,, \tag{16.19}$$

where η' is the membrane (point) efficiency. This relationship becomes

$$\eta = 1.386 \, \eta' \tag{16.20}$$

for an ideal cascade, in which $\theta = \frac{1}{2}$. However, the interrelationship becomes more complicated for countercurrent flow, and readers are referred to Weller and Steiner's treatise [18].

The following discussion is valid only for the case of perfect mixing, where the two efficiencies are equal. Using Eq. 16.4, for both components

$$\frac{F_L}{F} = \frac{y}{1-y} , \tag{16.21}$$

and substituting Eqs. 16.9 and 16.21 into Eq. 16.17, the stage-separation factor can be expressed as

$$\eta = \frac{1 - P_r}{1 + \dfrac{(1-P_r)(P_1 + P_2)/P_c}{a^* + (1-a^*)y}} , \tag{16.22}$$

where

$$P_r = \frac{P_2}{P_1}, \quad P_c = \frac{a\,\mu}{bM_H'^{\,\frac{1}{2}}} .$$

The latter is called the characteristic pressure.

Several special cases are of interest. First of all, for most isotopic mixtures $a^* - 1 \ll 1$, thus

$$\eta = \frac{1 - P_r}{1 + (1-P_r)(P_1 + P_2)/P_c} . \tag{16.23}$$

Second, when the dowstream pressure is negligibly small one obtains

$$\eta = \frac{1}{1 + P_1/P_2} . \tag{16.24}$$

Finally, when the ideal separation can be assumed, in other words, when the laminar flow contribution is zero, $b = 0$ or $P_c \to \infty$:

$$\eta = 1 - P_r . \tag{16.25}$$

It is interesting to note that Eqs. 16.22 through 16.25 can be interpreted as the membrane efficiency, and using Eq. 16.19, the overall stage efficiency may be obtained for cross flow. This is possible because the same results would be obtained if a small portion of membrane were to be taken as a limiting case of the perfect mixing model.

2 CASCADE OPERATION

The general principles of cascade operation presented in Chapter XIII are also valid for isotope separation. However, the special situation $a \cong 1$ results in greatly simplified mathematical expressions. It should be noted also that the separation factor a varies very little from stage to stage; therefore, it will be assumed as a constant throughout the entire chapter.

Minimum Number of Stages

For a cascade operation, in which the reject (undiffused) streams are recycled to the previous stages, the total number of stages depends on the amount of reflux. When there is no product taken off, that is, at total reflux, the number of stages would be a minimum for specified separation conditions. Setting $D = 0$ in Eq. 13.2 the operating line equation becomes

$$y_{n+1} = x_n \, . \tag{16.26}$$

Since the separation factor is constant, for any stage

$$a = \frac{y_n/(1-y_n)}{x_n/(1-x_n)} = \frac{R_n}{R_{n+1}} \, . \tag{16.27}$$

This equation is a recursion formula. Therefore, the ratio of the abundance ratios of the top product R_D and the bottom product $R_N = R_B$ can be related by

$$R_D = a^{N-1} R_N = a^{N-1} R_B \, , \tag{16.28}$$

where N is the total number of stages. Thus, the minimum number of stages is

$$N = \frac{\ln \dfrac{R_D}{R_B}}{\ln a} + 1 = \frac{\ln \left(\dfrac{y_D/(1-y_D)}{y_B/(1-y_B)} \right)}{\ln a} + 1$$

$$= \frac{\ln \left(\dfrac{y_D/(1-y_D)}{x_B/(1-x_B)} \right)}{\ln a} \, . \tag{16.29}$$

Since a is close to unity, the denominator $\ln a$ may be replaced by $a-1$. Thus, from Eq. 16.29 a great number of stages are required to enrich heavy isotopes from their mixture. This is well illustrated in the following example.

EXAMPLE 1

Calculate the minimum number of stages required in a UF_6 gaseous-diffusion plant producing 90% ^{235}U and a waste of 0.1% ^{235}U. The composition is in mole fraction. Assume $a=a^*$.

Solution.

$$a = \left(\frac{M_H}{M_L}\right)^{\frac{1}{2}} = \left(\frac{352}{349}\right)^{\frac{1}{2}} = 1.0043 \ .$$

From Eq. 16.29, the total minimum number of stages is

$$N = \frac{\ln\left[\left(\frac{0.9}{0.001}\right)\left(\frac{0.999}{0.1}\right)\right]}{\ln \quad 1.0043} = 2,122 \ .$$

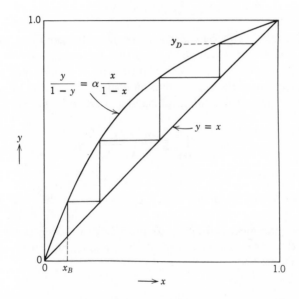

Fig. 16.2 Minimum number of stages.

In any actual separation plant, the reflux ratio is between the minimum and the maximum just as in the case of distillation. Therefore, the actual number of stages required will be greater than the minimum number of stages. In Fig. 16.2, the McCabe-Thiele type of procedure is used to obtain the minimum number of stages. In principle, the same type of procedure is involved in actual separation units, provided that the "operating line" is easily known. However, as discussed in Chapter XIII, the "operating line" is not even a single straight line; instead, it is a curve. Furthermore, no simple conditions exist in general, such as constant cut or constant overflow. However, if a new concept is introduced, namely, the *polytropic enriching* process, it will simplify the problem considerably. Also, the polytropic process can be reduced and identified with already-known special cases like the ideal cascade or operation with total reflux.

Polytropic Enriching

The separation factor a is the ratio of the abundance ratios of the two leaving streams from a stage, that is, the diffused and undiffused streams. In a cascade operation, a quantity β can be defined, similarly, that is the ratio of the abundance ratios of the two adjacent diffused streams:

$$R_n = \beta \, R_{n+1} .$$ (16.30)

This new quantity β is in general not a constant. Its value may change from stage to stage. However, it can be assumed that the quantity β does not change significantly from stage to stage. This will actually be true in special cases, that is, ideal cascade, total reflux (minimum number of stages), and infinite number of stages. Combining the definitions of a and β, Eqs. 16.13 and 16.30 yield

$$\frac{y_{n+1}}{1-y_{n+1}} = \frac{a}{\beta} \left(\frac{x_n}{1-x_n} \right) ,$$ (16.31)

which is the equation for the "operating line." The "operating line" is actually a locus of the operating points (x_n, y_{n+1}), which are the result of material balances.

In Fig. 16.3, the McCabe-Thiele-type procedure is shown. If the value of β is specified, the "operating line" can be drawn using Eq. 16.31. When $\beta = a$, Eq. 16.31 is reduced to

$$y_{n+1} = x_n ,$$ (16.32)

which is the 45° line, and the case is that of total reflux. When β approaches unity, Eq. 16.31 approaches Eq. 16.13. This is the case of infinite number of stages. Finally, when $\beta = a^{\frac{1}{2}}$, it is the case of the ideal cascade, which will be discussed later.

From these special cases, it is easily seen that β in general is limited to values between 1 and a:

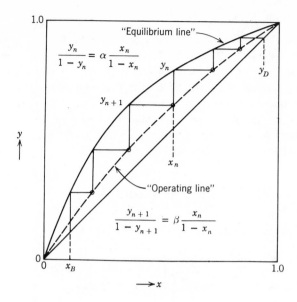

$$\frac{y_n}{1 - y_n} = \alpha \frac{x_n}{1 - x_n}$$

$$\frac{y_{n+1}}{1 - y_{n+1}} = \beta \frac{x_n}{1 - x_n}$$

Fig. 16.3 Polytropic enriching.

$$1 \leqslant \beta \leqslant a. \tag{16.33}$$

For such values of β, the number of stages required can be calculated using the recursion formula, Eq. 16.30, as follows:

$$N = \frac{\ln\left(\dfrac{R_D}{R_B}\right)}{\ln \beta} + 1 = \frac{\ln\left[\dfrac{y_D/(1-y_D)}{x_B/(1-x_B)}\right]}{\ln \beta} + 1 - \frac{\ln a}{\ln \beta}. \tag{16.34}$$

It is obvious that Eq. 16.34 reduces to Eq. 16.29 when $\beta = a$, where the minimum number of stages results. As the value of β decreases, the number of stages increases. Finally, it becomes infinity when β approaches to 1. In between these two extremes, the actual cascade operations are possible, and in particular $\beta = a^{\frac{1}{2}}$ gives the condition of the ideal cascade. These various cases are illustrated in Fig. 16.4 with β as the parameter.

Thus, it is well demonstrated that the treatment of the "operating line" is facilitated by introducing the parameter β. It should also be pointed out that when the compositions of two mixing streams are identical (the case of ideal cascade), β becomes the heads separation factor a' defined by Eq. 16.15.

Total Interstage Flow

To estimate the total power required in pumping, the total interstage flow rate must be calculated. Evidently the total interstage flow rate is a rough measure of the size of the gaseous-diffusion plant. It is also a gratifying result that the ideal cascade is proved to be the true optimum operation. This can be shown by minimizing the total interstage flow for a given separation process.

The derivation of the total interstage flow rate is quite a complicated procedure. Here only an outline, without detailed steps, will be presented. The total interstage flow J_E for the enriching section is

$$J_E = \sum_{n=1}^{N} (V_{n+1} + L_n) . \tag{16.35}$$

Using Eqs. 13.1 and 13.2, this can be re-expressed as

$$J_E = 2D \sum_{n=1}^{N} \left(\frac{y_D - y_{n+1}}{y_{n+1} - x_n} \right) + ND . \tag{16.36}$$

Substituting the recursion formula Eq. 16.30 and carrying out the summation of geometric series yield

$$J_E = \frac{2D}{a-\beta} \left[R_D (y_D - 1) \left(\frac{1 - \frac{1}{\beta^N}}{1 - \frac{1}{\beta}} \right) + (ay_D + \beta y_D - a)N \right.$$

$$\left. + \frac{a\beta \, y_D}{R_D} \left(\frac{1-\beta^N}{1-\beta} \right) \right] . \tag{16.37}$$

Simplifying this equation by using

$$\beta^N = \frac{R_D}{R_F} \tag{16.38}$$

results in the following expression:

$$J_E = \frac{2\beta(1 - a/R_F)D(y_D - 1)(R_D - R_F)}{(a-\beta)(\beta-1)}$$

$$+ \frac{(a+\beta) \, D \, (2y_D - 1)}{(a-\beta) \ln \beta} \ln \left(\frac{R_D}{R_F} \right)$$

Similarly, one can obtain the expression for the total interstage flow for the stripping section,

$$J_S = \frac{2\beta \ B \ (R_B - R_F)}{(a-\beta)(\beta-1)} \left[x_B - 1 + \frac{a \ x_B}{R_F R_B} \right]$$

$$+ \frac{(a+\beta) \ B \ (2x_B-1)}{(a-\beta) \ \ln \ \beta} \ \ln \left(\frac{R_B}{R_F} \right) \tag{16.40}$$

The total interstage flow for the entire cascade is the sum of Eqs. 16.39 and 16.40. However, this leads to a very complicated result, unless further approximations are made. For a large cascade the following approximation can be used:

$$R_B \simeq \frac{x_B}{1-x_B}. \tag{16.41}$$

Then Eq. 16.40 becomes a similar form to Eq. 16.39, and adding these two together, the total interstage flow equation results:

$$J = J_E + J_S = \frac{a + \beta}{(a-\beta) \ \ln \ \beta} K, \tag{16.42}$$

where

$$K = D(2y_D-1) \ \ln \left(\frac{y_D}{1-y_D} \right)$$

$$+ B(2x_D-1) \ \ln \left(\frac{x_B}{1-x_B} \right)$$

$$- F(2y_F-1) \ \ln \left(\frac{y_F}{1-y_F} \right). \tag{16.43}$$

The total flow equation is a product of two functional groups. The first involves only a and β, and represents the degree of difficulty in separation. The second, K, contains the specifications of the feed and two products, hence indicates the amount of the separation work. The second factor is also known as the *separative duty*.

The first factor becomes large when β approaches the value of a, which means the operation at total reflux (see Fig. 16.4); and, it becomes small when β approaches unity, in which case the total number of stages required will be infinite.

Ideal Cascade

When a is very close to unity, so is β. Therefore, the following approximations are valid:

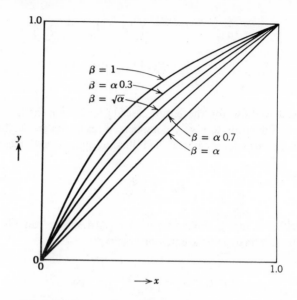

Fig. 16.4   Various "operating lines."

$$\ln \beta \simeq \beta - 1 , \tag{16.44}$$

$$\frac{2\beta}{\alpha + \beta} \simeq 1 . \tag{16.45}$$

After substituting Eqs. 16.44 and 16.45 into Eq. 16.42, the derivative with respect to β can be set equal to zero to obtain the optimum value for β, then J is the minimum total flow:

$$\frac{\partial J}{\partial \beta} = 0. \tag{16.46}$$

The above procedure gives

$$\beta = \alpha^{\frac{1}{2}} , \tag{16.47}$$

which means that the total interstage flow is a minimum when $\beta = \alpha^{\frac{1}{2}}$. For this case Eq. 16.30 becomes

$$R_n = \alpha^{\frac{1}{2}} R_{n+1}$$

$$= \alpha \ R_{n+2} . \tag{16.48}$$

Combining this with the definition of a,

$$x_n = y_{n+2} \ . \tag{16.49}$$

This is the requirement of matching compositions of two mixing (recycling) streams: This type of characteristic operation is called ideal cascade. Therefore, the ideal cascade is a special case of the polytropic enriching process with $\beta = a^{1/2}$, which gives the minimum interstage flow as well as the matching of recycle stream compositions. A graphical illustration of ideal cascade is shown in Fig. 13.10 of Chapter XIII.

The total number of stages for an ideal cascade can be obtained from Eq. 16.34 as

$$N = \frac{2 \ \ln\left(\dfrac{y_D(1-y_D)}{x_B/(1-x_B)}\right)}{\ln \ a} - 1 \ . \tag{16.50}$$

Comparing this with Eq. 16.29 shows that the number of stages required for an ideal cascade is roughly twice as many as the minimum number of stages.

Since the diffused streams must be recompressed before feeding to the next stages, it is important to have the expression in terms of the abundance ratio. For the enriching section

$$V_n = D\left(\frac{y_D - x_{n-1}}{y_n - x_{n-1}}\right) \tag{16.51}$$

becomes

$$V_n = \frac{D(aR_D - \beta R_n)(R_n + 1)}{(a-\beta) \ (R_D + 1) \ R_n} \ , \tag{16.52}$$

which is valid for any polytropic enriching process. Setting $\beta = a^{1/2}$ in Eq. 16.52.

$$V_n = \frac{D(a^{1/2}R_D - R_n) \ (R_n + 1)}{(a^{1/2}1) \ (R_D + 1) \ R_n} \ . \tag{16.53}$$

Similarly for the stripping section, the flow rate of the diffused stream at the mth stage is

$$V_m = \frac{B \ (R_m - a^{1/2}R_B) \ (R_m + 1)}{(a^{1/2} - 1) \ (R_B + 1) \ R_m} \ . \tag{16.54}$$

Finally, the total interstage flow rate for the entire cascade is

$$J = \frac{2(a^{\frac{1}{2}} + 1)}{(a^{\frac{1}{2}} - 1) \ln a} K ,$$ (16.55)

where K is the separative duty. Further simplification can be made with an approximation of

$$a \simeq 1 + \epsilon$$ (16.56)

such that Eq. 16.55 becomes

$$J = \frac{8}{\epsilon^2} K .$$ (16.57)

EXAMPLE 2

For the production of 1 (g)(mole)/hr of 90% $^{235}UF_6$ from the feed of 0.72% in Example 1, calculate the total number of stages and the total interstage flow rate using an ideal cascade. Sketch the distribution of the diffused gas flow rate as a function of stage. Also, plot the concentration of diffused gas versus stage number.

Stage. From material balance

$$F = \frac{D(y_D - x_B)}{(x_F - x_B)} = \frac{0.9 - 0.001}{0.0072 - 0.001} = 145 \frac{(g)(moles)}{hr} ;$$

$$B = F - D = 145 - 1 = 144 \frac{(g)(moles)}{hr} .$$

From Eq. 16.50,

$$N = 2 N_{min} - 1 = (2) (2122) - 1 = 4243,$$

and

$$R_F = \frac{0.0072}{1 - 0.0072} = 0.007252 .$$

The total interstage flow can now be calculated from Eq. 16.55:

$$J = \frac{2(a^{\frac{1}{2}}+1)}{(a^{\frac{1}{2}}-1)\ln a} [D(2y_D-1) \ln R_D + B(2x_B - 1) \ln R_B - F(2y_F-1) \ln R_F]$$

$$= \frac{2[(1.0043)^{\frac{1}{2}} + 1]}{[(1.0043)^{\frac{1}{2}} - 1]\ln 1.0043} \ (2 \times 0.9 - 1) \ln 9$$

$$+ \ 144 \ (2 \times 0.001 - 1)\ln 0.001001$$

$$- \ 145 \ (2 \times 0.0072 - 1)\ln 0.007252$$

$$= \ 1.2613 \times 10^8 \ \frac{(g)(moles)}{hr} \ .$$

The number of stages between the feed stage and the bottom will be

$$N_F = \frac{2 \ \ln(R_F/R_B)}{\ln \ a} = \frac{2 \ \ln\left(\dfrac{0.007252}{0.001001}\right)}{\ln \ 1.0043}$$

$$= \ 923.$$

Thus, the feed stage is located at the 923rd stage from the bottom. For the enriching section, the flow rate of diffused stream is calculated using Eq. 16.53:

$$V_n = \frac{D(a^{\frac{1}{2}}R_D - R_n)(R_n + 1)}{(a^{\frac{1}{2}}-1) \ (R_D + 1) \ R_n}$$

$$= \frac{D(1-a^{-n/2}) \ (R_D \ a^{(1-n)/2} + 1)}{(a^{\frac{1}{2}}-1) \ (R_D + 1) \ a^{-n/2}} \ ,$$

where the relationship

$$R_D = \beta^{n-1} \ R_n = a^{(n-1)/2} \ R_n$$

was used and $n < 3320$. Likewise, for the stripping section, from Eq. 16.54

$$V_m = \frac{B(R_m - a^{\frac{1}{2}}R_B)(R_m + 1)}{(a^{\frac{1}{2}} - 1)(R_B + 1) \ R_m}$$

$$= \frac{144[(1.0043)^{m/2} - 1.0043] \ [(1.0043)^{m/2+1} \times 0.001001 + 1]}{[(1.0043)^{\frac{1}{2}} - 1] \ (1.0043 \times 0.001001 + 1) \ (1.0043)^{m/2}} \ ,$$

where $m < 923$. Varying n and m for each section, the flow rate of diffused

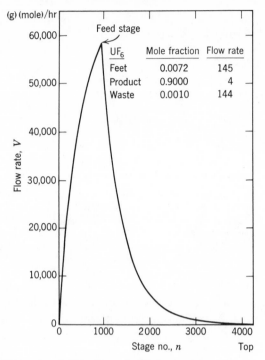

Fig. 16.5 Flow rate of diffused gas in ideal cascade.

gas from each stage can be calculated; the results are plotted in Fig. 16.5. The concentration y_n of the diffused stream is calculated by

$$y_n = \frac{R_n}{1 + R_n} \, ,$$

for the enriching section this becomes

$$y_n = \frac{R_D a^{(1-n)/2}}{1 + R_D a^{(1-n)/2}} ,$$

and for the stripping section,

$$y_m = \frac{R_B a^{(m-1)/2}}{1 + R_B a^{(m-1)/2}} \cdot$$

The results of computations are shown in Fig. 16.6. For the sake of simplicity, the stage numbers in both Fig. 16.5 and 16.6 are counted from the bottom.

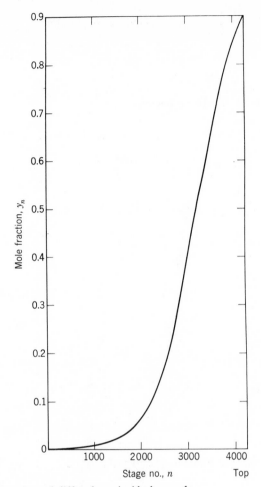

Fig. 16.6 Concentration of diffused gas in ideal cascade.

Continuous Cascade

From the above two examples, it is evident that a great number of stages are required in a cascade operation due to the closeness of the separation factor to unity, $0 < a - 1 = \epsilon \ll 1$. Furthermore, all of the physical quantities change very little from stage to stage. These facts permit us to treat such a cascade as if the entire cascade is one continuous unit. Then, the mathematical expressions can be simplified from a discrete symbol to a continuous one. The procedure is as follows:

$$y_n - y_{n-1} = \frac{dy}{dn} \, . \tag{16.58}$$

Here the symbols without subscript stand for the continuous quantity. Since

$$\frac{y_{n-1}}{1-y_{n-1}} = a^{\frac{1}{2}} \frac{y_n}{1-y_n} \ , \tag{16.59}$$

the left-hand side of Eq. 16.58 becomes

$$y_n - y_{n-1} = \frac{(1-a^{\frac{1}{2}})y(1-y)}{1-y + a^{\frac{1}{2}}y} \tag{16.60}$$

considering that $a^{\frac{1}{2}} \simeq 1 + \frac{\epsilon}{2}$:

$$\frac{dy}{dn} = - \frac{\epsilon}{2} y(1-y) \ . \tag{16.61}$$

Similarly,

$$\frac{dR}{dn} = R_n - R_{n-1}$$

$$= R_n - a^{\frac{1}{2}}R_n$$

$$= - \frac{\epsilon}{2} R \ , \tag{16.62}$$

which can be integrated immediately to yield

$$R = R_D \ \exp\left(- \frac{\epsilon}{2} n\right) . \tag{16.63}$$

or the number of stages can be expressed as

$$n = \frac{2}{\epsilon} \ \ln\left(\frac{R_D}{R}\right). \tag{16.64}$$

The flow expressions can be converted from the discrete form to a continuous one as follows. The flow rate of the diffused stream in the enriching section becomes

$$V = \frac{2D(y_D - y)}{\epsilon y(1 - y)} . \tag{16.65}$$

Therefore, the total interstage flow rate for the enriching section is

$$J_E = \sum_{n-1}^{N} (2V_n + D)$$

$$= \int_1^N (2 V + D) \, dn$$

$$= \int_{y_D}^{y_F} (2 V + D) \frac{dn}{dy} \, dy . \tag{16.66}$$

Substituting Eq. 16.61 into Eq. 16.66 and carrying out the integration yields

$$J_E = \frac{8D}{\epsilon^2} \left[(2y_D - 1) \ln \left(\frac{R_D}{R_F} \right) + \frac{(1 - 2y_F)(y_D - y_F)}{(1 - y_F) \, y_F} \right]$$

$$+ \frac{2D}{\epsilon} \ln \left(\frac{R_D}{R_F} \right). \tag{16.67}$$

Here the second term can be neglected. Likewise, the continuous formulas for the stripping section become

$$\frac{dR}{dm} = \frac{\epsilon}{2} R , \tag{16.68}$$

$$R = R_B \exp \left(\frac{\epsilon}{2} m \right), \tag{16.69}$$

$$m = \frac{2}{\epsilon} \ln \left(\frac{R}{R_B} \right), \tag{16.70}$$

$$V = \frac{2B (y - y_B)}{\epsilon \, y(1 - y)} . \tag{16.71}$$

Thus, the interstage flow is

$$J_S = \frac{8B}{\epsilon^2} \left[(2y_B - 1) \ln \left(\frac{R_B}{R_F} \right) + \frac{(1-2y_F)(y_B - y_F)}{(1-y_F)\, y_F} \right]$$

$$+ \frac{2B}{\epsilon} \ln \left(\frac{R_B}{R_F} \right). \tag{16.72}$$

Adding Eqs. 16.67 and 16.72, the total interstage flow for the entire cascade will be

$$J = J_E + J_S = \frac{8}{\epsilon^2} K, \tag{16.73}$$

where K is the separative duty defined by Eq. 16.43. Of course, this result is identical to Eq. 16.57. Now the total number of stages for the entire cascade is calculated from Eqs. 16.64 and 16.70:

$$N + M = \frac{2}{\epsilon} \ln \left(\frac{R_D}{R_B} \right), \tag{16.74}$$

which is the same result as Eq. 16.50 if $\ln a = \epsilon$ is used and the "minus one" is neglected.

In summary, it has been shown that all of the cascade equations can be made continuous when the separation factor is very close to unity. This simplifies the mathematical procedures, but yields the same practical results.

3 DESIGN CONSIDERATIONS

The separative duty defined by Eq. 16.43 holds for an entire cascade. Generalizing the concept of the separative duty for any stream of abundance ratio R_n, the separation potential [15-17] can be defined by

$$\phi_n = (2y_n - 1) \ln R_n. \tag{16.75}$$

The value of ϕ_n is always positive for all values of y_n, and it is concave upward. Using this separative potential, the separative duty can be re-expressed as

$$K = D\phi_D + B\phi_B - F\phi_F. \tag{16.76}$$

As mentioned previously, the separative duty indicates the magnitude of the amount of separation to be achieved. Many quantities in the cost estimate are directly proportional to this separative duty. For practical purposes, it is necessary to use equations derived for the ideal cascade with very small separation factor.

In order to estimate the compressor capacity, one may start with the calculation of the total diffused flow rate, which is roughly one-half of the total interstage flow. Utilizing Eqs. 16.18, 16.20, 16.23, and 16.57, the flow rate of the total diffused stream results as

$$\frac{J}{2} = \frac{2.082 \ K}{\epsilon *^2} \left[\frac{1 + (1-P_r)(P_1 + P_2)/P_c}{1 - P_r} \right]^2 . \tag{16.77}$$

Therefore, the total isothermal volumetric compressor capacity G is

$$G = \frac{JR'T}{2 \ P_2} = \frac{2.082 \ R'T \ K}{\epsilon *^2 \ P_2} \left[\frac{1}{1-P_r} + \frac{P_1+P_2}{P_c} \right]^2 . \tag{16.78}$$

This becomes a minimum when $P_2 = P_c/2$ and $P_1 = 1.207 \ P_c$ [17]. Thus, Eq. 16.78 reduces to

$$G_{min} = \frac{40.57 \ R'T \ K}{\epsilon *^2 \ P_c} . \tag{16.79}$$

The rate of power consumption may be estimated from thermodynamics as

$$W = \frac{J \ R'T}{2} \ \ln \left(\frac{P_1}{P_2} \right)$$

$$= - \frac{2.082 \ R'T \ K}{\epsilon *^2} \left[\frac{1}{1 - P_r} + \frac{P_1 + P_2}{P_c} \right] \ln P_r . \tag{16.80}$$

However, the actual power required in cascade operation will be greater than this value because of nonisothermal conditions, inefficiency, friction losses, and so on. The minimum power is obtained when $P_r = 0.285$, with both pressures approaching zero, as

$$W_{min} = \frac{5.112 \ R'T \ K}{\epsilon *^2} . \tag{16.81}$$

The total membrane area can be calculated from

$$S = \frac{J \ l}{2Q(P_1 - P_2)}$$

$$= \frac{2.082 \ K}{\epsilon^{*2} Q} \left[\frac{1}{P_1 - P_2} \left(\frac{1}{1 - P_r} + \frac{P_1 + P_2}{P_c} \right)^2 \right] . \qquad (16.82)$$

When P_2 approaches zero and P_1 becomes P_c, the membrane area is minimized to

$$S_{min} = \frac{8.328 \ Kl}{\epsilon^{*2} \ QP_c} . \qquad (16.83)$$

It should be noted that the above minimum quantities do not occur simultaneously at the same operating conditions. The optimum condition must be found after the total cost analysis has been made.

Now the cost study of isotope separation will be discussed. Assuming that the total operating cost is proportional to the separative duty, the total cost of product can be written as

$$D \ C_D = F \ C_F + K \ C_K , \qquad (16.84)$$

where C_D, C_F, and C_K are unit costs of product, feed, and separative work, respectively. Substituting Eq. 16.76 into Eq. 16.84,

$$C_D = \frac{F}{D} C_F + (\phi_D + \frac{B}{D} \phi_B - \frac{F}{D}) C_K . \qquad (16.85)$$

Noting the following relationships from material balances:

$$\frac{F}{D} = \frac{y_D - x_B}{y_F - x_B} , \qquad (16.86)$$

$$\frac{B}{D} = \frac{F}{D} - 1 , \qquad (16.87)$$

Eq. 16.85 can be rewritten as

$$C_D = (\phi_D - \phi_B)C_K + \left(\frac{y_D - x_B}{y_F - x_B} \right) [C_F + (\phi_B - \phi_F) C_K] . \qquad (16.88)$$

From this equation, the optimum waste composition is found by

$$\frac{dC_D}{dx_B} = 0 . \qquad (16.89)$$

The value of x_B that satisfies Eq. 16.89 will be called x_o. The final expression for the optimum unit cost becomes [17]

$$C_D = C_K \left((2y_D - 1) \ln\frac{y_D(1-x_o)}{(1-y_D)x_o} + \frac{(y_D - x_o)(1-2x_o)}{x_o(1-x_o)} \right). (16.90)$$

This is the formula that has been employed by the ABC for the price estimate of uranium.

The above discussions covered only the essential parts of the isotope separation by means of membrane diffusion. There are many aspects of cascade operation that could not be discussed in this limited space. In practice, it is almost impossible to have an ideal cascade, which must exhibit continuously changing flow rates of diffused and reject streams throughout the entire cascade. Thus, a so-called "squared-off cascade" is used in place of an ideal cascade. This involves the study of square-cascade theory, which is well discussed by Cohen [16]. Another important area, which has not been touched upon, is the control problem of cascade operation. The maintenance of the cascade variables at designated values presents a much higher degree of complexity than that encountered in ordinary plant operations. Numerous interdependent stage variables are fluctuating around the average values, even at steady-state operation. The equilibrium time is also a relevent topic in cascade theory. This is the time taken to reach a steady-state operation from the start, which may range from 50 to 100 days for a heavy-isotope separation plant. It is also an important economic factor because it causes serious delays as well as production losses. Finally, the economic study in design and the optimization problems are of utmost importance. An isotope separation plant requires not only a huge investment for installation cost but also high operating expenses. The equipment design and the optimum control are a continuous challenge in engineering problems.

Nomenclature

Symbol	Meaning
a	Constant in Knudsen-flow equation
B	Molar flow rate of bottom waste
b	Constant in Poiseuille equation
C	Unit cost
D	Molar flow rate of top product
d	Pore diameter
F	Molar flow rate of feed
G	Compressor capacity
J	Total molar flow rate in cascade
K	Separative duty
L	Molar flow rate of reject stream

Symbol	Meaning
l	Membrane thickness or capillary length
M	Total stage number of stripping section
M'	Molecular weight
m	Stage number counted from the bottom of stripping section
N	Total number of stages
n	Stage number counted from the top of enriching section
P	Pressure
P_c	Characteristic pressure
P_r	Pressure ratio (P_2/P_1)
Q	Permeability
R	Molecular abundance ratio of the light component
R'	Universal gas constant
S	Membrane area
T	Absolute temperature
V	Molar flow rate of the diffused stream
W	Rate of power consumption
x	Mole fraction of the light component in the reject stream
x^*	Mole fraction of the light component in the reject stream for ideal separation
x_o	Optimal mole fraction of the light component in waste
y	Mole fraction of the light component in the diffused stream
a	Stage separation factor
a'	Head separation factor
a^*	Ideal separation factor
β	Ratio of the abundance ratios of two adjacent diffused streams
ϵ	Small positive value
η	Stage separation efficiency
η'	Membrane efficiency
θ	Cut
λ	Mean free path
μ	Viscosity
ϕ	Separative potential

Subscripts	
B	Bottom stage
D	Top stage

E	Enriching section
F	Feed stage
H	Heavy element
L	Light element
M,m	Stage number in stripping section
N,n	Stage number in enriching section
S	Stripping section

References

1. J. G. Aston, *Phil. Mag.*, **39**, 449 (1920).
2. G. L. Hertz, *Physik. Z.*, **79**, 108 (1920).
3. G. L. Hertz, *Naturwiss.*, 21, 884 (1933).
4. H. Harmsen, G. L. Hertz, and W. Schutze, *Physik. Z.*, **90**, 703 (1934).
5. D. E. Woolderidge and F. A. Jenkins, *Phys. Rev.*, **49**, 404 (1936).
6. D. E. Woolderidge and W. R. Smyth, *Phys. Rev.*, **50**, 233 (1936).
7. H. D. Smythe, *Atomic Energy for Miltary Purposes,* Princeton Univ. Press, Princeton, N.J., 1945.
8. J. F. Hogerton, *Chem. Eng.*, **52**, (12) 98 (1945).
9. P. C. Keith, *Chem. Eng.*, **53**, (2) 112 (1946).
10. R. B. Bird, W. E. Stewart, and E. N. Lightfoot, *Transport Phenomena,* Wiley, New York, 1960, p. 19.
11. C. Frejacques, O. Bilous, J. Dizmier, D. Massignon, and P. Plurien, "Peaceful Uses of Atomic Energy," Proceedings of the Second U.N. International Conference, Vol. 4, p. 418, Geneva (1958).
12. R. D. Present and W. G. Pollard, *Phys. Rev.*, **73**, 762 (1948).
13. S-T. Hwang and K. Kammermeyer, *Canad. J. Chem. Eng.*, **44**, (2) 82 (1966).
14. S-T. Hwang and K. Kammermermeyer, *Prog. Separat. Purification,* **4**, 1 (1971).
15. H. London, *Separation of Isotopes,* George Newnes, London, 1961, Chap. 8.
16. K. Cohen, *Theory of Isotope Separation as Applied to the Large-scale Production of* ^{235}U, National Nuclear Energy Series III-1B, McGraw-Hill, New York, 1951.
17. M. Benedict and T. H. Pigford, *Nuclear Chemical Engineering,* McGraw-Hill, New York, 1957.
18. S. Weller and W. A. Steiner, *J. Appl. Phys.*, **21**, 279 (1950).

Chapter **XVII**

APPLICATION OF MEMBRANES IN ANALYTICAL CHEMISTRY

One of the most important problems in analytical chemistry is the separation of materials into pure species. Therefore, any separation processes are potential analytical means, or at least can serve as a preparatory procedure. All of the membrane processes discussed hitherto are separation processes. Thus, it is natural to conceive the membrane-separation processes as analytical tools. However, in practice, a rather limited number of such processes is found in analytical applications.

Membranes are also used in many cases as isolating barriers to protect sensitive systems from the ambient environment. Combination of these two roles of membranes are well utilized in ion-selective membrane electrodes and specific gas probes. In both cases, special chemicals and electrodes are surrounded by ion-specific or permselective membranes so that specific ions or gases may permeate through the membranes while the whole electrode system is protected by the membrane. These types of probing techniques will become a major means of analysis.

1   ION-SELECTIVE MEMBRANE ELECTRODES

So far, membranes have been considered as barriers for mass transport. Transported constituents were molecules, atoms, and infrequently ions. However, membranes are also found in use primarily to transmit electrons after the desired ion-exchange processes have taken place.

For most of the ion-selective membrane electrodes, the role of the membrane is seldom to transmit specific ions. Rather, the selective uptake of particular ions is considered to be responsible in the measurement of ion activity in solution. Thus, the situation is somewhat different in regard to membrane characterization and classification. The purpose of having a membrane in an ion-selective electrode is to give the specific response only to the selected ionic species.

Classifications of Membrane Electrodes

According to the types of membranes, the membrane electrodes can be classified as follows:

1. Homogeneous Membrane Electrodes
   A. Solid-state electrodes
       Glass electrodes
       Single-crystal electrodes
       Polycrystal electrodes
       Mixed-crystal electrodes
       Metal electrodes
   B. Liquid-membrane electrodes
2. Heterogeneous Membrane Electrodes
   A. Precipitate electrodes
   B. Solid ion-exchange membrane electrodes
   C. Enzyme-substrate electrodes

Alternatively, the membrane electrodes may be simply classified according to their functions:

1. pH electrodes
2. Cation-selective electrodes
3. Anion-selective electrodes

In either case, the membrane is the most important element. Depending on how the membrane is constructed or to what specific ions the membrane responds, one of the above classifications may be used. For all electrodes, a particular membrane permits only a particular kind of ion to penetrate and exchange with on-site ions. Also, it should be noted here that most electrode measurements are extremely rapid, thus allowing continuous monitoring and on-line installation.

Glass Electrodes

Various types of glasses have been used as membranes in ion-selective glass electrodes. When these electrodes were developed originally, they were primarily used in measuring pH values. However, the scope of application has been widened to include many of the univalent cations such as Na^+, K^+, Li^+, NH_4^+, Ag^+, Rb^+, Cs^+.

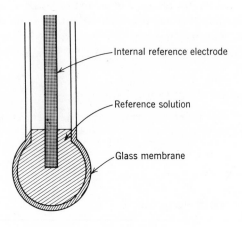

Fig. 17.1   Schematic of glass electrode.

A schematic view of a glass electrode is shown in Fig. 17.1. Thus a thin membrane of special glass encloses a reference solution (usually $0.1M$ HCl), in which a reference electrode is immersed. Generally, nonhygroscopic glasses display little electrode function. A variety of soft glasses has been successfully used to produce desired selectivity to various univalent cations.

The mechanism of glass-electrode function is not completely understood. However, it has been established that the glass surface must possess a capability of undergoing hydration upon contact with an aqueous solution. As hydration takes place, water molecules penetrate the silicate structure to form an inner and outer hydrated layer. If a cross section is taken through a glass electrode, it might look as follows:

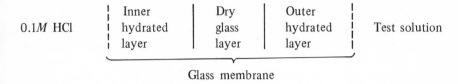

	Inner hydrated layer	Dry glass layer	Outer hydrated layer	
0.1M HCl	Inner hydrated layer	Dry glass layer	Outer hydrated layer	Test solution

Glass membrane

In a functioning glass electrode, the bulk of the glass membrane consists of the dry glass layer. Only a very thin (50-10,000 Å) region close to either surface is hydrated to give a gel like structure. This hydrated glass layer exhibits distinctly different physical characteristics from those for the dry layer. Besides the fact that the hydrated layer experiences some swelling, it enhances the diffusivity of univalent cations about 1000 times from the value in dry glass.

When the dry-glass electrode is immersed in an aqueous solution initially, one or more hours of soaking are necessary to ensure the formation of the hydrated layer. Thereafter, the hydrated layer continues to dissolve into solution on one side, while further hydration of additional dry glass takes place on the other side. These countereffects maintain roughly a constant thickness of the hydrated layer. The lifetime of a glass membrane depends upon the nature of the glass and the type of the solution used.

As the hydration process proceeds, cations move into the swollen layers and an ion exchange takes place within the inner and outer layers. The amount of cation occupying the exchange sites on either side of the membrane will be proportional to the activity of the cations in the adjacent solution. Therefore, the overall performance of a glass electrode depends upon two factors: one is how fast cations diffuse into the hydrated layers, and the other is how many cations be exchanged in the membrane. Thus, the total electrode potential consists of the diffusion potential and the phase-boundary (ion-exchange equilibrium) potential. This view was proposed originally by Teorell [1, 2] and later developed further by Eisenman [3, 4]. The total electrode potential can well be described by the Nernst equation, or its modified version

$$E = E_G^{\circ} - \frac{RT}{\mathcal{F}} \ln a_{M^+} , \qquad (17.1)$$

where E_G° is the standard potential for the particular electrode and a_{M^+} is the cation activity.

Different compositions of glass yield different characteristics of the electrode. By changing the glass composition, the degree of hydration will change, and the sensitivity to any specific cation will vary. However, no evidence has been found that the electrical resistance of a glass electrode has any bearing on the electrode function.

There are two kinds of possible glass-electrode errors: one is the alkaline error, and the other is the acid error. At high pH (greater than 10), most electrodes yield observed pH values that are lower than the actual values. On the other hand, at low pH (below 1), the glass electrode shows an acid error, so that the observed pH value is higher than the true value. Besides strong alkalinity or acidity, high temperatures and prolonged exposure cause false readings. However, corrections can be made for alkaline error and acid error if calibration curves are obtained. In fact, many commercial electrodes are furnished with pertinent correction tables.

Finally, it should be pointed out that the glass electrodes can be also used in nonaqueous media, such as liquid ammonia or molten salts.

In conclusion, glass electrodes are versatile and rapid in response. They measure the ionic acitvity, but not the concentration in solution. Although the earlier pH glass electrodes were developed a long time ago [5-7], more sophisticated ion-selective electrodes are now widely used in many areas of research such as biomedical investigations and pollution monitoring.

Other Solid-State Electrodes

Various solid-state membrane electrodes can be fashioned by replacing the glass membrane with single crystals, polycrystallines, mixed crystals, or even pure metals. However, metal-membrane electrodes will be treated separately, since they involve an entirely different mechanism of detecting selected ions. A schematic is depicted in Fig. 17.2. These membranes must exhibit selective

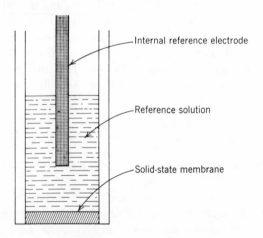

Fig. 17.2 Schematic of solid-state electrode.

ionic conductivity besides mechanical stability and chemical inertness, including low solubility in the sample solution.

The ionic conductivity is created by the mobility of the smallest lattice ions in the crystal. The lattice-defect mechanism is believed to be responsible in the transport of ions. That is, a vacancy is filled by a mobile ion, thus leaving a new vacancy behind, and so on. Because of the geometric requirement and charge distribution of the vacancy, only a particular ion can be transmitted by this vacancy mechanism. Any other ions cannot go into the vacancy; therefore, they do not contribute to the ionic conductivity. Thus, solid-state electrodes become ion selective by limiting the movement of all ions except the one to be detected.

The rare-earth fluoride membrane electrode is an excellent example of a single-crystal electrode. Especially, lanthanum fluoride is known as a fluoride ion conductor [8]. By treating with Eu^{2+}, the electrical resistance can be lowered further. The performance of such an electrode shows Nernstian behavior in a wide range, covering from saturated solutions to $10^{-7}M$ fluoride ion acitvity.

Many of the silver salts are good ionic conductors and thus are used as membrane materials. In such membranes, the silver ions are the mobile ones. However, they can be used not only to measure the silver ionic activities but also to detect the counterion levels, which are determined from the solubility product. Commercial electrodes are available for chloride, bromide, iodide, and sulfide ions using solid silver halides or sulfide as membranes.

Also, mixed crystals of various silver salts and silver sulfide have been fabricated to form electrode membranes. These mixed crystals show much lower resistances than the pure silver salts.

Since silver sulfide was so successful as a matrix material for silver salts, it was a logical extension to try other metal sulfides to form membranes selective to that metal ions. Indeed, many metal sulfides were found to be useful, such as CuS, CdS, and PbS [9].

Metal-Membrane Electrodes

One electrochemical way or viewing ion-selective membrane electrodes is to interpret the electrode as one-half of a concentration cell. The membrane separates a reference solution from a test solution. The concentration difference generates an electropotential, which is measured. Any membrane that can develop the potential and transmit electrons can be considered as a candidate, provided that the response is ion selective.

Various metal membranes have been tried [10], all of which involved an oxidation-reduction reaction at the surface. Since the reversibility of the oxidation-reduction reaction is required to ensure proper function, a rather stringent restraint is imposed upon the selection of metals. Very rapid steady-state potentials were obtained and reproduced for silver, copper, cadmium, and thallium membrane electrodes [10].

Liquid-Membrane Electrodes

The mobility of ions would be greatly enhanced if the ion exchanger were in a liquid state. There are many organic liquid ion exchangers that possess excellent selectivity for specific ions. The only problem with liquid membranes is to maintain both mechanical and chemical stability while in use. The liquid ion exchanger should be in electrolytic contact with the sample solution, but it should not be soluble in it. One way of providing a stable liquid membrane between two aqueous solutions is to intersperse a porous hydrophobic membrane as shown in Fig. 17.3. The liquid ion exchanger wicks into the pore space from the annular reservoir, which is in contact with the porous membrane. The inner tube holds the reference solution and the reference electrode.

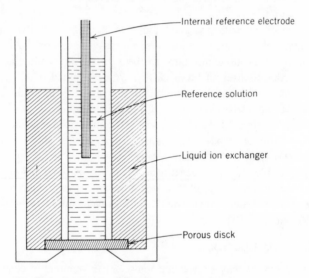

Internal reference electrode

Reference solution

Liquid ion exchanger

Porous disck

Fig. 17.3 Schematic of liquid-membrane electrode.

When a sample solution is brought into contact with the liquid ion exchanger at the phase interface, a selective ion-exchange process takes place. This is where the selectivity of an electrode arises. Once the selected ions are combined with the ion-exchange molecules, the site and the ion travel together through the membrane phase according to its own potential gradient. Then, the ions are released on the other side of the membrane into the reference solution. This mechanism is quite similar to the "carrier" transport, which was discussed previously.

Various phosphate esters are used as complexing agents for the calcium electrode, which is especially valuable in biological research [9]. Another example of a liquid-membrane electrode is the potassium electrode that utilizes valinomycin as the carrier [11]. This electrode is also useful for studies of biological systems. The main feature of this type of potassium electrode is in its sharp selectivity for K^+ over Na^+ of better than 5000 to 1. It is also possible to fabricate anion-selective liquid-membrane electrodes. Positively charged transition-metal complexes have been employed as anion exchangers [9]. These electrodes were successful in measuring anions such as NO_3^-, ClO_4^-, and BF_4^-. A chloride electrode was made using a dimethyl-distearyl ammonium cation. This type of liquid-membrane electrode can detect the chloride ions without much interference by sulfide, bromide, or iodide ions.

Precipitate Electrodes

This type of electrode employs membranes fabricated from inert binding materials, in which various insoluble salts of desired ions are impregnated. Pungor [12-14] used silicone rubber as a membrane matrix, which was imbedded with silver halides or sulfide, and barium sulfate. These electrodes showed selective responses to chloride, bromide, iodide, and sulfate ions. Binder materials, such as paraffin wax, collodion, and various plastics also have been utilized by other investigators [15-19]. Fischer and Babcock [18] used parafin to impregnate barium sulfate, and using a radioactive tracer technique showed that barium ions were immobile, but counterions could migrate through the membrane.

In the preparation of membranes, the grain size, conditions of precipitation, and the proportion of active material to inert components are all-important. The particles of active materials should touch each other to provide the ionic conduction. Usually, a half and half mixture of inert and active material gives the optimum proportion. The optimum grain size is in the order of several microns.

One simple method of making an impregnated silicone-rubber membrane is described by Buchanan and Seago [19]. They used a room-temperature vulcanized silicone rubber as a binder. A finely powdered sample of the insoluble metal compound was added and mixed with silicone rubber to obtain a coherent slurry. The slurry was pressed to give a 0.5-mm-thick sheet between a heavy polyethylene plane and a polyvinyl chloride sheet. After curing this membrane in air, it was conditioned in a hot (90°C), concentrated (1-2M) solution of the appropriate metal salt for a week. These authors did not find any effect of the crystalline form, the degree of hydration, and the associated anion of the imbedded salt on the membrane electrode response or selectivity. They also reported that such electrodes were nonspecific in their responses to cations.

Enzyme-Substrate Electrodes

Even though there are no commercial electrodes of this type as yet, the potential is great and promising in the future for the enzyme-substrate electrodes in biochemical and biomedical studies. Guilbaut et al. [20-22] reported how to couple an electrode sensor with an enzyme.

Their technique is basically a modification of an ammonium ion-sensitive glass electrode with a coating of enzyme substrate layer. The enzyme-catalyzed hydrolysis liberates the ammonium ions, which then diffuse through the membrane and are detected by the electrode. Thus, the potential developed by this type of electrode is a measure of the substrate concentration in the sample solution.

Based on the same principle, Montalov [23] developed a new electrode that can detect enzyme activity simply by inverting the positions of enzyme and sample solution. The ammonium ion-sensitive electrode was covered with a thin layer of concentrated urea solution held by a cellophane membrane. When this electrode is placed in a solution of the enzyme urease, the urea will diffuse out to react with the enzyme urease. The reaction product is ammonium ion, which will diffuse back through the membrane to the electrode. By measuring this ammonium ino activity, the activity of the enzyme in the sample solution is ascertained. Thus, the electrode can be called an enzyme electrode. This type of enzyme electrode is likely to see considerable development in biological studies.

2 SPECIFIC GAS PROBES

With the advent of biomedical research and pollution studies, continuous, rapid, and accurate devices of measuring both gaseous and dissolved gases became important. Indeed, numerous probes and other devices have been developed and are on the market. Perhaps, the most successful example is the oxygen probe. Others are still in their development stages, and much room exists for further improvements.

Although many different principles of analysis were involved for various probes, some kind of membranes are invariably employed. The main purpose of having these membranes is to isolate the inside content from the test solution and to allow selective gas permeation into the detector probes. Usually, these probes can be used in measuring dissolved gas concentrations as well as the gas-phase partial pressures.

Oxygen Probe

The oxygen probe is basically an amperometric cathode of any metal (usually gold or platinum), which is connected through an electrolytic gel (silver chloride) to an anode (silver). The whole system is protected by an oxygen-permeable membrane. The membrane can be Teflon, polyethylene, nylon, or cellophane.

Oxygen diffuses through the membrane according to its concentration gradient. As oxygen reaches the cathode, which is under a considerable overvoltage, the cathodic reduction of oxygen takes place. This produces a current that is proportional to the rate of oxygen consumption. By measuring the electric current, the oxygen concentration of the sample solution (or gas) can be determined. The oxygen-diffusion rate is directly proportional to the concentration difference across the membrane. However, the inside concentration of oxygen is zero, because as soon as the oxygen molecules reach the cathode they are consumed by the chemical reaction. Therefore, the electric current, which is proportional to the oxygen diffusion rate, is also proportional to the absolute oxygen concentration on the outside of the membrane. It is, thus, easily seen that the proportionality constant depends upon factors of membrane permeability, membrane area, membrane thickness, and temperature. As temperature changes, the permeability of membrane to oxygen varies. Therefore, the proportionality factor changes also. Most commercial oxygen probes are temperature compensated using a built-in thermistor.

When the oxygen probe is used in a liquid solution to measure the dissolved oxygen concentration, care should be exercised so as not to have erroneous readings due to the stagnant boundary layer of liquid on the membrane surface. As discussed in Chapter II, the overall permeability of dissolved gas through a membrane depends not only on the membrane properties but also on the boundary layer of liquid. The boundary layer cannot be completely removed no matter how much agitation is applied. Thus, a practical solution to the boundary resistance can be obtained by maintaining a constant resistance of the boundary layer. This can be achieved by providing constant thickness of the boundary layer through the same degree of agitation.

Carbon Dioxide Probe

There are no commercial carbon dioxide probes on the market that are reliable and accurate as of today, although many companies offer various types of sensors. Most probes are essentially pH electrodes with a slight modification that allows a reaction space for the reaction of carbon dioxide and carbonic acid. The activity of hydrogen ions generated by this reaction is measured by the pH electrode.

Thus, the probe is made of a pH-sensitive glass electrode and a reference electrode (silver-silver chloride) housed together in a medium of aqueous electrolyte gel solution. The whole unit is covered with a silicone-rubber membrane through which carbon dioxide permeates readily. When the probe is placed in a test solution, carbon dioxide permeates through the membrane into the gel. Since the gel consists mostly of water, the following equilibria will be established:

$$CO_2 + H_2O \rightleftharpoons H_2CO_3 \rightleftharpoons H^+ + HCO_3^- \rightleftharpoons 2H^+ + CO_3^{2-}.$$

Then, the pH-sensitive electrode measures the potential developed by the hydrogen ions. If the concentration of carbon dioxide changes in the test solution, the concentration of hydrogen ions in the gel will change accordingly to maintain equilibrium, causing a variation in the electromotive force,

The silicone-rubber membrane protects the electrode system from convection, yet allows easy permeation of carbon dioxide. When measuring the dissolved carbon dioxide in a liquid solution, the boundary layer of stagnant liquid at the membrane surface presents an extra resistance to the permeation of carbon dioxide similar to that encountered in the oxygen probe. This boundary-layer effect is especially pronounced for the carbon dioxide and silicone-rubber system due to the unusually high permeability [24] of CO_2.

Sulfur Dioxide Probe

Based on a similar principle as the oxygen probe, a sulfur dioxide probe was produced and marketed. A special electrolyte solution and an electrode system are covered with a permselective membrane. As sulfur dioxide permeates through the membrane preferentially, an electrochemical reaction takes place, which produces an electric current that can be sensed by the electrode. Therefore, the current is directly proportional to the concentration of the sulfur dioxide present outside of the membrane. The membrane protects the electrode system and prevents any ionic and high-molecular compounds from interfering. As in other similar situations the boundary-layer phenomenon in the liquid phase will present the usual difficulties.

3 MEMBRANE METHODS OF ANALYTICAL SEPARATIONS

In sample preparation and in actual analysis, the use of separation techniques is the most important aspect of modern analytical chemistry. When a specific measurement technique is applied to a particular compound, the presence of extraneous substances in the sample usually causes interference with the determination of the desired quantity. Therefore, no analytical method is completely selective. Thus, the problem of separating and isolating a particular compound is always the central concern in chemical analysis.

All of the conventional techniques of separations have been employed in various analytical problems. The separations based on phase changes include distillation, extraction, crystallization, sublimation, precipitation, centrifugation.

leaching, and others. However, the recent developments of various chromato-graphic techniques and electrophoresis have become even more important in analytical chemistry.

Since membrane processes are basically separation techniques, all of such processes discussed previously may, in principle, be applied to analytical chemistry. Nevertheless, only a limited number of membrane methods have been utilized and those mainly in biochemical analysis. Ultrafiltration is used to concentrate the sample solution or to remove large particles. Elimination of solutes is achieved by dialysis, and electrodialysis is often used to desalinate the sample. Since each of these processes has already been discussed in detail, only a brief comment will be given here on how they are applied to some analytical problems.

Ultrafiltration

Most protein molecules are large enough to be ultrafiltered by appropriate pore-size membranes. Choosing different pore sizes, all molecules of proteins to be concentrated are filtered, and smaller molecules together with some solutions are eliminated. It is quite interesting to note that a considerable effort was made by Elford [25-27] to determine molecular weights of proteins by graduated ultrafiltration techniques using collodion membranes.

Ultrafiltration is also used as a standard technique for the microbiological analysis of water. By counting the number of organisms that are filtered, the bacterial analysis of water can be performed.

Another use of ultrafiltration is in the preparation of plasma from whole blood by removing particulates.

Dialysis

Since every species has its own dialysis rate for a given membrane, a mixture of two compounds in a solution may be separated. An example would be the separation of glucose from sucrose by selective dialysis [28]. However, dialysis is rarely used to determine the molecular sizes of different solutes. The most widely used application of dialysis is in eliminating small solute molecules. Removal of salts from a protein solution is frequently done by using cellophane membranes.

Dialysis is also effective in concentrating protein solutions by employing a highly concentrated solution of a polymer such as polyoxyethylene or polyvinylpyrrolidone [29]. The polymer solution is placed on the opposite side of the membrane from the protein solution. Because of the osmotic pressure, water and smaller solutes will permeate through the membrane to the polymer solution. Therefore, this procedure is called "osmotic concentration."

Electrodialysis

Electrodialysis is also an excellent method to desalinate a protein solution [29]. Separation of amino acid can be done in this way. However, this process has been utilized to a lesser extent than any of the other membrane processes, perhaps because of the extra need for electrical equipment.

The foregoing discussions of the use of membranes in analytical instruments is of necessity limited in scope. Although the most important basic types of devices have been covered, there is a multitude of variations and potential methods of application already in existence or in a state of development. Consequently, a host of new uses is likely to develop at an accelerating rate.

References

1. T. Teorell, *Proc. Natl. Acad. Sci.*, **21**, 152 (1935).
2. T. Teorell, *Proc. Soc. Exp. Biol. Med.*, **23**, 282 (1935).
3. G. Eisenman, "The Electrochemistry of Cation-Sensitive Glass Electrodes," in *The Glass Electrode*, Wiley-Interscience, New York, 1965.
4. G. Eisenman, Ed., *Glass Electrodes for Hydrogen and Other Cations: Principles and Practice*, Marcel Dekker, New York, 1967.
5. M. Cremer, *Z. Biol.*, **47**, 562 (1906).
6. F. Haber and Z. Klemensiewicz, *Z. Physik. Chem.*, **67**, 385 (1909).
7. D. A. MacInnes and M. Dole, *Ind. Eng. Chem., Anal. Ed.*, **1**, 57 (1929).
8. M. S. Frant and J. W. Ross, Jr., *Science*, **154**, 1533 (1966).
9. J. W. Ross, Jr., "Solid-State and Liquid Membrane Ion-Selective Electrodes," in *Ion-Selective Electrodes*, R. A. Durst, Ed., NBS Special Publ. 314, U.S. Government Printing Office, Washington, D.C., 1969.
10. E. B. Buchanan and J. L. Seago, *J. Electrochem. Soc.*, **114**, 595 (1967).
11. M. S. Frant and J. W. Ross, Jr., *Science*, **167**, 987 (1970).
12. E. Pungor, J. Havas, and K. Toth, *Acta Chim. Acad. Sci. Hung.*, **41**, 239 (1964).
13. E. Pungar, J. Havas, and K. Toth, *Z. Chem.*, **5**, 9 (1965).
14. E. Pungar, J. Havas, and K. Toth, *Acta Chim. Acad. Sci. Hung.*, **48**, 17 (1966).
15. H. P. Gregor and K. Sollner, *J. Phys. Chem.*, **58**, 409 (1954).
16. P. Cloos and J. J. Fripiat, *Bull. Soc. Chim. Fr.*, 423 (1960).
17. A. Shatkay, *Anal. Chem.*, **39**, 1056 (1967).
18. R. B. Fischer and R. F. Babcock, *Anal. Chem.*, **30**, 1732 (1958).
19. E. B. Buchanan, Jr. and J. L. Seago, *Anal. Chem.*, **40**, 517 (1968).
20. G. G. Guilbault, R. K. Smith, and J. G. Montalvo, Jr., *Anal. Chem.*, **41**, 600 (1969).
21. G. G. Guilbault and J. G. Montalvo, Jr., *J. Amer. Chem. Soc.*, **91**, 2164 (1969).

22. .G. G. Guilbault and J. G. Montalvo, *Anal. Lett.*, **2**, 283 (1969).
23. J. G. Montalvo, *Anal. Chem.*, **41**, 2093 (1969).
24. S. T. Hwang and G. D. Strong, *J. Polymer Sci. Symposium*, **41** (1973).
25. W. J. Elford, *Proc. Roy. Soc. (London), B*, **106**, 216 (1930).
26. W. J. Elford, *Proc. Roy. Soc. (London), B*, **112**, 384 (1933).
27. W. J. Elford, *Trans. Faraday Soc.*, **33**, 1094 (1937).
28. L. C. Craig and A. O. Pulley, *Biochemistry*, **1**, 89 (1962).
29. C. J. Van Oss, "Separation and Purification of Plasma Proteins," in *Progress in Separation and Purification*, Vol. 1, E. S. Perry, Ed., Wiley-Interscience, New York, 1968.

Appendix A

COMPUTER PROGRAMS FOR PERVAPORATION*

All computer programs are written in FORTRAN.

The individual programs cover model situations as delineated in particular for gaseous diffusion because these models in principle hold for any membrane-separation process. Necessary modifications concern primarily the concentration dependence of the transfer coefficients.

PV-1 COMPLETE-MIXING CASE – VARIABLE PERMEABILITIES

1. *Given:* $XI, X\emptyset$

   $PHAI = O$

   Relative permeabilities of A, QAS $[=Q_A(x)/Q_A(x=1)]$, and permeability ratios, $SFS = OAS/QBS$, in terms of mole fraction of A in the high-pressure (or liquid) phase, XS, in a tabulated form.

   To find: $Y1, Z1, THETA$

2. *Equations used:*

$$y^p = \frac{\dfrac{a^* x^0}{(1-x^0)}}{1+\dfrac{a^* x^0}{(1-x^0)}}, \quad a^* = a^*(x^v),$$

$$\theta = \frac{x^f - x^0}{x^p - x^0},$$

$$z_1 = \frac{y^p \theta}{\dfrac{Q_A(x^0)}{Q_A(x=1)} x^0}.$$

3. *Note:* x^0 must be larger than the minimum possible stripping concentration x_m, *where*

---

*This Appendix especially relates to Chapter VII.

$$x_m = 1 - \frac{(1-x^f)}{1 - (1-\frac{1}{a*})x^f} \quad , \quad a = a*(x^f).$$

4. *Method:* The Lagrange-Aitken method is used to find a and QA in terms of x.

PV-2 PLUG-FLOW CASE – VARIABLE PERMEABILITIES

1. *Given:*   XI, XØ
   RHAI = O
   Relative permeabilities of A, QAS$[=Q_A(x)/Q_A(x=1)]$, and permeability ratios, SFS=QAS/QBS, in terms of mole fraction of A in the high-pressure (or liquid) phase, XS, in a tabulated form.
   To find: Y1, Z1, THETA
2. *Equations used:*

$$\frac{dq}{dx} = \frac{q\left[x + \frac{(1-x)}{a*(x)}\right]}{x(1-x)\left[1- \frac{1}{a*(x)}\right]}$$

$$\frac{dz}{dx} = \frac{q}{\frac{Q_A(x)}{Q_A(1)}\left\{x(1-x)\left[1- \frac{1}{a*(x)}\right]\right\}} \quad \text{at } x=x^f, \quad q=1, \quad z=0$$

To find:

$$q=1-\theta \quad \text{and} \quad z=z_1 \quad \text{at } x=x^0.$$

3. *Method:*
   Integration: Runge-Kutta-Gill method
   Interpolation: Lagrange-Aitken method

Appendix B

COMPUTATIONAL METHOD — GASEOUS DIFFUSION*

In this Appendix the following two design problems will be demonstrated:

1. *Design I.* To find the dimensionless area s_T, the permeate mole fraction y_p, and cut θ for given feed mole fraction x_f, reject mole fraction x_0, ideal separation factor a^*, and pressure ratio Pr.

2. *Design II.* To find s_T, x_0, y_p for given x_f, θ, a^*, and Pr.

Except for the model of complete mixing, computations for various models involve solving a system of ordinary differential equations based on the differential material balance. When the design is an initial-value problem, the Runge-Kutta-Gill method of numerical integration [1] is exclusively used. In a boundary-value problem, the method for an initial-value problem is also employed with the second-order iteration of the Newton-Raphson scheme [2]. The corresponding computer programs in FORTRAN IV are listed in Table B.1. The notation of the symbols used in the computer program is as follows:

XF	x_f	Concentration of the feed stream in mole fraction
XO	x_0	Concentration of the reject stream in mole fraction
YP	y_p	Concentration of the permeate stream in mole fraction
ALPHA	a^*	Separation factor
PR	Pr	Pressure ratio of the low pressure to the higher pressure
THETA	θ	Cut
ST	s_T	Dimensionless diffusional area defined by Eq. 13.80

The convention for the ideal separation factor a^* is such that all mole fractions in a binary mixture refer to the higher-permeable species A, and a^* is the ratio of the permeability Q_A of the higher-permeable species A to that Q_B of the lower species B, that is, $a^* = Q_A/Q_B > 1$.

WELLER-STEINER CASE I

Program WS11 — Design I

*This Appendix especially relates to Chapter XIII, Section 7.

Table B.1 Summary of Programs for Models of Gaseous Diffusion

Models	Name of Program	Variables	
		Given	To Find
Weller-Steiner I	WS11	x_f, x_o; a^*, Pr	θ ,y_p,s_T
	WS12	x_f, θ ; a^*, Pr	x_o, y_p, s_T
Weller-Steiner II	WS2P1	x_f, x_o; a^*, Pr	θ ,y_p,s_T
	WS2P2	x_f, θ ; a^*, Pr	x_o, y_p, s_T
Cocurrent	COCRP1	x_f, x_o; a^*, Pr	θ ,y_p,s_T
	COCRP2	x_f, θ ; a^*, Pr	x_o, y_p, s_T
Countercurrent	CTCRP1	x_f, x_o; a^*, Pr	θ ,y_p,s_T
	CTCRP2	x_f, θ ; a^*, Pr	x_o, y_p, s_T

1. Given: XF, XO; ALPHA, PR
   To Find: YP, ST, THETA
2. Formulation of working equations (see Eqs. 13.74, 13.76, and 13.81):

$$A = -(a^*-1)\text{Pr} ,$$

$$C = -a^* x_o ,$$

$$D = 1-x_o-A-C ,$$

$$y_p = \frac{D + (D^2-4AC)^{1/2}}{(2A)} ,$$

$$\theta = \frac{(x_o - x_f)}{(x_o - y_p)} ,$$

$$s_T = \frac{y_p \theta}{x_o-\text{Pr } y_p} . \tag{B.1}$$

3. Note: x_o must be larger than the minimum possible stripping concentration x_{oM}, which is given by Eq. 13.85.

Program WS12 – Design II
1. Given: XF, THETA, ALPHA, PR
   To Find: XO, VP, ST
2. Formulation of working equations (see Eqs. 13.74, 13.79, and 13.81):

$$A = (a^*-1) [\theta+\text{Pr} (1-\theta)] ,$$

$$B = 0.5 [(a^*-1)\left\{ \theta+\text{Pr}(1-\theta) + x_f \right\}+1] ,$$

$$C = a^*x_f ,$$

$$y_p = \frac{B - (B^2-AC)^{\frac{1}{2}}}{A} ,$$

$$x_0 = \frac{x_f - y_p \theta}{1 - \theta} ,$$

$$s_T = \frac{y_p \theta}{x_0 - \text{Pr} y_p} . \tag{B.2}$$

WELLER-STEINER CASE II – CROSS FLOW

Program WS2P1 – Design I
 1. Given: XF, XO, ALPHA, Pr
     To find: YP, ST, THETA
 2. Formulation of working equations: Rearrangement of Eqs. 13.89 and 13.90 yields

$$\frac{dq^{\text{I}}}{dx_A} = \frac{-q^{\text{I}}}{x_A + \dfrac{a^*}{(1-a^*) + \dfrac{\text{Pr}-1}{(x_A-\text{Pr} y_A)}}}$$

$$= f_1 (x_A, y_A, q^{\text{I}}; a^*, \text{Pr}) \tag{B.3}$$

and

$$\frac{ds}{dx_A} = \frac{f_1(x_A, y_A, q^{\text{I}}; a^*, \text{Pr})}{x_A - \text{Pr} y_A + [1-x_A-\text{Pr}(1-y_A)]/a^*}$$

$$= f_2 (x_A, y_A, q; a^*, \text{Pr}) \tag{B.4}$$

where y_A is given by Eq. 13.91. The system of differential equations is numerically integrated with initial conditions:

$$q^{\text{I}} = 1, \quad s = 0 \quad \text{at } x_A = x_f$$

to find

$$q^I = 1-\theta, \quad s = s_T \quad,$$

$$y_p = \frac{x_f - (1-\theta)x_o}{\theta} \text{ at } x_A = x_0 \;.$$

The integration is performed by the Runge-Kutta-Gill method.

Program WS2P2 – Design II
1. Given: XF, THETA, ALPHA, PR
   To find: XO, YP, ST
2. Formulation of working equations: Eq. 13.89 and 13.90 can be reduced to

$$\frac{dx_A}{dq^I} = -\frac{(x_A-\text{Pr } y_A)f_4(x_A,y_A; \; a^*, \; \text{Pr}) + x_A}{q^I}$$

$$= f_3(x_A,y_A, \; q^I; \; a^*, \; \text{Pr}) \quad, \tag{B.5}$$

$$\frac{ds}{dq^I} = -\frac{1}{x_A-\text{Pr } y_A + [(1-x_A)-\text{Pr}(1-y_A)]/a^*}$$

$$= f_4 \; (x_A, \; y_A; \; a^*, \; \text{Pr}), \tag{B.6}$$

where y_A is given by Eq. 13.91. The system of differential equations is integrated by the Runge-Kutta-Gill method with inital conditions

$$s = 0, \quad x_A = x_f \quad \text{at } q^I = 1$$

to find

$$s = s_T, \quad x_A = x_o, \quad y_p = x_f - \frac{(1-\theta)x_o}{\theta} \text{ at } q^I = 1-\theta \;.$$

BLAISDELL COCURRENT FLOW

Program COCRP1 – Design I
1. Given: XF, XO, ALPHA, PR
   To find: THETA, YP ST
2. Formulation of working equations: The system of differential equations, Eqs. B.3 and B.4 with Eqs. 13.106 and 13.107 is integrated by the Runge-Kutta-

Gill method with initial conditions of

$$s = 0, \quad q^{\mathrm{I}} = 1 \quad \text{at } x_A = x_f$$

to find

$$s = s_T, \quad q^{\mathrm{I}} = 1-\theta \quad \text{at } x_A = x_o \ .$$

Here, x_o must be larger than the minimum possible stripping concentration x_{oM} given by Eq. 13.111.

3. Note: This computer program requires a subroutine in the program WS2P1.

Program COCRP2 – Design II

1. Given: XF, THETA, ALPHA, PR
   To find: XO, YP, ST

2. Formulation of working equations: The system of differential equations, Eqs. B.5 and B.6, where y is given by Eqs. 13.106 and 13.107, is integrated by the Runge-Kutta-Gill method with initial conditions

$$s = 0, \quad x_A = x_f \quad \text{at } q^{\mathrm{I}} = 1$$

to find

$$s = s_T, \quad x_A = x_o \quad \text{at } q^{\mathrm{I}} = 1-\theta \ .$$

BLAISDELL COUNTERCURRENT FLOW

Program CTCRP1 – Design I

1. Given: XF, XO, ALPHA, PR
   To Find: THETA, YP, ST

2. Formulation of working equations: The system of differential equations, Eqs. B.3 and B.4, is transformed by substituting

$$\overline{q} = \frac{q^{\mathrm{I}}}{1-\theta} \quad \text{and} \quad \overline{s} = \frac{s-s_T}{1-\theta}$$

for q and s, respectively. Thus, Eqs. B.3 and B.4 remain the same, but Eq. 13.113 is transformed into

$$y_A = \frac{x_o - \overline{q}x_A}{1-\overline{q}}, \quad \overline{q} \neq 1 \ ,$$

$$y_A = y_o \, , \qquad \qquad \overline{q} = 1 \, . \qquad \qquad \text{(B.7)}$$

Therefore, the system of equations, Eqs. B.3 and B.4, where y_A is given by Eq. Eq. 13.7, and integrated with initial conditions

$$\overline{s} = 0, \qquad \overline{q} = 1 \qquad \text{at } x_A = x_o$$

to find

$$\overline{s} = - \frac{s_T}{1-\theta} \, , \qquad \overline{q} = \frac{1}{1-\theta} \qquad \text{at } x_A = x_f \, .$$

This transformation thus avoids the unnecessary trial-and-error procedure and allows a straightforward solution of the initial-value problem. It has already been mentioned that, without such transformation, the backward integration with a trial-and-error method is necessary in order to avoid the instability due to the singularity in Eq. 13.113 at $x_A = x_o$.

Here, x_o must be larger than the minimum possible stripping concentration $x_{oM} = \text{Pr } y_f$.

Program CTCRP2 – Design II
 1. Given: XF, THETA, ALPHA, PR
      To find: XO, YP, ST
 2. Formulation of working equations: In order to avoid the instability of the forward numerical integration, the following backward integration is performed utilizing the second-order iteration scheme of the Newton-Raphson method. Differential equations, Eqs. B.5 and B.6, where y is given by Eq. 13.113, are used in combination with the following additional differential equation:

$$\frac{dx_k}{dq^I} = - \frac{1}{q^I} \left[(x_k - \text{Pr } y_k) f_4 \left\{ 1 - (x_A - \text{Pr } y_A) f_4 (1 - \frac{1}{a^*}) \right\} + x_k \right], \qquad \text{(B.8)}$$

where y_k is given by

$$y_k = \frac{q^I x_k - 1 + \theta}{q^I - 1 + \theta} \, . \qquad \qquad \text{(B.9)}$$

Actually, Eq. B.8 is obtained by differentiating Eq. B.5 with respect to x_o, and x_k and y_k are the corresponding derivatives of x_A and y_A, respectively. The solution of the problem requires that an initial value x_o is guessed, and that an iteration procedure is applied to the integration of the system of differential equations, Eqs. B.5, B.6, and B.8 in conjunction with the algebraic equations,

Eqs. 13.113 and B.9. Then, the initial conditions become

$$s = 0, \quad x_k = 1, \quad x_A = x_o^{(r)} \quad \text{at } q = 1 - \theta.$$

The iteration is repeated until

$$\left| \frac{x_o^{(r+1)} - x_o^{(r)}}{x_o^{(r+1)}} \right| < 0.0005, \quad r = \text{iteration number},$$

where

$$x_o^{(r+1)} = x_o^{(r)} + \frac{x_f - x_A^{(r)}}{x_k^{(r)}} \tag{B.10}$$

and $x_A^{(r)}$ and $x_k^{(r)}$ are values of x_A and x_k, respectively, at $q = 1$, and are obtained with $x_o = x_o^{(r)}$.

References

1. L.Lapidus, *Digital Computation for Chemical Engineers,* McGraw-Hill, New York, 1962.
2. W. F. Ames, *Non-linear Differential Equations in Transport Processes,* Academic, New York, 1968.

Appendix C

COMPUTATIONAL METHOD – OSMOTIC PROCESS*

The following computational methods are essentially the same as those used in gaseous diffusion. The computation concerns only the design problem: to find s_T, Γ_o, Γ_p, for given a_A, π_f, θ and a_l or A_p. The corresponding computer programs are summarized in Table C.1. The notations used in the programs are as follows:

AA	a_A	Separation factor defined by Eq. 13.145
AL	a_l	Relative mass-transfer coefficient defined by Eq. 13.145
AP	A_p	Parameter defined by Eq. 13.167
GI	Γ^I	Relative concentration in the reject phase
GII	Γ^{II}	Relative concentration in the permeate phase
GO	Γ_o	Relative reject concentration
GP	Γ_p	Relative permeate concentration
GW	Γ_W	Relative concentration at the membrane wall
P	P	Dimensionless pressure drop
PAI	π_f	Dimensionless osmotic pressure in the feed
QI	q^I	Dimensionless flow rate in the reject phase
ST	s_T	Dimensionless membrane area
THETA	θ	Fractional feed recovery or cut

Table C.1 Summary of Computer Programs – Reverse Osmosis

Name of Program	Description	Variables Given	To Find
CMPLR	Complete mixing with concentration polarization	θ, a_A, a_l, π_f	Γ_p, Γ_o, s_T
PLUGR	Plug flow with fully developed boundary layers	θ, a_A, a_l, π_f	Γ_p, Γ_o, s_T
COCRPR	Cocurrent plug flow with a pressure loss but without concentration polarization	θ, a_A, A_p, π_f	Γ_p, Γ_o, s_T
CTCRPR	Countercurrent plug flow with a pressure loss but without concentration polarization	θ, a_A, A_p, π_f	Γ_p, Γ_o, s_T

*This Appendix relates especially to Chapter XIII, Section 8.

Program CMPLR

1. Given: THETA, AA, AL, PAI
   To find: GP, GO, ST
2. Rearrangement of Eqs. 13.141 through 13.144 yields

$$f(\Gamma_p) = \frac{\Gamma_p (1 - \theta)}{(a_A + \pi_f \Gamma_p)(1 - \Gamma_p)} - \exp\left[\frac{1}{a_l} - \frac{\pi_f \Gamma_p}{a_l(a_A + \pi_f \Gamma_p)}\right] = 0. \qquad \text{(C.1)}$$

This algebraic equation is solved for Γ_p using the Newton-Raphson method. Then, Γ_o and S_T are evaluated from the following equations:

$$\Gamma_o = \frac{1 - \Gamma_p \theta}{1 - \theta}, \qquad \text{(C.2)}$$

$$\Gamma_W = \Gamma_p \left(1 + \frac{1}{a_A + \pi_f \Gamma_p}\right), \qquad \text{(C.3)}$$

$$s_T = \frac{\theta}{1 - \pi_f(\Gamma_W - \Gamma_p)}. \qquad \text{(C.4)}$$

Program PLUGR

1. Given: THETA, AA, AL, PAI
   To find: GP, GO, ST
2. For a given Γ^I, the algebraic equation

$$f(\Gamma^{II}) = \ln \frac{\Gamma^{II}}{(\Gamma^I - \Gamma^{II})(a_A + \pi_f \Gamma^{II})} - \frac{a_A}{a_l(a_A + \pi_f \Gamma^{II})} = 0 \qquad \text{(C.5)}$$

is solved for Γ^{II} using the Newton-Raphson method. Then the following system of differential equations is integrated by the Runge-Kutta-Gill method (see Eqs. 13.152 to 13.155):

$$\frac{d\Gamma^I}{dq^I} = -\frac{\Gamma^I - \Gamma^{II}}{q^I}, \qquad \text{(C.6)}$$

$$\frac{ds}{dq^I} = -\frac{1}{1 - \pi_f(\Gamma_W - \Gamma^{II})}, \qquad \text{(C.7)}$$

where

$$\Gamma_W = \Gamma^{II} \left[1 + \frac{1}{(a_A + \pi_f \Gamma^{II})}\right]. \qquad \text{(C.8)}$$

The initial condition at $q^I = 1$ is

$$\Gamma^I = 0, \quad s = 0.$$

Such integration is performed to obtain

$$\Gamma^I = \Gamma_o \quad \text{and} \quad s = s_T \quad \text{at} \quad q^I = 1 - \theta,$$

and then Γ_p is evaluated from Eq. C.2.

Program COCRPR

1. Given: THETA, AA, AP, PAI
   To find: GP, GO, ST

2. Rearrangement of Eqs. 13.162 to 13.164 yields

$$\frac{ds}{dq^I} = -\frac{1}{P - \pi_f(\Gamma^I - \Gamma^{II})}, \tag{C.9}$$

$$\frac{d\Gamma^I}{dq^I} = \frac{1}{q^I}\left(\frac{a_A(\Gamma^I - \Gamma^{II})}{P - \pi_f(\Gamma^I - \Gamma^{II})} - \Gamma^I\right), \tag{C.10}$$

$$\frac{dP}{dq^I} = -\frac{A_p\,(1-q^I)}{P - \pi_f(\Gamma^I - \Gamma^{II})}, \tag{C.11}$$

where

$$\Gamma^{II} = \frac{a_A(\Gamma^I - \Gamma^{II})}{P - \pi_f(\Gamma^I - \Gamma^{II})}. \tag{C.12}$$

This system of differential material balances is integrated simultaneously with the additional differential equations:

$$\frac{d\Gamma_p^I}{dq^I} = \frac{1}{q^I}\left[\frac{a_A(\Gamma^I - \Gamma^{II})}{P - \pi_f(\Gamma^I - \Gamma^{II})}\left(\frac{\Gamma_p^I - \Gamma_p^{II}}{\Gamma^I - \Gamma^{II}} - \frac{P_p - \pi_f(\Gamma_p^I - \Gamma_p^{II})}{P - \pi_f(\Gamma^I - \Gamma^{II})}\right) - \Gamma_p^I\right], \tag{C.13}$$

$$\frac{dP_p}{dq^I} = \frac{A_p(1-q^I)\,[P_p - \pi_f(\Gamma_p^I - \Gamma_p^{II})]}{[P - \pi_f(\Gamma^I - \Gamma^{II})]^2}, \tag{C.14}$$

where

$$\Gamma_p^{II} = \frac{a_A\Gamma_p^I - \Gamma^{II}(P_p - \pi_f\Gamma_p^I)}{2\pi_f\Gamma^{II} + P - \pi_f\Gamma^I + a_A}. \tag{C.15}$$

Actually, Eqs. C.13 to C.15 are obtained by differentiating Eqs. C.10 to C.12 with respect to the unknown initial value P_0 of the dimensionless pressure drop P at $q^I = 1$. And Γ_p^I, Γ_p^{II}, and P_p are the corresponding derivatives of Γ^I, Γ^{II}, and P, respectively. The initial conditions are

$$s = 0, \quad \Gamma^I = 1, \quad P = P_0^{(r)}, \quad \Gamma_p^I = 0, \quad P_p = 1 \quad \text{at } q^I = 1.$$

Such integration is initiated with a reasonable guess for P_0 and then iterated until the following prescribed error criterion is satisfied.

$$\left| \frac{P_0^{(r+1)} - P_0^{(r)}}{P_0^{(r)}} \right| < 10^{-4}, \quad r = \text{iteration number.}$$

Here

$$P_0^{(r+1)} = P_0^{(r)} - \frac{P^{(r)} - 1}{P_p^{(r)}}. \tag{C.16}$$

$P^{(r)}$ and $P_p^{(r)}$ are values of P and P_p, respectively, at $q^I = 1 - \theta$, which are obtained with $P_0 = P_0^{(r)}$. Thus, the boundary-value problem is reduced to the initial-value problem with the Newton-Raphson method. The numerical integration is done by the Runge-Kutta-Gill method. It should be noted that this iteration procedure does not converge when the fractional feed recovery θ is specified larger than the maximum attainable recovery for a given A_p, or θ is too close to the limiting value.

Program CTCRPR
   1. Given: THETA, AA, AP, PAI
       To find: GP, GO, ST
   2. This program integrates the differential material balances, Eqs. C.9, C.10, and

$$\frac{dP}{dq^I} = - \frac{A_p(1 - \theta - q^I)}{P - \pi_f(\Gamma^I - \Gamma^{II})} \tag{C.17}$$

incorporated with Eq. C.12. The numerical integration is performed with the Runge-Kutta-Gill method. If the fractional feed recovery θ is specified beyond the maximum attainable recovery for a given A_p, an error message is produced.

INDEX